Lovell Reeve

The Land and Freshwater Mollusks Indigenous To, Or Naturalized In, the British Isles

Lovell Reeve

The Land and Freshwater Mollusks Indigenous To, Or Naturalized In, the British Isles

ISBN/EAN: 9783337411091

Printed in Europe, USA, Canada, Australia, Japan

Cover: Foto ©berggeist007 / pixelio.de

More available books at **www.hansebooks.com**

THE
LAND AND FRESHWATER
MOLLUSKS

INDIGENOUS TO, OR NATURALIZED IN,

The British Isles.

BY

LOVELL REEVE, F.L.S.

LONDON:
REEVE AND CO., 5, HENRIETTA STREET, COVENT GARDEN.
1863.

Dedication.

TO JOHN EDWARD GRAY, Ph.D., F.R.S.

Dear Dr. Gray,

It is forty-two years since you published, as a student, in the 'London Medical Repository,' your first memoir on some of our Land and Freshwater Mollusks. Ten years later, you were good enough to encourage me, as a youth, in the pursuit of our favourite science, and I enjoyed many pleasant days over your collection of shells in the society of yourself and Mrs. Gray.

Another ten years passed, and I had the honour of reading, at a meeting of the Academy of Sciences of Paris, my first crude notions of a 'System of Conchology.' The volumes in which they were embodied, published by Messrs. Longman and Co., in 1841, were costly, and the only chance left to me of pursuing the subject was to turn printer and publisher myself. With a lithographic press, a staff of print-colourers, a stock of types, and a printing-press, the means of production became comparatively easy. During the twenty-two years elapsed since, I have worked unremittingly on species, considering them more with reference to the phenomena of distribution than of classification. My 'Conchologia Iconica' has reached its fourteenth volume, and eighteen-hundredth plate.

Impressed with the feeling that it would be useful to show how our British molluscan fauna is represented in other parts of the globe, I have commenced an attempt to bring this experience to bear on the subject. Our opinions on what constitutes a genus and a species have been greatly at variance; but with a maturer knowledge of the general organization of the animal, and a clearer perception of the variableness of specific characters of shells, I have come to entertain many of the views which I formerly disregarded. No man has contributed so assiduously, through a long course of years, to our knowledge of the Land and Freshwater Mollusks of the British Isles, as yourself. It is in recognition of this service, and in grateful remembrance of some of the happiest days of my life, that I do myself the pleasure of dedicating to you the present volume.

I am, dear Dr. Gray,

Your very obedient Servant,

LOVELL REEVE.

Hutton, Brentwood.
January, 1863.

PREFACE.

The Land and Freshwater Mollusks of the British Isles have been ably described by Turton, Gray, Forbes and Hanley, Jeffreys, and others, as a fauna *per se ;* and great attention has been lavished on details of form and colour, of local habitation, and of parts of the anatomy. French conchologists, commencing at the opening of the present century with the terse and philosophic Draparnaud, and terminating with the accomplished Moquin-Tandon, have done even more for the natural history of our mollusks, while treating of them as members of the fauna of France, than has been done by British authors. But neither the conchologists of France nor of Britain have worked out the distribution or representation of the European genera and species, and the resulting phenomena, in other parts of the world. It is on this ground that I venture to add another to the already numerous manuals on the subject.

The bent of a long experience in foreign conchology, has led me to the study of our native Land and Freshwater Mollusks with especial reference to their relation with those of other countries. The genera of our own latitude are in this Work collated with the similar or representative genera of other latitudes. The species are regarded in their true character of an outlying fragment of that great province of distribution, elsewhere called the Caucasian Province, which has its centre of creation on the confines of Europe and Asia Minor, and extends on either side from Finland to North Africa, and from Arctic Siberia to the Himalayas. I have not, however, attempted to carry this kind of research beyond its application to the distribution of British generic and specific molluscan forms in space. The inquiry as to its bearings on their distribution in duration of time, I commend to the researches of the palæontologist. The views which I had instinctively formed on the origin of species have gathered strength from the present study. Reasoning from the facts before me, apart from any considerations of geology, if such a mode of reasoning in the present state of science may be allowed to have any weight, the conclusions at which I have ventured to arrive do not accord with the theories of Edward Forbes or of Darwin, but seem to point to another solution.

With the hope of making the work useful to collectors of shells, as well as to observers of the external characters and habits of the mollusk, figures are given of the shell of each species (the upper one when in outline representing the natural size), engraved from drawings on wood by Mr. G. B. Sowerby; and a vignette of a living animal of each genus, drawn and engraved in most instances from nature by Mr. O. Jewitt. Three years have been devoted to collecting living specimens. Where these could not be obtained or delineated, recourse has been had to some figures of M. Moquin-Tandon, and to some unpublished drawings obligingly placed at my disposal by the Rev. M. J. Berkeley.

The very characteristic portrait of Citoyen Draparnaud prefixed to the volume, has been engraved on steel by Mr. W. Holl from a photograph of a scarce print in Paris, which, I have reason to believe, is the only likeness extant of the renowned but short-lived Professor of Montpellier.

CONTENTS.

	PAGE
DEDICATION .	v
PREFACE	vii
ALPHABETICAL LIST OF SPECIES	x
SYSTEMATIC LIST OF SPECIES	xii
ANALYTICAL KEY	xiv
EXPLANATION OF MAP	xix
MAP OF DISTRIBUTION OF CAUCASIAN FORMS	1
SYNOPTICAL TABLE OF CLASSIFICATION	2
CLASS I. CEPHALA	3
FAMILY LIMACINEA .	5
—— COLIMACEA	33
—— AURICULACEA	124
—— LYMNÆACEA	133
—— CYCLOSTOMACEA	174
—— LITTORACEA	181
—— PERISTOMATA	184
—— NERITACEA	200
CLASS II. ACEPHALA	204
FAMILY MYTILACEA	206
—— NAIADES	211
—— CARDIACEA . .	225
TABLES OF DISTRIBUTION	242
DISTRIBUTION AND ORIGIN OF SPECIES	252
LIST OF WORKS REFERRED TO IN THE SYNONYMY	259
INDEX . .	267

ALPHABETICAL LIST OF SPECIES.

		Page			Page
1.	Achatina acicula	97	37.	Helix Carthusiana	67
2.	Acme lineata	179	38.	—— ericetorum	71
3.	Ancylus fluviatilis	171	39.	—— fasciolata	70
4.	—— lacustris	172	40.	—— fulva	80
5.	Anodonta cygnea	215	41.	—— fusca	79
6.	Arion ater	9	42.	—— hispida	76
7.	—— hortensis	11	43.	—— lamellata	81
8.	Assiminea Grayana	183	44.	—— lapicida	74
9.	Azeca tridens	95	45.	—— nemoralis	64
10.	Balea perversa	106	46.	—— obvoluta	73
11.	Bulimus acutus	88	47.	—— Pisana	68
12.	—— montanus	90	48.	—— pomatia	61
13.	—— obscurus	91	49.	—— pulchella	82
14.	Bythinia Leachii	190	50.	—— pygmæa	86
15.	—— similis	188	51.	—— revelata	78
16.	—— tentaculata	189	52.	—— rotundata	83
17.	Carychium minimum	127	53.	—— rufescens	75
18.	Clausilia biplicata	101	54.	—— rupestris	84
19.	—— laminata	100	55.	—— sericea	77
20.	—— perversa	103	56.	—— virgata	69
21.	—— Rolphii	102	57.	Limax agrestis	20
22.	Conovulus bidentatus	131	58.	—— brunneus	22
23.	—— denticulatus	129	59.	—— cinereus	25
24.	—— myosotis	130	60.	—— flavus	24
25.	Cyclas cornea	238	61.	—— gagates	19
26.	—— lacustris	241	62.	—— marginatus	21
27.	—— ovalis	240	63.	—— Sowerbyi	17
28.	—— Pisidioides	239	64.	—— tenellus	23
29.	—— rivicola	237	65.	Lymnæa auricularia	159
30.	Cyclostoma elegans	177	66.	—— glabra	165
31.	Dreissena polymorpha	209	67.	—— glutinosa	167
32.	Geomalacus maculosus	13	68.	—— involuta	168
33.	Helix aculeata	81	69.	—— limosa	157
34.	—— arbustorum	63	70.	—— palustris	162
35.	—— aspersa	59	71.	—— stagnalis	160
36.	—— Cantiana	66	72.	—— truncatula	164

ALPHABETICAL LIST OF SPECIES.

	Page		Page
73. Neritina fluviatilis	203	101. Succinea oblonga	44
74. Paludina contecta	194	102. —— putris	42
75. —— vivipara	195	103. Testacella haliotidea	30
76. Physa fontinalis	151	104. —— Maugei	32
77. —— hypnorum	153	105. Unio margaritifer	223
78. Pisidium amnicum	228	106. —— pictorum	221
79. —— Casertanum	232	107. —— tumidus	219
80. —— Henslowianum	234	108. Valvata cristata	200
81. —— obtusale	229	109. —— piscinalis	198
82. —— nitidum	231	110. Vertigo alpestris	119
83. —— pulchellum	233	111. —— antivertigo	116
84. —— pusillum	230	112. —— edentula	122
85. Planorbis albus	139	113. —— minuta	123
86. —— carinatus	142	114. —— Moulinsiana	117
87. —— complanatus	143	115. —— pusilla	120
88. —— contortus	146	116. —— pygmæa	118
89. —— corneus	138	117. —— striata	119
90. —— crista	141	118. —— vertigo	121
91. —— fontanus	147	119. Vitrina pellucida	39
92. —— glaber	140	120. Zonites alliaria	48
93. —— nitidus	148	121. —— cellarius	47
94. —— spirorbis	145	122. —— crystallinus	53
95. —— vortex	144	123. —— excavatus	52
96. Pupa Anglica	112	124. —— nitidulus	49
97. —— cylindracea	111	125. —— nitidus	51
98. —— muscorum	110	126. —— purus	49
99. —— secale	109	127. —— radiatulus	50
100. Succinea elegans	43	128. Zua subcylindrica	93

SYSTEMATIC LIST OF SPECIES.

Limacinea.

		Page
1.	Arion ater	9
2.	—— hortensis	11
1.	Geomalacus maculosus	13
1.	Limax Sowerbyi	17
2.	—— gagates	19
3.	—— agrestis	20
4.	—— marginatus	21
5.	—— brunneus	22
6.	—— tenellus	23
7.	—— flavus	24
8.	—— cinereus	25
1.	Testacella haliotidea	30
2.	—— Maugei	32

Colimacea.

		Page
1.	Vitrina pellucida	39
1.	Succinea putris	42
2.	—— elegans	43
3.	—— oblonga	44
1.	Zonites cellarius	47
2.	—— alliaria	48
3.	—— nitidulus	49
4.	—— purus	49
5.	—— radiatulus	50
6.	—— nitidus	51
7.	—— excavatus	52
8.	—— crystallinus	53
1.	Helix aspersa	59
2.	—— pomatia	61
3.	—— arbustorum	63
4.	—— nemoralis	64
5.	—— Cantiana	66
6.	—— Carthusiana	67
7.	—— Pisana	68
8.	—— virgata	69

		Page
9.	Helix fasciolata	70
10.	—— cricetorum	71
11.	—— obvoluta	73
12.	—— lapicida	74
13.	—— rufescens	75
14.	—— hispida	76
15.	—— sericea	77
16.	—— revelata	78
17.	—— fusca	79
18.	—— fulva	80
19.	—— lamellata	81
20.	—— aculeata	81
21.	—— pulchella	82
22.	—— rotundata	83
23.	—— rupestris	84
24.	—— pygmæa	86
1.	Bulimus acutus	88
2.	—— montanus	90
3.	—— obscurus	91
1.	Zua subcylindrica	93
1.	Azeca tridens	95
1.	Achatina acicula	97
1.	Clausilia laminata	100
2.	—— biplicata	101
3.	—— Rolphii	102
4.	—— perversa	103
1.	Balea perversa	106
1.	Pupa secale	109
2.	—— muscorum	110
3.	—— cylindracea	111
4.	—— Anglica	112
1.	Vertigo antivertigo	116
2.	—— Moulinsiana	117
3.	—— pygmæa	118
4.	—— alpestris	119
5.	—— striata	119

SYSTEMATIC LIST OF SPECIES.

	Page
6. Vertigo pusilla	120
7. —— vertigo	121
8. —— edentula	122
9. —— minuta	123

Auriculacea.

1. Carychium minimum	127
1. Conovulus denticulatus	129
2. —— myosotis	130
3. —— bidentatus	131

Lymnæacea.

1. Planorbis cornens	138
2. —— albus	139
3. —— glaber	140
4. —— crista	141
5. —— carinatus	142
6. —— complanatus	143
7. —— vortex	144
8. —— spirorbis	145
9. —— contortus	146
10. —— fontanus	147
11. —— nitidus	148
1. Physa fontinalis	151
2. —— hypnorum	153
1. Lymnæa limosa	157
2. —— auricularia	159
3. —— stagnalis	160
4. —— palustris	162
5. —— truncatula	164
6. —— glabra	165
7. —— glutinosa	167
8. —— involuta	168
1. Ancylus fluviatilis	171
2. —— lacustris	172

Cyclostomacea.

1. Cyclostoma elegans	177
1. Acme lineata	179

	Page
Littoracea.	
1. Assiminea Grayana	183

Peristomata.

1. Bythinia similis	188
2. —— tentaculata	189
3. —— Leachii	190
1. Paludina contecta	194
2. —— vivipara	195
1. Valvata piscinalis	198
2. —— cristata	200

Nerilacea.

1. Neritina fluviatilis	203

Mytilacea.

1. Dreissena polymorpha	209

Naiades.

1. Anodonta cygnea	215
1. Unio tumidus	219
2. —— pictorum	221
3. —— margaritifer	223

Cardiacea.

1. Pisidium amnicum	228
2. —— obtusale	229
3. —— pusillum	230
4. —— nitidum	231
5. —— Casertanum	232
6. —— pulchellum	233
7. —— Henslowianum	234
1. Cyclas rivicula	237
2. —— cornea	238
3. —— Pisidioides	239
4. —— ovalis	240
5. —— lacustris	241

ANALYTICAL KEY.

—◆—

MOLLUSKS WITH A HEAD. RESPIRING AIR OR WATER.

Inoperculated. Respiring air.
Vital organs not coiled. Shell rudimentary; simple or subspiral.

Respiratory chamber near the head, orifice towards the front	ARION.
Body stout, wrinkled with tuberosities, shell a few grains........................	— *ater*.
Body slender, wrinkled with leaflets, shell a few aggregated grains	— *hortensis*.
Respiratory chamber near the head, orifice quite in front..........................	GEOMALACUS.
Body cylindrically attenuated, shell a solid flat ovate shield....................	— *maculosus*.
Respiratory chamber near the head, orifice towards the hinder part...........	LIMAX.
Body stoutly oblong, keeled throughout, shell thick or obscure...............	— *Sowerbii*.
Body tapering, veined, keeled throughout, shell thick, convex.................	— *gagates*.
Body cylindrically oblong, keeled towards the tail, shell very small..........	— *agrestis*.
Body gelatinous, striped, keeled towards the tail, shell thin....................	— *marginatus*.
Body tapering, slightly keeled towards the tail, shell unknown................	— *brunneus*.
Body slender, compressed towards the tail, transparent, shell unknown.....	— *tenellus*.
Body cylindrically elongated, keeled towards the tail, shell nucleated.......	— *flavus*.
Body elongately tapering, keeled towards the tail, shell large, nucleated...	— *cinereus*.
Respiratory chamber near the tail, covered by an external subspiral shell ...	TESTACELLA.
Body yellowish, finely veined, shell small, triangularly ear-shaped............	— *haliotidea*.
Body brown-mottled, rugosely veined, shell larger, squarely oblong..........	— *Maugei*.

Vital organs coiled within a spiral shell.
Inhabiting land. Tentacles retractile, with the eyes at their summit.

Carrying a depressly globose glassy shell, with shield in front, .obed...........	VITRINA.
Shell of three imperforate pellucid whorls smooth and shining,..................	— *pellucida*.
Carrying a membranaceous inflated shell. Inhabiting wet places	SUCCINEA.
Shell large, amber-coloured, yellowish, ovately-oblong, inflated...............	— *putris*.
Shell smaller, reddish-amber, constrictedly fusiformly inflated................	— *elegans*.
Shell small, yellowish, oblong turbinated, but little inflated....................	— *oblonga*.
Carrying an orbicular shell of few colourless shining whorls, lobed.............	ZONITES.
Shell comparatively large, rather narrowly umbilicated, of six whorls........	— *cellarius*.
Shell smaller, narrowly umbilicated, of five glassy whorls.......................	— *alliarius*.
Shell rather larger, openly umbilicated, of five dull opake whorls.............	— *nitidulus*.
Shell very small, rather largely umbilicated, of four greenish glassy whorls	— *purus*.
Shell very small, moderately umbilicated, of four stria-rayed whorls..........	— *radiatulus*.

ANALYTICAL KEY.

Shell larger, moderately umbilicated, of four shining fulvous whorls......... *Z. nitidus.*
Shell rather larger, excavately umbilicated, of four fulvous whorls............ — *excavatus.*
Shell very small, narrowly umbilicated, of four crystalline whorls .... — *crystallinus.*
Carrying an orbicular shell, painted or horny, lip sometimes reflected.......... HELIX.
Shell large, imperforate, of four to four and a half whorls, lip reflected..... — *aspersa.*
Shell very large, subumbilicated, four and a half whorls, lip dilated.......... — *pomatia.*
Shell rather large, subumbilicated, of five whorls, lip expanded............... — *arbustorum.*
Shell rather large, imperforate, of five painted whorls, lip expanded......... — *nemoralis.*
Shell moderate, narrowly umbilicated, of six rufous horny whorls............. — *Cantiana.*
Shell smaller, minutely umbilicated, of six yellowish horny whorls............ — *Carthusiana.*
Shell rather large, narrowly umbilicated, of five black-pencilled whorls ..... — *Pisana.*
Shell moderate, slightly umbilicated, of five rust-banded whorls .... — *virgata.*
Shell smaller larger umbilicated, of five wrinkled banded whorls.............. — *fasciolata.*
Shell rather large, broadly umbilicated, of five rather narrow whorls........ — *ericetorum.*
Shell moderate, largely umbilicated, of six whorls, aperture triangular...... — *obvoluta.*
Shell rather large, lens-shaped, brown, of five whorls, lip expanded........... — *lapicida.*
Shell small, deeply umbilicated, of six yellowish or rufous whorls. .... — *rufescens.*
Shell small, openly umbilicated, of six downy horny whorls .... — *hispida.*
Shell small, minutely umbilicated, of six silky hairy whorls..................... — *sericea.*
Shell small, of six greenish wrinkly membranaceous whorls..................... — *revelata.*
Shell small, of five glassy fuscous-olive membranaceous whorls................ — *fusca.*
Shell smaller, trochiform, of six thin glossy fulvous whorls .... — *fulva.*
Shell very small, subglobose, of six horny lamellated whorls.................... — *lamellata.*
Shell very small, of four horny whorls, lamellated, edged with lashes .... — *aculeata.*
Shell minute, largely umbilicated, of four subhyaline tubular whorls .... — *pulchella.*
Shell small, largely umbilicated, of six depressed striated whorls.............. — *rotundatus.*
Shell minute, largely umbilicated, of five brown horny whorls................. — *rupestris.*
Shell more minute, largely umbilicated, of four horny whorls................... — *pygmæa.*
Carrying a minutely umbilicated turriculate shell of eight to nine whorls..... BULIMUS.
Shell horny, marbled with opake white. Inhabiting chalk soil .... — *acutus.*
Shell stouter, dark olive, shagreened. Inhabiting wooded districts .... — *montanus.*
Shell smaller, constricted, fuscous. Inhabiting wooded districts............. — *obscurus.*
Carrying an imperforate cylindrical shell. Inhabiting wet places .... ZUA.
Shell of five smooth shining horny whorls. Columella slightly truncated... — *subcylindrica.*
Carrying an imperforate cylindrical shell. Inhabiting wooded districts....... AZECA.
Shell of seven smooth whorls, with an ear-shaped toothed aperture .... — *tridens.*
Carrying a smooth acicular hyaline shell. Burrowing in earth. Blind........ ACHATINA.
Shell minute, imperforate, of six whorls. Columella involute, truncated... — *acicula.*
Carrying a slender sinistral shell, enclosing itself by a clausilium............. CLAUSILIA.
Shell moderately large, nearly smooth, aperture two-plaited................... — *laminata.*
Shell rather larger, densely ridge-wrinkled, aperture two-plaited .... — *biplicata.*
Shell smaller, swollen, finely wrinkled, aperture five-plaited.................. — *Rolphii.*
Shell small finely wrinkled, dark brown, aperture two-plaited .... — *perversa.*
Carrying a conically turreted sinistral transparent horny shell................... BALEA.
Shell minutely umbilicated, of seven glossy finely-striated whorls............. — *perversa.*

ANALYTICAL KEY.

Carrying a horny narrow-whorled shell. Lower tentacles short .... *PUPA*.
    Shell rather large, of eight to nine whorls, aperture seven-toothed.......... — *secale*.
    Shell small, of seven glossy whorls, aperture toothless or one-toothed....... — *muscorum*.
    Shell very small, of six rather swollen whorls, aperture one-toothed......... — *cylindracea*.
    Shell small, of six narrow whorls, aperture triangular, five-toothed .... — *Anglica*.
Carrying a minute shell. No lower tentacles. Inhabiting wet places.......... *VERTIGO*.
    Shell tumidly cylindrical, of five whorls, aperture six to nine-toothed ...... — *antivertigo*.
    Shell tumidly ovate, of five whorls, aperture two to four-toothed .... — *Moulinsiana*.
    Shell ovately cylindrical, of five whorls, aperture four to five-toothed ...... — *pygmæa*.
    Shell oblong-cylindrical, of five whorls, aperture four-toothed .... — *alpestris*.
    Shell shortly cylindrical, of four tumid whorls, aperture six-toothed — *striata*.
    Shell sinistral, stout, of five whorls, aperture seven-toothed..................... — *pusilla*.
    Shell sinistral, attenuated at base, of five whorls, aperture five-toothed ... — *vertigo*.
    Shell straightly cylindrical, of six to seven whorls, aperture toothless ...... — *edentula*.
    Shell oblong-cylindrical, of five narrow tumid whorls, aperture toothless — *minuta*.

Inhabiting land. Tentacles not retractile, eyes at their base.

Carrying a minute oblong shell. Inhabiting wet places .... *CARYCHIUM*.
    Shell of five glassy whorls, aperture small, three-toothed....................... — *minimum*.
Carrying a much larger fusiform shell. Inhabiting places near the sea........ *CONOVULUS*.
    Shell ovate, of six to seven whorls, aperture toothed on both sides .... — *denticulatus*.
    Shell oblong, of six to seven whorls, columella four-toothed .... — *myosotis*.
    Shell ovate, of six to seven whorls, columella two-toothed .... — *bidentatus*.

Inoperculated. Respiring both air and water.
Inhabiting water. Head with muzzle, tentacles various, eyes at their base.

Carrying a discoid shell of three to seven whorls, tentacles bristle-like ...... *PLANORBIS*.
    Shell large, ventricose, of five to five and a half striated whorls .... — *corneus*.
    Shell small, of four and a half longitudinally striated whorls .... — *albus*.
    Shell very small, smooth, of four and a half rounded whorls .... — *glaber*.
    Shell very small, nautiloid, of three whorls, epidermis edged with lashes... — *crista*.
    Shell moderate, depressed, of five whorls, keeled below the centre .... — *carinatus*.
    Shell larger, depressed, of five whorls, keeled at the basal edge ... . — *complanatus*.
    Shell rather small, depressed, of seven very slowly increasing whorls ...... — *vortex*.
    Shell small, very depressed, of six faintly two-angled whorls .... — *spirorbis*.
    Shell rather small, depressed, of seven broad closely coiled whorls .... — *contortus*.
    Shell minute, amber-horny, of four convexly sloping whorls .... — *fontanus*.
    Shell minute, amber-horny, of four radiately segmented whorls .... — *nitidus*.
Carrying an inflated sinistral shell. Mantle reflected. Tentacles slender... *PHYSA*.
    Shell ovate, of four glassy whorls, convoluted ventricosely .... — *fontinalis*.
    Shell oblong, of five transparent whorls, convoluted attenuately............... — *hypnorum*.
Carrying a horny shell. Mantle rarely reflected. Tentacles flatly triangular *LYMNÆA*.
    Shell obliquely ovate, of four whorls, the last ventricosely inflated............ — *limosa*.
    Shell orbicular, of three whorls, the last enormously expanded............... — *auricularia*.

Shell ovately turreted, of six whorls, the last globosely inflated .... *L. stagnalis.*
Shell oblong-ovate, of five to six rough convex whorls .... *— palustris.*
Shell smaller, acuminately ovate, of six rough convex whorls.................... *— truncatula.*
Shell small, elongately turreted, of eight closely coiled whorls .... *— glabra.*
Shell globose, extremely thin, of three transparent inflated whorls .... *— glutinosa.*
Shell truncately ovate, thin, of three whorls, spire immersed.................... *— involuta.*
Carrying a limpet-shaped shell. Tentacles short, slenderly triangular ...... *ANCYLUS.*
Shell rotundately ovate, vertex posterior, incurved to the right............... *— fluviatilis.*
Shell oblong-ovate, depressed, vertex more central, incurved to the left ... *— lacustris.*

Operculated. Respiring air.

Head produced into a ringed proboscis. Tentacles bristle-like.

Carrying a solid turbinated shell. Operculum shelly, spiral .... *CYCLOSTOMA.*
Shell conically ovate, of five spirally corded whorls .... *— elegans.*
Carrying a thin minute elongated shell. Operculum horny, subspiral ...... *ACME.*
Shell of six transparent horny whorls, mostly linearly striated .... *— lineata.*

Operculated. Respiring water.

Head produced into a ringed muzzle. Tentacles various.

Tentacle and eye-stalk united in one, with the eye at their summit .... *ASSIMINEA.*
Shell conical, of six sloping whorls. Operculum horny, few-whorled......... *— Grayana.*

Carrying a turbinated shell with the aperture entire.

Tentacles filiform, eyes on sessile swellings at their base .... *BYTHINIA.*
Shell very small, of six whorls. Operculum horny, two-whorled............ *— similis.*
Shell of five ventricose whorls. Operculum subtestaceous, concentric ... *— tentaculata.*
Shell of five scalariform whorls. Operculum subtestaceous, concentric... *— Leachii.*
Tentacles cylindrical, eyes on short stalks. Neck lobed on each side .... *PALUDINA.*
Shell scalariformly turbinated of five to six whorls, umbilicated............... *— contecta.*
Shell obtusely turbinated, of four to five whorls, scarcely umbilicated ...... *— vivipara.*
Tentacles cylindrical, eyes at inner base. Branchiæ an external plume...... *VALVATA.*
Shell small, of five globose whorls. Operculum concentric...................... *— piscinalis.*
Shell minute, of four discoid whorls. Operculum concentric .... *— cristata.*

Eyes on slender stalks detached from the tentacles.

Carrying an obliquely ovate shell. Operculum paucispiral, hooked .... *NERITINA.*
Shell rather solid, of three whorls, generally painted with network .... *— fluviatilis.*

MOLLUSKS WITHOUT HEAD. RESPIRING WATER.

One pair of adductor muscles.

Bearing an oblong fan-shaped shell. Affixed by a byssus.

Mantle closed, except for the pedal, branchial, and excretory orifices .... *DREISSENA.*
Shell triangularly trapezoid, very gibbous, painted with olive waves ...... *— polymorpha.*

ANALYTICAL KEY.

Two pairs of adductor muscles.

Bearing a pearly shell, mantle freely open except behind.

Shell equivalve, very inequilateral, auriculated behind, thin, toothless......	*ANODON.*
Shell ovate or oblong, compressed or ventricose, green or fulvous-olive ...	— *cygnea.*
Shell equivalve, very inequilateral, slopingly produced behind, toothed ...	*UNIO.*
Shell ovately wedge-shaped, olive-green, rayed, umboes wrinkled	— *tumidus.*
Shell elongately oblong, fulvous olive, umboes faintly wrinkled	— *pictorum.*
Shell oblong kidney-shaped, black, umboes more or less eroded	— *margaritifer.*

Bearing a thin horny shell, mantle closed at each end.

Branchial and excretory siphons prolonged into tubes united in one	*PISIDIUM.*
Shell large, triangular, posteriorly slopingly produced	— *amnicum.*
Shell very small, suborbicularly ovate, umboes obtuse, prominent............	— *obtusale.*
Shell very small, obliquely oval, umboes moderately tumid	— *pusillum.*
Shell very small, suborbicular, glossy, umboes obtuse, grooved	— *nitidum.*
Shell small, triangularly orbicular, posteriorly obliquely produced............	— *Casertanum.*
Shell minute, obliquely orbicular, anteriorly obliquely produced	— *pulchellum.*
Shell minute, obliquely subtriangular, umboes calyculate........................	— *Henslowianum.*
Branchial and excretory siphons prolonged into tubes, partially united.......	*CYCLAS.*
Shell large, oval-globose, striated, slightly lunuled, ligament apparent.....	— *rivicola.*
Shell small, suborbicular, ligament scarcely apparent............................	— *cornea.*
Shell moderate, oblong-oval, posteriorly subtriangularly sloped...............	— *Pisidioides.*
Shell moderate, oblong-oval, pale, umboes nearly central.......................	— *ovalis.*
Shell small, squarely orbicular, rhombic, thin, umboes calyculate............	— *lacustris.*

EXPLANATION OF MAP.

The Map, facing page 1, is intended to show the boundary of the Caucasian province of molluscan distribution over which the British species range, and the position which it occupies on the surface of the globe with reference to the general system. The student is supposed to be acquainted with the geography of the two hemispheres, or to have a geographical atlas before him for reference. The lines of latitude and longitude, and the names of places, are excluded, in order to exhibit more clearly the physical features to which it is desired to call attention; namely, to the mountains and rivers, the configuration of the land and water, and to the parallels of equality of temperature, copied from Humboldt's system of isothermal lines. An attempt has been made by the lithographer to indicate the specific centre of the province by deepening the colour of that part of it in which the most characteristic of the genera and species congregate, and are most prolific in individuals; the part, in short, in which the genera and species obtain their *maximum* of development.

The primary division of the terrestrial portion of the globe into broadly-defined provinces of molluscan distribution, is as follows:—

Eastern Hemisphere.	*Western Hemisphere.*
The Caucasian Province.	The North American Province.
The West African Province.	The Columbian Province.
The South African Province.	The Brazilian Province.
The Malayan Province.	The Bolivian Province.
The Australian Province.	The Chilian Province.

The Caucasian Province is divided to some extent in its western part by the mountain chain of the Pyrenees, Alps, and Carpathians, but this division does not affect the range of the species of land and freshwater mollusks which inhabit Britain. The region north of that apparently natural barrier, extending to the utmost limit of molluscan life towards the pole, is termed the Germanic Region. The southern portion, reaching to the boundary of the province in the opposite direction, at about the 15th parallel of north latitude, on the west side of Africa, and extending on the east side to Abyssinia and the island of Socotra, is called the Lusitanian Region. But the whole of Britain on the Continental side is surrounded by Germanic forms, and all the species inhabiting the Continent,—all, that is to say, excepting *Geomalacus maculosus, Assiminea Grayana,* and *Lymnæa involuta,* which, so far as we know at present, are peculiar to Britain, are, if not of Germanic origin, Germanic now. The most obvious subdivision of the British land and freshwater mollusks in their Continental range, is to separate, as shown in our Tables of Distribution (pp. 242 to 249), those inhabiting Europe throughout, from those which inhabit the Centre and North, and the Centre and South. The species furthest removed from the Centre of Europe on the northern side is *Helix lamellata,* and it is the most northerly species in Britain. The species furthest removed from the Centre on the southern side is *Helix Pisana,* and it is the most southerly species in Britain.

Molluscan life increases vastly in its development towards the Equator, and it follows that there are many more Germanic species common to the Lusitanian region, than there are Lusitanian species common to the Germanic. The Azores and Canary Islands are outlying fragments of the Lusitanian region, inhabited chiefly by continental species. Madeira is also Lusitanian in type, but it is inhabited to a very limited extent by continental species, having a very characteristic inland molluscan fauna of its own.

Of the Asiatic range of the British species little is known beyond the information supplied by Gerstfeldt of those inhabiting the valley of the Amoor, and by Drs. Hooker and Thomson and M. Jacquemont of those inhabiting Cashmere and Thibet. It is certain, however, that Malayan species, having a copious and highly developed specific centre in the Indian Archipelago, inhabit Burmah and Siam, while they mingle with Caucasian species in the North-western provinces of India, in the valley of the Amoor, and in Japan.

Of the remaining nine Provinces, divisible into twenty-five regions, the North American is the only one in which there are any species of land or freshwater mollusks common to Britain, for particulars of which see 'Distribution and Origin of Species,' p. 252.

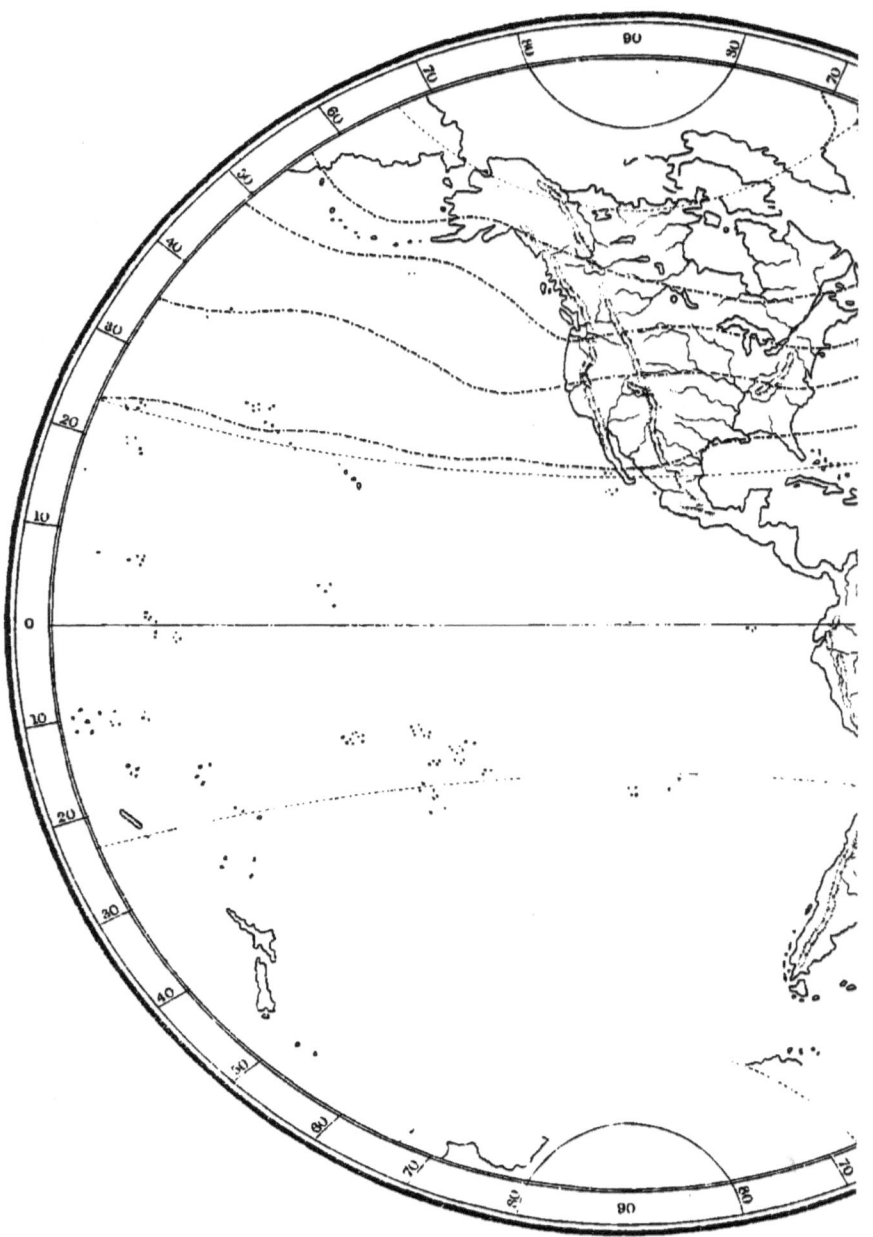

BRITISH
LAND AND FRESHWATER
MOLLUSKS.

THE inland molluscan fauna of the British Isles is very limited in kind, compared with that of more southern European countries, or with that of America within the same isothermal latitude as Britain. The warmer temperate districts of Europe, marked by the olive, and even the vine, show a material increase in the development of molluscan life; while in the intertropical region of the palms, embracing all that is most productive in the conditions of light, heat, and vegetation, its abundance is prodigious. Upwards of seven thousand land and freshwater shells have been described from all parts of the world, of which scarcely more than a hundred and twenty are British. To the casual observer, a few slugs and snails in the garden or on the chalk down, a few water-snails and little horny bivalves in our ponds and ditches, and a pearl-mussel or two in our running streams, are all that present themselves to his attention. The remainder are minute forms, living more or less concealed in moss, or among the decaying roots of shrubs and bushes, only to be procured with diligent search at the proper times and seasons. Our molluscan fauna, too, is simply European in its character; it possesses no local typical speciality like the fauna of Madeira or of the Sandwich Islands. Our groves do not glow beneath the same tropical sun as the woods and forests of the Philippine Islands, of Bolivia, or of Venezuela; and our running streams are small indeed compared with the mighty rivers of America. But although our land and freshwater mollusks are small, and singularly limited in kind, they present equally interesting features for comparison and study. With these few introductory remarks, we pass to our tabular exposition of the classification.

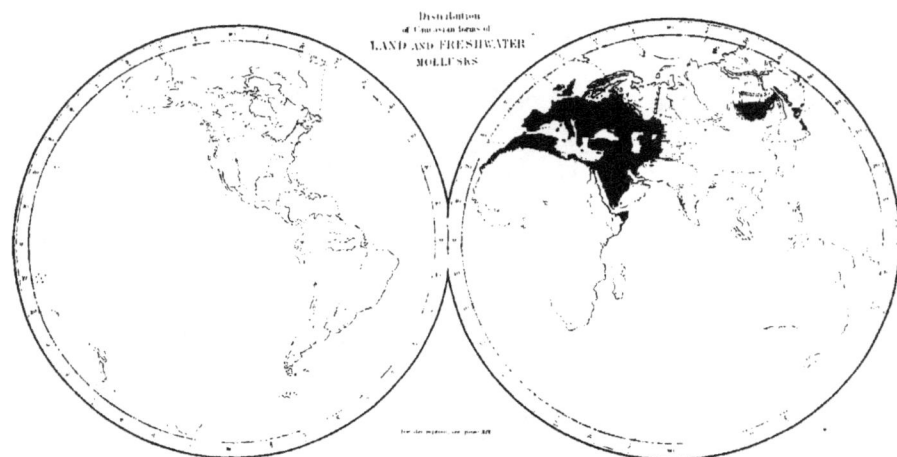

Distribution
of Caucasian fauna of
LAND AND FRESHWATER
MOLLUSKS

CLASS I.—CEPHALA (with head).

Tribes.	Orders.	Families.	Genera.
1. INOPERCULATA (without operculum).	1. PULMONIFERA (breathing air only).	1. *Limacinea.*	1. Arion. 2. Geomalacus. 3. Limax. 4. Testacella.
		2. *Colimacea.*	1. Vitrina. 2. Succinea. 3. Zonites. 4. Helix. 5. Bulimus. 6. Zua. 7. Azeca. 8. Achatina. 9. Clausilia. 10. Balea. 11. Pupa. 12. Vertigo.
		3. *Auriculacea.*	1. Carychium. 2. Conovulus.
	2. PULMOBRANCHIATA (breathing both air and water).	1. *Lymnæacea.*	1. Planorbis. 2. Physa. 3. Lymnæa. 4. Ancylus.
2. OPERCULATA (with operculum).	1. PULMONIFERA (breathing air only).	1. *Cyclostomacea*	1. Cyclostoma. 2. Acme.
	2. BRANCHIFERA (breathing water only).	1. *Littoracea.*	1. Assiminea.
		2. *Peristomata.*	1. Bythinia. 2. Paludina. 3. Valvata.
		3. *Neritacea.*	1. Neritina.

CLASS II.—ACEPHALA (without head).

Tribes.	Orders.	Families.	Genera.
LAMELLIBRANCHIATA (lamella-gilled).	1. UNIMUSCULOSA (one adductor muscle).	1. *Mytilacea.*	1. Dreissena.
	2. BIMUSCULOSA (two adductor muscles).	1. *Naiades.*	1. Anodonta. 2. Unio.
		2. *Cardiacea.*	1. Pisidium. 2. Cyclas.

Class I. CEPHALA—WITH HEAD.

The two Classes of Cephala and Acephala, signifying Headed and Headless, into which Land and Freshwater, and indeed all mollusks, are primarily divided, correspond as near as may be with the old Linnæan designations of Univalve and Bivalve. The Acephala, which are all water-dwelling mollusks, are always provided with a shell; but the Cephala, or Gastropods, as they are also called, are sometimes, as in *Arion*, a garden slug, without any shell; the term Univalve is then inappropriate. There exist about a hundred and twenty species of land and freshwater mollusks in Britain, fifteen of which—the Mussels and Cyclads—are acephalous, without a head. The remaining hundred and five, representing twenty-nine genera, belong to the Class under consideration. They possess a well-developed head, with either two or four tentacles, having a pair of eyes, sometimes at their extremity, sometimes at the base; and the mouth is provided with a jaw and minute palate-teeth which present some curious varieties of structure. The body of the Cephals is elongated and rather depressed, the upper surface developing a mantle which is the calcifying organ, secreting, in most instances, a shell covering or enclosing the respiratory and visceral organs of the animal. The lower surface of the body forms a grasping fleshy disk, by the contraction and dilatation of which the creature crawls.

Tribe I. INOPERCULATA—WITHOUT OPERCULUM.

Attention may now be directed to the Synoptical Table opposite. The class Cephala, it will be seen, is subdivided into Inoperculated and Operculated. The Inoperculated Cephals, those unprovided with any operculum, comprise the pulmoniferous or air-breathing families, *Limacinea, Colimacea, Auriculacea*, including about half of the entire series; and the pulmobranchiate or lung-gilled family *Lymnæacea* breathing both air and water. The Operculated Cephals comprise the pulmoniferous family *Cyclostomacea*, breathing air only, and the branchiferous or gilled families *Peristomata* and *Neritacea*, breathing water only. There is, however, no absolute line of demarcation between them. All are modified and run into each other by gradations of character. The *Ampullariæ*, a foreign genus of *Peristomata*, though breathing water as their natural element of respiration, possess an apparatus which gives them the faculty of breathing air in cases of emergency, such, for example, as when the ponds in which they are dwelling become dried up.

Dr. Gray is of opinion that the shell and its operculum are homologous to the two valves of a bivalve, and that the normal form of a mollusk is to be protected by two valves or shells. This is an ingenious theory, in which few participate to the extent of the author. The operculum affords fair subsidiary characters for the distinction of groups. The foreign *Cyclostomacea*, numbering eight hundred species, are subdivided on this method into as many as forty genera. The arrangement is almost universally adopted, but it embraces, in some instances, a rather anomalous association of forms.

The Inoperculated Cephals are subdivided into two Orders:—

1. **Pulmonifera.** Respiratory organ a vascular sac for the respiration of air.
2. **Pulmobranchiata.** Respiratory organ a vascular sac for the respiration of air, with the addition of branchial lamellæ (gills), for the respiration also of water.

Order I. **PULMONIFERA**—Air-breathing.

Respiratory organ a vascular sac for the respiration of air.

One-half of our entire series of Land and Freshwater mollusks belong to this Order. They include the Slugs, *Limacinea*, the Snails, properly so called, *Colimacea*, all of which are terrestrial, and the *Auriculacea*, which live near the sea, within reach of the spray, or on the banks of rivers or of estuaries. The pulmonary organ or lung is a sac over the surface of which the blood flows in minute vessels for the purpose of being aërated.

Although breathing air as their natural element, the Pulmonifera love damp and moisture, and sustain life with comparatively little inconvenience for a long time under water. A slug may not uncommonly be seen attached to the inner side of a waterbutt, in which it has become immersed by the rising of the water, without evincing any disposition to move; and a snail, on being dropped in a glass of water, will put forth its tentacles with little less than its accustomed ease, and crawl up leisurely to the brim, with an air-bubble at its respiratory orifice, like that of the water-beetle. It may, however, be observed that the slug, when submerged, secretes a more copious discharge of mucus, and forms for itself a kind of viscid envelope.

The Pulmoniferous Inoperculated Cephals are distributed into three Families:—

1. **Limacinea.** Respiratory and visceral organs incorporated with the main contractile mass of the body. Eyes at the extremity of the tentacles. Shell wanting, or rudimentary.
2. **Colimacea.** Respiratory and visceral organs distinct from the main contractile mass of the body, coiled within a spiral shell. Eyes at the extremity of the tentacles.
3. **Auriculacea.** Respiratory and visceral organs distinct from the main contractile mass of the body, coiled within a spiral shell. Eyes at the base of the tentacles.

Family I. LIMACINEA.

Respiratory and visceral organs incorporated with the main contractile mass of the body. Eyes at the extremity of the tentacles. Shell wanting, or rudimentary.

The typical character of the Family of *Limacinea* is expressed by the respiratory and visceral organs being incorporated with the main contractile body of the animal, as distinguished from that of the *Colimacea* and *Auriculacea*, in which they are coiled within a whorled shell. In the simplest terrestrial form of mollusk, *Arion*, *Geomalax* and *Limax*, the pulmonary sac is situated near the head, with slight variations in the position of the respiratory orifice. In *Testacella* it is situated at the tail end of the animal. There is an intermediate form of slug, *Parmacella*, in which the pulmonary sac is situated about the middle of the body; but this is a stranger to our islands, not having reached nearer to us than the south of France, at Arles in the department of the Mouths of the Rhone.

In these three genera of *Limacinea* there is no visible shell, but the calcifying function commences to be developed in the shield covering the pulmonary sac, with the secretion of a few calcareous grains. In *Arion* they are at first isolated, then agglomerated in a rude mis-shapen manner. In *Limax* an embryo umbonal nucleus is formed; and the shell, in a further stage of development, increases in symmetrical order by regular concentric layers of growth, covered by a membranous epidermis. In the fourth genus, *Testacella*, the shell is secreted externally, and the umbonal nucleus commences to coil, on the spiral plan of growth which is developed and matured in the whorled shell of the succeeding Family.

The *Limacinea* shun the light of day, rarely indulging their voracious appetites except at night. They inhabit gardens and roadside hedges in damp places, and congregate in cellars and outhouses, and under planks and stones around old walls, pumps and wells. These remarks apply chiefly to the genera *Arion*, *Geomalax*, and *Limax*, which feed on vegetable matter, though not entirely abstaining from flesh. *Testacella* burrows into the ground to the depth of from two to three feet, and feeds, or rather gorges, upon worms. The feebleness of the shell-producing function in the *Limacinea* is largely compensated by the faculty of secreting mucus of a particularly viscid kind, from all parts of the body. The slug will lower itself to the ground from a tree or shrub—and even from a shelf, when brought into the room—by the mere accumulation of mucus at the extremity of the tail hardening into a gelatinous thread. The animal functions are not suspended during hybernation, and at other periods, as in the snail; and the animal is at all times more tenacious of life. The continued secretion of mucus is necessary to the slug's existence; when this faculty ceases and the integuments dry, the animal dies.

The geographical distribution of the *Limacinea* is almost worldwide; they are by no means so confined to temperate climates as has been supposed. More observations have been made on them in Europe than in other countries, but they have been collected by M. Quoy and others, at New Zealand, New Hebrides, Australia, South Africa, and Ascension Island; and Mr. Cuming saw numbers at the Philippine Islands, although he did not preserve specimens. In South America the family is plentifully represented by a genus unknown in Europe, *Vaginulus;* and in North America a genus is known abundantly under the name *Tebennephorus* in every part of the country between Lake Erie and the Gulf of Mexico. Some of our European *Limacinea* have been transported to the United States, and become naturalized there to a distance of a hundred miles from the coast. The home of *Testacella* is in the Canary Islands. Its appearance in the south of Europe and the British Isles is partly due to transportation with exotic plants. It has rarely been observed except in gardens, and mostly about the hothouse or conservatory. *T. haliotidea* has long established itself in Guernsey. Its presence in that island was detected sixty years ago by Mr. Lukis, and I have received specimens from him, while this sheet is passing through the press, accompanied by a letter, written with his wonted enthusiasm, informing me that it has been most abundant in his garden during the present year. *Testacella* has also been observed in the West Indies at Guadaloupe and Martinique, but hardly sufficiently to prove that it is indigenous to those islands.

The genus *Onchidium* included in this Family by authors is a sea slug. It is a native chiefly of New Zealand and the Friendly Islands; a small species occurs on the shores of Brittany and Cornwall, congregated in little groups on the rocks, where the waves break over them.

The genera of British *Limacinea* are : —

1. **Arion.** Body rounded on the back, a terminal pore at the tail extremity, pulmonary sac near the head, respiratory orifice in front, shield granulated. Shell wanting, or represented by a few isolated or agglomerated grains.
2. **Geomalacus.** Body rounded on the back, pulmonary sac near the head, respiratory orifice quite in front, shield smooth. Shell, a solid oval plate.
3. **Limax.** Body more or less keeled on the back, pulmonary sac near the head, respiratory orifice more or less behind, shield sometimes granulated, sometimes concentrically striately wrinkled. Shell, a solid oval plate, passing into a valve developed in symmetrical order from an umbonal nucleus.
4. **Testacella.** Body rounded, tapering towards the head, pulmonary sac near the tail extremity, covered by a lobed mantle secreting an external shell. Shell car-shaped, enlarged from a sub-spiral nucleus.

Genus I. **ARION**, *Férussac*.

Animal; body cylindrical, elongated, rounded on the back, and furnished with a mucus-pore at the posterior extremity, integuments crowded with oblong tuberosities or with leaflets; head with two pairs of tentacles, the upper pair being much the longer, with minute eyes at their bulbous extremity; shield at the front part of the body, covered with fine granulations, respiratory orifice at the right margin, a little anterior to the centre; shell wanting, or represented by a few isolated or agglomerated grains.

The slugs of this genus are distinguished from those of *Limax* proper by the presence of a pore or gland, for the more copious secretion of mucus, at the extremity of the tail, and in having the pulmonary sac and overlapping shield nearer the head, with the respiratory orifice in front; and the shield has no internally developed shell, its place being occupied by merely a few calcareous grains, which are sometimes isolated, sometimes aggregated into a rude irregular mass. But there are good generic characters in *Arion*, as distinguished from *Limax*, apart from the position of the pulmonary sac and the want of a symmetrically-formed shell. The body is enveloped by integuments of more considerable density, rising into wrinkle-like tuberosities or leaflets, and there is no dorsal keel. Important anatomical differences are also recorded.

We have only two undoubted species of *Arion* in this country *A. ater* and *hortensis*. A third, *A. flavus*, has been described, and I have living specimens before me, obligingly communicated by Mr. E. J. Lowe, from the vicinity of Nottingham, as well as from Mr. Bridgman of Norwich; but M. Moquin-Tandon includes the species in his list of 'Espèces Incertaines,' and I cannot bring my mind to the conviction that it is anything more than a yellow dwarfed variety of *A. hortensis*. Notwithstanding the record given by Mr. Jeffreys in his 'Gleanings,' that Mr. Bridgman finds *A. flavus* on horse-chestnut leaves near Norwich, I am favoured with that gentleman's opinion, that it has always been a question with him whether *A. flavus* may not be the young of *A. hortensis*. Species are much too readily named. As many as ten species of *Arion* have been described in France alone, four of them within the last few years by M. Millet and M. Normand; but the only species since recognized by Moquin-Tandon in addition to our *A. ater* and *hortensis* are *A. albus* and *subfuscus*, which are apparently varieties of the first-named. Three new species have also been described by M. Morelet, from Portugal. The only *Arion* observed in the New World as indigenous is a species collected near Discovery Harbour, Puget Sound, *A. foliolatus*, Gould. *A. hortensis* has been transported to the United States, and has become acclimatized within a limited range in the vicinity of Boston.

The British species of *Arion* are :—

1. **ater.** Animal stout, three to five inches in length, mostly of a dark chocolate colour inclining to black, integuments rising into oblong tuberosities. Shell represented by calcareous granules, sometimes rudely agglomerated.

2. **hortensis.** Animal slender, one to two inches long, mostly grey or yellowish, more or less striped with blue-black, integuments rising into leaflets. Shell represented by calcareous granules, agglomerated into a subhemispherical form.

1. **Arion ater.** *Black Arion.*

Animal; body stout, from three to five inches in length, extremely variable in colour, but mostly of a deep chocolate-red or olive-black, integuments tessellately wrinkled throughout with short swollen ridges or tuberosities; margin of the foot expanded, generally yellowish or orange, radiately lineated with black or fuscous red; head stout, marked with two dark lines between the tentacles, which are livid blue-black; shield minutely granulated.

Shell; represented by calcareous grains, sometimes isolated, sometimes rudely agglomerated.

Limax ater (1674), Lister, *Phil. Trans.* vol. ix. p. 96. *Hist. Anim. Angl.* p. 131. pl. ii. f. 17.
Limax ater and *rufus*, Linnæus (1746), *Faun. Suec.* p. 507.
Limax albus, Linnæus (1767), *Syst. Nat.* 12th edit. p. 1081.
Limax succineus, Müller (1774), *Hist. Term.* vol. ii. p. 7.
Limax luteus, Razoumowsky (1789), *Hist. Nat. Jorat.* vol. i. p. 269.
Limax subfuscus, Draparnaud (1805), *Hist. Moll.* p. 123. pl. ix. f. 8.
Arion empiricorum, albus, fuscatus, subfuscus, and *melanocephalus,* Férussac, *Hist. Moll.* vol. i. p. 60 to 65 ; vol. ii. p. 17, 18. pl. i. to iii.
Arion ater, Turton (1831), *Man.* p. 104.
Limax virescens, Millet (1854), *Moll. Maine et Loire*, p. 11.
Arion (Lochea) rufus, albus, and *subfuscus*, Moquin-Tandon (1855), *Hist. Moll.* vol. ii. p. 10, 13. pl. i. f. 1 to 27.

Hab.—Throughout Europe, (in woods and gardens, in shady places, under hedges and leaves, and about wells or pumps).

This majestic richly-diapered slug may be recognized by its large swollen dimensions and characteristic oblong tuberosities. In colour it is mostly of a shining dark chocolate-red or olive-black, but it is extremely variable, and, as may be seen by the list of synonyms, has received at the hands of different authors the names of black, red, brown, white, and amber. De Férussac, thinking to avoid confusion, introduced the name *empiricorum* (of the quack doctors) in allusion to its alleged medicinal properties, and it has been adopted by Forbes and Hanley. The Linnæan names *ater* or *rufus*, the first of which is most in use in Britain, the second mostly on the Continent, either express the prevailing colour of the species, but Lister's name *ater* had a well-established priority in this country long before the time of Linnæus or Müller; and it has been ruled that the names of natural objects should not be altered according to the fancy or judgment of any author as to their appropriateness.

Whatever the colour of the body may be, the tentacles of *A. ater* are always of a livid blue-black, the eyes at the bulbous extremity of the larger pair of tentacles being minute and often difficult to distinguish; it is even doubted by some authors whether these specks are organs of vision at all. As in the rest of the family the head and tentacles are wonderfully retractile, being completely drawn within the skin when the animal is at rest, with the body foreshortened into a lump, protected almost vertically in front by the shield. The shield is finely granulated, and encloses beneath the posterior portion a few calcareous grains, which are sometimes isolated, sometimes rudely agglomerated. The edge of the foot, when expanded, shows a neatly-defined yellowish or orange border, rayed with black lines.

Arion ater has a wide range of habitation throughout the temperate and north temperate regions of the Old World. It has not as yet, according to Binney, been transported to America; but a species very nearly allied to it, *A. foliolatus*, Gould, has been collected, near Discovery Harbour, Puget Sound. Though nocturnal in habit, *A. ater* may be occasionally seen in the daytime. Plants are its natural food, but it has not unfrequently been observed to devour earthworms.

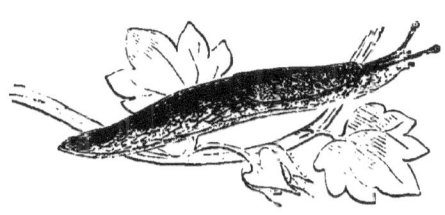

2. **Arion hortensis.** *Garden Arion.*

Animal; body slenderly elongated, from an inch to two inches in length, yellowish, sometimes grey or greenish-grey, dark-banded or deep blue-black, faintly banded at the sides, integuments wrinkled longitudinally with close-set leaflets, margin of the foot light grey or warm yellow; head dingy leaden blue; shield elongately oblong, rather coarsely granulated, faintly striped.

Shell; represented by calcareous grains more or less aggregated.

Limax flavus, Müller (1774), *Term. Hist.* vol. ii. p. 10, (not of Linnæus).
Arion hortensis, Férussac (1819), *Hist. Moll.* p. 65, pl. ii. f. 4, 6.
Limax fasciatus, Nillson (1822), *Moll. Suec.* p. 3.
Arion circumscriptus, Johnston (1828), *Edin. New Phil. Journ.* vol. v. p. 76.
Arion flavus, Férussac (1840), *Hist. Moll.* Supp. p. 96 B.
Limax intermedius, Normand (1852), *Descr. Lim.* p. 16.
Arion (*Prolepis*) *fuscus*, Moquin-Tandon (1855), *Hist. Moll.* vol. ii. p. 14. pl. 1. f. 28 to 30.

Hab.—Nearly throughout Europe. Boston, United States. (In gardens and roadside meadows, in damp places, under stones, or among dead leaves.)

A very much smaller species than the preceding, more delicately and more prettily coloured, occurring chiefly in the spring. The body is more slenderly elongated than that of *Arion ater*, and the wrinkled surface of the integuments is finer and partakes more of the form of crowded leaflets. The shield is proportionately lengthened, and it is rather more coarsely granulated. The typical colour of *Arion hortensis* is a dark greenish-grey banded at the sides with black, but it is extremely variable. M. Moquin-Tandon describes and names as many as eleven varieties inhabiting France, adopting the specific name of *fuscus*, on the ground that the species is the *Limax fuscus* of Müller. The descriptions of the Danish author are extremely short, and, unaccompanied with figures, are not easy to identify. The yellow variety (*Limax flavus*, Müller, *L. aureus?* Gmelin, *Arion flavus*, Férussac) referred to in our remarks on the genus is believed by some collectors to be a distinct species. Mr. E. J. Lowe informs me, that in specimens collected by him in the vicinity of Nottingham he finds the back more rigidly set, and the mucus orange-coloured.

Arion hortensis has been transported to Boston, United States, and become regularly acclimatized in the gardens and neighbourhood of that city. As an instance of the density of its integuments, this little *Arion* may be distinguished from *Limax* when trodden underfoot, a common fate of these mollusks, by its more leathery toughness.

Genus II. **GEOMALACUS**, *Allman*.

Animal; body lanceolate, subcylindrical, rather narrowly attenuated posteriorly, margin of the foot brown, transversely grooved; head with two pairs of tentacles, the upper pair much the longer, with eyes at their extremity; shield near the head, with the respiratory orifice on the right side in front.
Shell; solid, flat, subovate.

This interesting form of slug was discovered about twenty years since in comparative plenty in county Kerry, Ireland, and it has not been collected in the British Isles in any other locality. It differs in several anatomical particulars from its congeners. Its more obvious peculiarities consist in the pulmonary sac and overlapping shield being situated near the head, as in *Arion*, with the respiratory orifice still more in front; while a solid flat subovate shell is developed within the shield, as in *Limax*. The body, including the shield, is comparatively smooth. Forbes and Hanley, and Mr. Jeffreys, describe the back as being keeled, but one of the most important characters of a negative kind, in which *Geomalacus* resembles *Arion* more than *Limax*, is, that it is not keeled.

A year previous to the discovery of *Geomalacus* in Ireland, M. Morelet described under the name *Limax anguiformis*, a slug of very similar character collected by him in Portugal. He makes no mention of any shell.

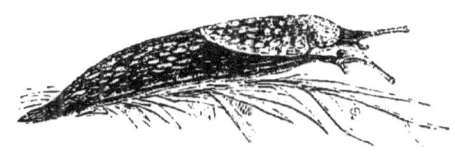

1. Geomalacus maculosus. *Spotted Geomalacus.*

Animal; about two inches in length, copiously mottled with yellowish or white spots.

Shell; solid, flat, subovate.

Geomalacus maculosus, Allman (1846), *Ann. and Mag. Nat. Hist.* vol. xvii. p. 297. pl. 9. f. 1 to 3.

Hab.—County Kerry, west Ireland (among moss and other plants, at the shady bases of moist rocks); Andrews.

The specific characters of *G. maculosus*, consists in its being everywhere conspicuously mottled with yellow or white spots upon a dark ground. Respecting its habits, Mr. Andrews, the discoverer, says:—

"On the rocks of Oulough, near Lake Carogh, to the south of Castlemain Bay, within a limited circuit, and at a distance of about fifty yards from the water, the *Geomalaci*, on a misty or showery day, may be noticed quiescently stretched, their richly maculated character being strikingly conspicuous. On what they feed I know not; I never could detect them in the eating mood. At the little Glen of Limnavar, on similar rocks at the same range from the water, I again met with the *Geomalaci*, particularly a white variety, but more sparingly than at Oulough."

Professor Allman, to whom we are indebted for the first description of *Geomalax*, remarks that the habits of this mollusk are somewhat curious. "It possesses a singular power of elongating itself, so as at times to assume the appearance of a worm. By this means it can insinuate itself into apertures which we could scarcely conceive it possible to enter. This curious property indeed was very nearly the cause of my losing the first, and only specimen I had seen. I had placed the mollusk as I supposed securely in a botanical collecting box, when to my surprise, I found shortly after, that it had transgressed the limits I had assigned it. The creature not liking its confinement, had insinuated itself beneath the lid."

Genus III. **LIMAX**, *Linnæus*.

Animal; body cylindrically elongated, more or less tapering, rounded on the back, sometimes keeled throughout, sometimes keeled only towards the tail, integuments variously wrinkled, furrowed, and veined; head with two pairs of tentacles, of which the upper pair is much the longer, bearing minute eyes at their bulbous extremity; shield sometimes finely granulated, sometimes concentrically striately wrinkled, enclosing a rudimentary shell, respiratory orifice at the right margin, more or less posterior to the centre.

Shell; an oval agglomerate, or a symmetrically-developed valve, increasing in concentric order from an umbonal nucleus, covered by a thin membranaceous epidermis.

The slugs of the *Limax* type differ from those of *Arion*, in having the pulmonary sac situated rather more removed from the head, with the respiratory orifice behind, instead of being in front, of the centre of the right margin of the shield; and sometimes it is quite at the posterior extremity. Among other generic differences it may be noted that the body of the creature is always more cylindrical and tapering, it is always more or less keeled, and the rudimentary agglomerate of calcareous particles assumes the form of an oval plate passing into an oblong square shell, developed in regular concentric order from an umbonal nucleus, and covered by a membranaceous epidermis. In colour the *Limax* is as inconstant as the *Arion*, and presents greater variety of marking.

We have eight species of *Limax* recorded as British, but two of them, *L. brunneus* and *tenellus*, are doubtful. The first is not improbably the young of a dark variety of *L. agrestis*, while the second is founded on a single specimen collected many years ago, of which the specific value has not been confirmed by subsequent research. *L. brunneus*, though reported by M. Bouchard-Chantereaux in 1838 to be breeding abundantly in the vicinity of Boulogne, is included by M. Moquin-Tandon, the highest and latest authority on the subject, along with *L. tenellus* in his list of 'Espèces Incertaines.'

There are, then, six undoubted species of *Limax* in the British Isles, all continental types, affording good distinguishing peculiarities of form, general marking, and habit. The series embrace two very obvious divisions, *L. Sowerbyi* and *gagates*, constituting the sub-genus *Milax* of Gray, *Amalia*, of Moquin-Tandon, have the shield granulated like that of *Arion;* and, like *Arion* the rudimentary shell which it encloses is little more than an irregular agglomerate of calcareous particles. It is, however, more constant, of a more regular ovate form, and in developing a rudimentary nucleus, presents a nearer approach to shell-structure. Another peculiarity of this division of the *Limaces*, more conspicuous in appearance, but of less physiological value, is the presence of an erect keel, along the whole length of the back, reaching from the shield to the extremity of the tail; while, in the species which follow, the back is keeled in a less prominent manner only towards the tail. The mantle, here designated the shield from the peculiarity of its form and office, is that part of the animal to which the conchologist will probably direct his attention—the shell, of which we have so nicely graduated a development in the genera of this family, being no unimportant part of the animal.

In the remaining four species of the genus, *L. agrestis, marginatus, flavus*, and *cinereus*, forming the subgenus *Eulimax* of Moquin-Tandon and Gray, the shield is striated with wrinkles which are disposed in regular concentric order; and the shell becomes, in its fullest development in the last-named species, an oblong square valve, constructed of symmetrical additions of growth around an umbonal nucleus. An epidermis of thin horny tissue is also formed.

We have said that our six undoubted species of *Limax* are all continental types. They are the only satisfactorily established species in Europe. Not that the continental conchologists have been backward in describing species. No less than forty have been recorded as natives of Europe, the greater portion of them, according to Dr. Grateloup, living in France. Yet M. Moquin-Tandon, in his recent work on the terrestrial mollusks of that country, reduces the number to eight, and two even of these are at least doubtful, *L. Corsicus*, which is said to differ from *L. cinereus* in a character by no means constant, and *L. Alpinus*, which M. Moquin-Tandon says he never saw but once, and which De Blainville pronounced to be also *L. cinereus*.

Not many extra-European habitats have been recorded. *L. gagates* and *flavus* have been collected in Madeira, and *L. agrestis, flavus,* and *cinereus* in Syria. *L. agrestis* and *flavus* have been transported to Boston, New York, Philadelphia, and other maritime cities of the United States, and are breeding abundantly to a distance of nearly a hundred miles inland. The North Americans have two species of their own; and two species have been described by D'Orbigny from South America. Two species have also been described by the same author, from the Canary Islands. Not many intertropical *Limaces* have been described, though doubtless many exist. Mr. Cuming informs me that he saw many at the Philippine Islands, but did not preserve specimens. MM. Quoy and Gaimard describe a species from New Zealand, one from Mauritius, and one from Ascension Island; and two are described by Dr. Krauss from South Africa.

To the remarks already made on the habits of the *Limacinea* generally, it may be added that *L. agrestis* has a partiality for the flower and kitchen gardens, *L. marginatus* for trees, and *L. flavus* for cellars, caves, and walls. All more or less spin mucous threads, and will let themselves down from shrubs or other objects not agreeable to them for food or habitation.

The species of *Limax* are :—

1. **Sowerbyi.** Animal tawny yellow, tessellated with dusky-brown, back keeled throughout, shield granulated. Shell, an unsymmetrical concretion.

2. **gagates.** Animal dull grey, back keeled throughout, radiately furrowed and veined, shield granulated. Shell, an unsymmetrical concretion.

3. **agrestis.** Animal yellowish-grey, red-mottled, back keeled towards the tail, shield concentrically striated. Shell flat.

4. **marginatus.** Animal gelatinous, livid grey, black loop-banded, back keeled towards the tail, shield concentrically striated. Shell oval, concave.

5. **brunneus.** Animal dark dingy brown, back keeled towards the tail, shield concentrically striated.
6. **tenellus.** Animal pale dull yellow, transparent, shield concentrically striated.
7. **flavus.** Animal yellowish, speckled with grey, keeled towards the tail, shield concentrically striated, spotted. Shell concentrically enlarged from an umbonal nucleus.
8. **cinereus.** Animal yellowish-ash, streaked and spotted with black, body tapering, keeled towards the tail, shield concentrically striated. Shell concentrically enlarged from an umbonal nucleus.

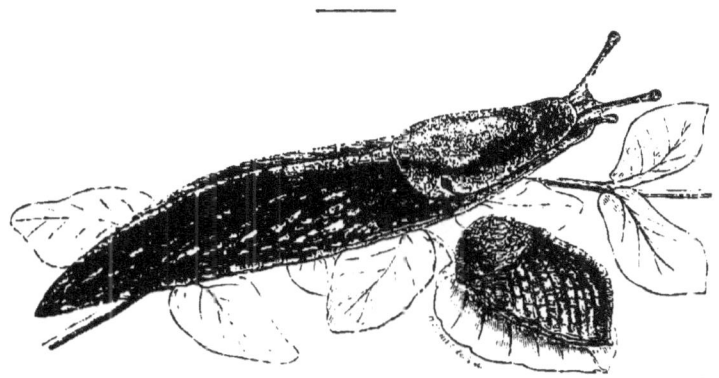

1. **Limax Sowerbyi.** *Sowerby's Limax.*

Animal; body oblong, linearly grooved, conspicuously keeled throughout the back, tawny yellow, tessellated with dusky brown, keel pale bright yellow, head ash-grey, tentacles livid blue; shield moderate-sized, oblong, widened and truncated behind, granulated, speckled, respiratory orifice subposterior.

Shell; small, often rather obscure, rude, irregular, sometimes flat, sometimes thickened, and convex.

Limax marginatus, Draparnaud (1805), *Hist. Moll.* p. 124. pl. 9. f. 7. (not of Müller).
Limax Sowerbii, Férussac (1821), *Hist. Moll.* p. 96. pl. 8 D. f. 7, 8.
Limax carinatus, Leach, (1831), Turton, *Man.* p. 115. pl. 3. f. 17.
Milax Sowerbii, Gray (1855), *Cat. Brit. Mus.* p. 175.
Hab.—Central Europe. Britain (in gardens).

This and the following species, constituting the subgenus *Milax* of Gray, *Amalia* of Moquin-Tandon, are distinguished from the other *Limaces* by two very obvious characters. The back of the creature is conspicuously keeled throughout its entire length, and the shield is finely granulated, like that of *Arion*, while it encloses merely an irregular calcareous button, for it can scarcely be called a shell, which is sometimes flat and partially membranaceous, sometimes rudely thickened and convex. The general colour of the body of *L. Sowerbyi*, which is linearly grooved, is a tawny-yellow, faintly tessellated with dusky brown, and the keel, which stands erect upon the back, especially when the animal is in a state of rest as represented in our vignette, is a pale bright amber. The head is ash-grey with the usual dark lines upon the neck, and the tentacles are of a dull livid blue. The shield is of a moderate size, almost short in comparison with that of *L. gagates*, being rather abruptly widened and truncated behind.

Limax Sowerbyi is not uncommon in gardens and shady places in the neighbourhood of London, and some other cities. The specimen from which our figure is drawn was taken by the artist, Mr. Jewitt, in the neighbourhood of Hampstead. Dr. Gray mentions that like many of its congeners, it will feed on animal food and even devour the dead remains of an individual of its own species, leaving only the skin of the back.

Mr. Jeffreys refers this species to Müller's *Limax marginatus*. Draparnaud, it is true, has acted on that opinion, but only in doubt. It is, as correctly indicated by Forbes and Hanley, the *L. marginatus* of Draparnaud but not of Müller. The true *L. marginatus* of Müller, it can hardly be doubted, is the slug described by M. Bouchard Chantereaux as *L. arborum*. It cannot be supposed that *L. arborum* was not known until he collected it in comparative plenty in 1838, in the neighbourhood of Calais. M. Bouchard Chantereaux named it after its habit of living on trees. Mr. Clarke, who collected it abundantly in Ireland, and has figured it in different stages of growth, says "I have seldom found this slug elsewhere than on trees." Müller, when describing the habits of his *L. marginatus*, says, "In Fago, vulgaris," and dwells particularly on the dark loop band which is characteristic of the species, "cinereus clypeo utrinque striga obscura. Striga clypei in omnibus nota constans. Juniores et adulti iisdem coloribus,"—as shown in Mr. Clarke's figures.

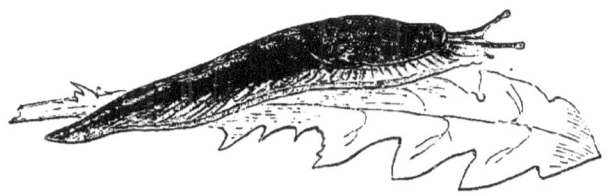

2. Limax gagates. *Agate Limax.*

Animal; body oblong, tapering towards the tail, conspicuously keeled throughout the back, dull grey, becoming colourless towards the margin, radiately furrowed with anastomosing veins, head and tentacles grey; shield oblong, continuously ridged, truncated, granulated, dark, respiratory orifice sub-central.

Shell; oval, small, thick, irregularly convex.

Limax gagates, Draparnaud (1801), *Tabl. Moll.* p. 100. *Hist. Moll.* pl. 9. f. 1, 2.

Milax gagates, Gray (1855), *Cat. Brit. Mus.* p. 174.

Limax (Amalia) gagates, Moquin-Tandon (1855), *Hist. Moll.* vol. ii. p. 19. pl. 2, f. 1 to 3.

Hab. Central, Southern, and Western Europe, Madeira, Ireland, Isle of Man, Portland Island, Isle of Wight, Torquay, Guernsey, Scotland, (by the road-side or at the base of old walls).

The *Agate Limax*, so named by Draparnaud from its dark agate-like shield, has been long known on the Continent, but in Britain, excepting a single record of its discovery lately in Scotland, it has only been observed in Ireland, in the Isle of Man, and in the few southern English localities recorded above. Dr. Gray remarks in the late edition of his Manual that *L. gagates* is very likely only a variety of *L. Sowerbyi*, but the characters by which it is distinguished from that species are by no means unimportant. The body, though as conspicuously keeled, is more slender and tapering towards the tail, and the integuments instead of being linearly grooved are smooth and rayed at the sides with linear veins. The shield is more ovate and continuously ridged, while the respiratory orifice is nearer the centre. In colour the body is of a dingy leaden hue, fading almost to white at the margins, and often speckled, and the shield is darker still.

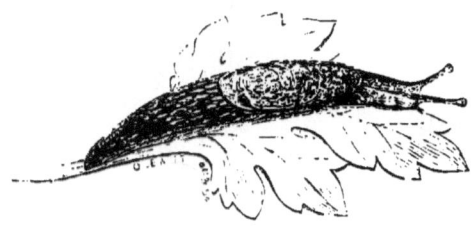

3. Limax agrestis. *Field Limax.*

Animal; body cylindrically oblong, rounded on the back, keeled towards the tail, reticulately wrinkled, yellowish grey, sometimes faintly, sometimes darkly mottled with red-brown, pale semitransparent at the margin, head yellowish with a dusky line running towards each tentacle; shield oblong, rather large, concentrically striately wrinkled, respiratory orifice posterior.
Shell; oval, very small, almost flat.

Limax agrestis, Linnæus (1758), *Syst. Nat.* 10th edit. p. 652.
Limax filans, Hoy (1790), *Trans. Linn. Soc.* vol. i. p. 183.
Limax sylvaticus, Draparnaud (1805), *Hist. Moll.* p. 126. pl. 9. f. 11.
Limax bilobatus, Férussac (1819–21), *Hist. Moll.* p. 74. pl. 5. f. 10.
Limax salicium, Bouillet (1836), *Moll. terr. et fluv. Auvergne*, p. 18.
Limax tunicata, Gould (1841), *Report Invert. Massachusetts*, p. 3.
Limax (Eulimax) agrestis, Moquin-Tandon (1855), *Hist. Moll.* vol. ii. p. 22. pl. 2. f. 18 to 22, and pl. 3. f. 1, 2.

Hab. Europe, Madeira, United States (in fields, woods, and gardens).

This little species of *Limax*, often called from the opake character of its mucus, the Milky Slug, is well known as the most prolific and mischievous of its class. It propagates with marvellous rapidity and abundance, feeding voraciously on the leaves of vegetables, and proving especially destructive among the newly bedded-out plants of the florist. An hour or so after sunrise *Limax agrestis* may be seen in the flower and kitchen gardens by hundreds, but by the time the sun is fairly up, not one is to be found; all have disappeared until twilight, and the dews rouse them to life and reaction. It is about an inch to two inches in length, keeled towards the tail, of a yellowish-grey colour, more or less darkly mottled with reddish-brown. The shield in this and the remaining species of the genus, loses the granulated chagrined character. When the animal is in motion, the shield first smoothens, and then contracts into concentric stria-like wrinkles with the dilatation and contraction of the body; and the shell is more symmetrically developed.

Limax agrestis has been transported to the United States, and breeds abundantly in the vicinity of the principal maritime cities.

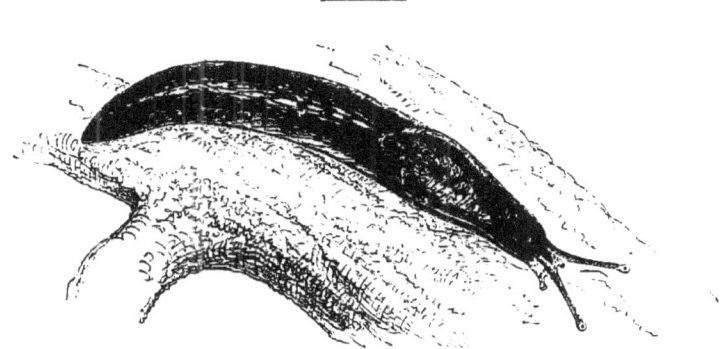

4. **Limax marginatus.** *Bordered Limax.*

Animal; body gelatinous, moderately elongated, rounded on the back, keeled towards the tail, livid grey, marked with pale anastomosing wrinkles, striped with a loop of two distant longitudinal bands; shield oblong, rounded in front, widening and then sloping to a point behind, finely concentrically striately wrinkled, banded with a dark loop like the body; respiratory orifice small, posterior, arched over by the dark band.

Shell; oval, thin, slightly concave, silvery white.

Limax marginatus, Müller (1774), *Verm. Hist.* p. 10. (not of Draparnaud).
Limax hortensis, Férussac (1822), pl. 8A. f. 4.
Limax arborum, Bouchard-Chantereaux (1838), *Cat. Moll. terr. et fluv. Pas-de-Calais*, p. 164.
Limax salicetum, Bouillet (1836), *Cat. Moll. terr. et fluv. Auvergne*, p. 18.
Limax glaucus and *arboreus*, Clarke (1843), *Ann. and Mag. Nat. Hist.* vol. xii. p. 334. pl. 11. f. 4 to 10.
Limax scandens, Normand (1852), *Desc. Lim.* p. 6.
Limax (Eulimax) arborum, Moquin-Tandon (1855), *Hist. Moll. terr. et fluv.* vol. ii. p. 24.

Hab. Central Europe, Britain (mostly on trees).

The most obvious distinguishing characters of *Limax marginatus* are its livid glaucous colour and its gelatinous substance, often so translucent that the internal organization may be traced. Typical specimens, such as the one represented in our figure, are moreover characterized in all stages of growth by two conspicuous loops of black bands, one on the shield, the other on the back reaching to the extremity of the tail. In examples of the species in a young state, with drawings of which, made as long back as 1828, Mr. Berkeley has obligingly favoured me, the black loop-like bands are still a prominent feature. The shield is rounded in front, and widening over the pulmonary sac, terminates posteriorly in an angular slope.

The habits of *L. marginatus* are variously described by different observers. It is most frequent on beech, ash, and walnut trees. Mr. Thompson says: "After rain, I have seen them in numbers, gliding down the smooth bark of the beech from feeding on the higher foliage, their bodies appearing between the light like pellucid jelly." Forbes and Hanley mention having collected the species near Connor Cliffs in the west of Ireland, in a rocky locality entirely devoid of trees; and Mr. Clarke notices having met with it among the ruins of a chapel in Connemara, crawling in considerable numbers on the walls and tombstones.

The shell is variously developed: sometimes it is of a thin silvery white substance partially membranaceous, sometimes rather thick, slightly convex above, concave below.

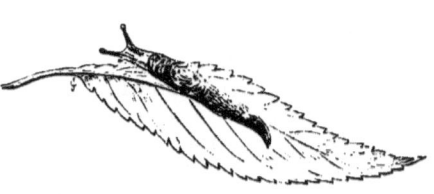

5. Limax brunneus. *Brown Limax.*

Animal; body tapering, rounded on the back, slightly keeled towards the tail, dark dingy brown throughout; shield oblong-oval, large, concentrically ridged and striated.
Shell; —— ?

Limax brunneus, Draparnaud (1801), *Tabl. Moll.* p. 104.

Hab. North-east of England (in damp woods); Johnston, Alder. Pas-de-Calais, France (common on the banks of rivers and brooks, and under moss and stones); Bouchard-Chantereaux.

M. Bouchard-Chantereaux ascribed to *Limax brunneus* of Draparnaud, a small slug varying from an inch to an inch and a half in length, living in comparative plenty in the neighbourhood of Boulogne on the banks of running streams, and in damp places among moss and under stones. Dr. Johnston, of Berwick-on-Tweed, and Mr. Alder, of Newcastle, observed the species also in their own locality, and it is from Mr. Alder's drawing, reduced to the natural size, that our figure is taken. Since the publication of these details, ten years ago, by Forbes and Hanley, no further observation has been recorded, and M. Moquin-Tandon includes *Limax brunneus* in his list of 'Uncertain Species.' The question presents itself,—Is *L. brunneus* anything more than the young of a dark variety of *L. agrestis*? It is said to differ from that species not only in colour, but in having a transparent mucus. These differences are not of much specific value, and it is doubtful whether *Limax brunneus* can be retained as a species.

6. **Limax tenellus.** *Tender Limax.*

Animal; body slender, rounder on the back, compressed towards the tail, pale dull yellow, very transparent and lubricous, with an obscure band on each side of the shield and back, the tentacles being black; shield rounded behind, covered with fine concentric wrinkles.

Shell; ——?

Limax tenellus, Müller (1774), vol. ii. p. 11.

Hab. Northumberland (in a wood at Allansford, near Shortly Bridge); Blacklock. South of France, Férussac.

The foregoing characters, published in 1853, by Forbes and Hanley, are derived from a single specimen, collected in the above recorded locality by Mr. Blacklock, and drawn by Mr. Alder, of whose figure our wood-engraving is a copy reduced to the natural size. This is all that is known in England of the slug ascribed to Müller's *Limax tenellus*. It has been observed, according to Férussac, in the South of France, at Quercy, and in the neighbourhood of Montpellier; but nothing has been added to that observation, and *Limax tenellus* is included by Moquin-Tandon, along with *L. brunneus* in his list of 'Uncertain Species.'

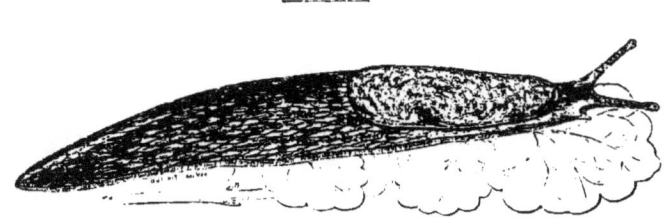

7. Limax flavus. *Yellow Limax.*

Animal; body cylindrically elongated, rounded on the back, keeled towards the tail, yellowish, speckled with grey, wrinkled and furrowed; head rather small, bluish, tentacles rather short; shield oblong-ovate, rounded at the extremities, concentrically striately wrinkled, ash-grey, variegated in concentric order with yellow spots, respiratory orifice posterior to central.

Shell; somewhat squarely ovate, slightly concave, striated concentrically from an umbonal nucleus.

Limax flavus, Linnæus (1758), *Syst. Nat.* 10th edit. p. 652.
Limax variegatus, Draparnaud (1801), *Tabl. Moll.* p. 103.
Limacella unguicula, Brard (1815), *Hist. Coq.* p. 115. pl. 4. f. 3, 4, 11.
Hab. Europe, Syria, Madeira (in caves and damp places in woods, and about houses in cellars and wells).

The ordinary colouring of this moderately keeled and strikingly marked species is a dull yellow, variegated on the back with ash grey, so as to leave a copious sprinkling of yellow dots and spots. On the shield the spots range in the concentric order of the striæ, and present a characteristic feature by which the species may be readily recognized. In habit *L. variegatus* inclines rather to damp situations about houses, in cellars and wells, and in woods; it prefers to dwell on the damp ground or in caves.

The shell is slight, and somewhat squarely ovate, inclining to the more developed form of that of *L. cinereus*, the lines of growth being added concentrically from an umbonal nucleus.

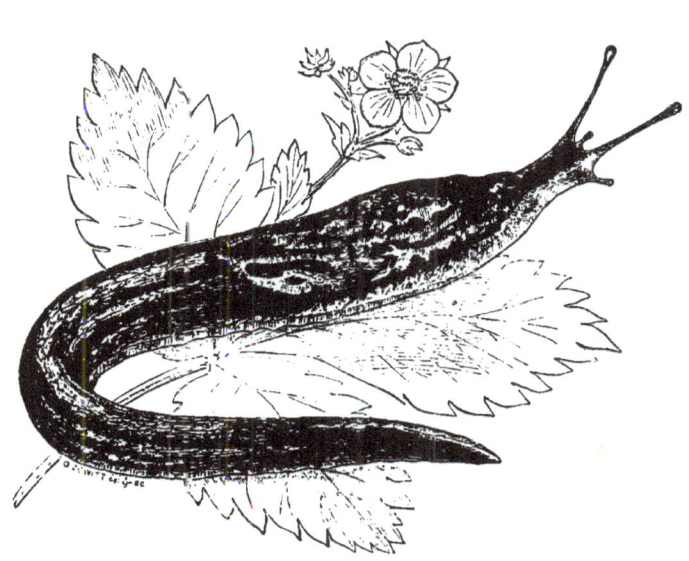

8 **Limax cinereus.** *Ash Limax.*

Animal; body elongated, rounded on the back, tapering and keeled towards the tail, yellow-ash, conspicuously streaked and leopard-spotted with black, head pinkish-grey mottled, tentacles yellowish-grey, long, slender; shield oblong, rounded in front, inclined to triangular behind; respiratory orifice at the posterior end of right margin, striated concentrically from an umbonal nucleus, outer edge membranaceous.

 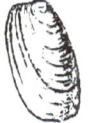

Shell; oblong-square, large, depressly concave.

Limax cinereus, Lister (1674), *Phil. Trans.* vol. ix. p. 96. *Hist. Anim. Angl.* p. 131. pl. 2. f. 17.
Limax maximus, Linnæus (1758), *Syst. Nat.* p. 108.

Limax cinereus, Müller (1774), *Verm. Hist.* vol. ii. p. 5.
Limax maculatus, Leach (1815), MS. in *Brit. Mus.*
Limax antiquorum, Férussac (1819), *Hist. Moll.* p. 68, pl. 4. f. 1 to 8.
Limax Alpinus, Férussac (1822), *Tabl. Syst.* p. 21. *Hist. Moll.* pl. 4A f. 5 to 7.
Limax Valentiana, Férussac (1822), *Tabl. Syst.* p. 21. and *Hist. Moll.* pl. 8A. f. 5, 6.
Limax Cyrenæus, Companyo (1837), *Rapport Moll. Pyren.* in *Bull. Phil. Perpignan*, vol. iii. p. 88.
Limax (Eulimax) maximus, Moquin-Tandon (1855), *Hist. Moll.* vol. ii. p. 28. pl. 4. f. 1 to 8.
Limacella parma, Brard (1815), p. 110. pl. 4. f. 1, 2, 9, 10. Shell.

Hab.—Europe, Syria, Madeira (about outhouses and in cellars, in gardens and on walls, under hedges and logs of wood).

This very handsome leopard-spotted slug was the first to attract the attention, nearly two centuries ago, of our famous countryman Dr. Martin Lister, of Oxford, in whose 'Historia Animalium Angliæ' it is accurately figured and described. It is the largest of the tribe, conspicuous for its elongately tapering eel-like form, and is plentifully distributed throughout Europe about outhouses and in cellars, under hedges and logs of wood, in gardens and on walls. The shield, it will be observed, is rather ample, rounded in front, and terminating in a sloping angle behind. The surface of the shield is striately wrinkled in a concentric manner, and the respiratory orifice is at the posterior extremity of the right margin. Below the shield, the body is longitudinally wave-wrinkled throughout, being at first rounded on the back, and then gradually keeled. The tentacles are unusually long, partaking of the slenderness of the body.

The spotted variety, from which our figure is drawn, a specimen collected for its size and beauty, by Mr. Metcalfe, at the little island of Herm, may be regarded as the type of the species; I have, however, collected it as large and as beautifully spotted in my own garden. While examining some living specimens, I met with a curious instance of mucus-spinning. In the course of the evening, one escaped from a glass on the mantelpiece, and having made its way to the edge of the marble, was observed to let itself down by a thread of mucus a distance of from three to four feet, into the fender. The time occupied was about five minutes.

The shell of *L. cinereus* is the largest and most symmetrically developed of the genus; it is of an oblong square form, regularly concentrically enlarged from an umbonal nucleus, depressly concave, tinged with a blush of pink, membranaceous at the outer edge.

M. Moquin-Tandon characterizes the species as having the concentric striæ of the shield disposed into two separate masses, a mass on each half of the shield, and remarks that De Férussac's *L. alpinus* differs from it in having the concentric striæ in one mass only, like the other striated-shield Limaces. I have not been able to detect this difference, nor is it noticed by any other writer. He gives sixteen varieties occurring in France, and they are little less numerous in Britain. The animal is rather sluggish in its movements, and when irritated dilates its shield, and secretes abundantly a whitish mucus.

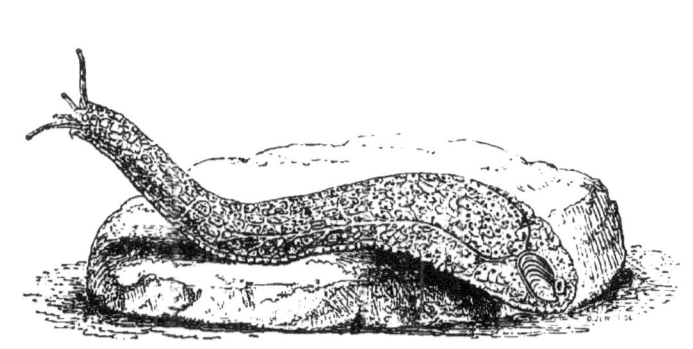

GENUS IV. **TESTACELLA**, *Cuvier*.

Animal; body semi-cylindrical, swollen posteriorly, tapering and susceptible of elongation towards the head, integuments wrinkled or linearly veined; head with two pairs of tentacles, the upper pair much the longer with eyes just below the extremities, tentacles rather approximating at the base, then divergent; shield combining with the general integuments; pulmonary sac at the posterior end of the body, covered by a lobed slightly extensile mantle, which secretes an external shell, beneath the right side of which is the respiratory orifice.
Shell; ear-shaped, with a subspiral umbonal nucleus, covered by a thin horny epidermis.

In the year 1740, M. Dugué, a gentleman residing at Dieppe, communicated to the Academy of Sciences of Paris, the following interesting discovery of the mollusk which has received the name of *Testacella* :—"There is in this city a garden in which a slug has been found that is unknown to the gardeners of the country. It is from eighteen to twenty lines long, and is somewhat of the form of the red slugs which crawl upon the earth and have no shell. It burrows in the soil like the earthworm, and goes out only at night. It carries upon its tail a shell like a nail at the end of the finger, and is as hard. The entire animal is so hard that one can scarcely cut it with a knife. It has been kept in a pot with some earthworms three or four inches in length and as stout as a pen; it feeds on them though much less strong in appearance. It occupied from four to five hours in swallowing one entirely, but did not risk during this long time losing its prey; when once it has seized a worm by one end, there is no chance of escape. It deposits its eggs in the ground perfectly round at first, and nothing more than a little pellicule filled with a viscous humour, but at the end of fifteen days, or a little more, the humour thickens, the round form changes to an oval, and the slug hatches it like a chicken."

Linnæus had not yet matured his system of nomenclature, and it was not until sixty years later, that the *Testacella* was named by Cuvier from specimens collected by M. Faure Bignet, at Crest in the department of Drôme. Subsequently it has been collected in the centre and South of England and Ireland, in the Channel Islands, in Spain, Portugal, Syria, Algeria, the Canary Islands, and the West Indies; and it has been abundantly bred at Bordeaux for experimental purposes, by M. Gassies, who in conjunction with M. Fischer, has published a very admirable monograph of the genus.

The interest attaching to *Testacella* in a systematic point of view, arises from the circumstance of its being an intermediate link between the two families of *Limacinea* and *Colimacea*. The pulmonary sac which in *Limax* is in the front part of the body, and in *Parmacella*, a genus inhabiting the South of France, is in the middle, is in *Testacella* situated at the hinder extremity; and in place of an integumentary shield it is provided with a delicate mantle secreting an external ear-shaped shell. The shell is composed of regular concentric layers, commencing from a subspiral umbonal nucleus, and presents the first indication of the convoluted plan of growth, which is developed in *Vitrina* and *Helix*, and becomes the typical structure of the class. But *Testacella* possesses essential characters of its own. It is a ground slug of strictly carnivorous habits, penetrating the soil to the depth of two to three feet, or more, and preying voraciously upon earthworms. There is, moreover, as already noticed, no shield in *Testacella*, although in *Vitrina* a modification of the *Limax* shield appears along with a more advanced convolution of the shell.

It was stated by De Férussac that the mantle of *Testacella* though ordinarily concealed beneath the shell, is susceptible of being extended over the whole body of the animal. "When the *Testacellæ* are surprised by drought, they envelope themselves entirely with their mantle. The mantle, which is extremely gelatinous, preserves the animal in the dry soil in a state of freshness and humidity, which appears indispensable for it to live. The body is, it is true, contracted, but it augments in thickness although it diminishes in length." This statement is regarded with incredulity by Moquin-Tandon, but Mr. Woodward has given a wood-engraving of this singular phenomenon as observed by Mr. Cunnington in a field near Devizes, in a supplementary note to his 'Manual of the Mollusca,' accompanied by the following interesting description:—" During winter and dry weather the *Testacella* forms a sort of cocoon in the ground by the exudation of its mucus. If this cell is broken, the animal may be seen completely shrouded in its thin opaque white mantle, which rapidly contracts until it extends but a little way beyond the margin of the shell." There is, however, reason to believe that this thin white cocoon is not an extension of the mantle, but a pellicle of slimy mucus. Another phenomenon in the structure of the mantle may be seen in our figure of *T. Maugei*, engraved from a drawing by the Rev. M. J. Berkeley, in which it is lobed externally on either side for the lateral embrace of the shell.

Testacella is recorded from the West Indies, at Guadaloupe and Martinique, but we have not been able to procure specimens of *T. Antillarum* and *Matheronii* to test their specific value. The genus is not known in the United States or in any other part of America. The shell occurs in a fossil state in the lacustrine deposits of the South of France, chiefly in the vicinity of Montpellier.

The *Testacellæ* produce calcareous eggs of large size, as symmetrically formed as those of a bird. A specimen before me at this moment, from the collection of Mr. Cuming, was taken by him from the living animal while in the act of ejecting it from the ovary.

The species are :—

1. **haliotidea.** Animal comparatively smooth, linearly grooved in a veined manner, yellowish. Shell small, subtriangular, depressed.
2. **Maugei.** Animal more rugose, linearly grooved at the sides, olive mottled with brown. Shell larger, subquadrately ovate, convex.

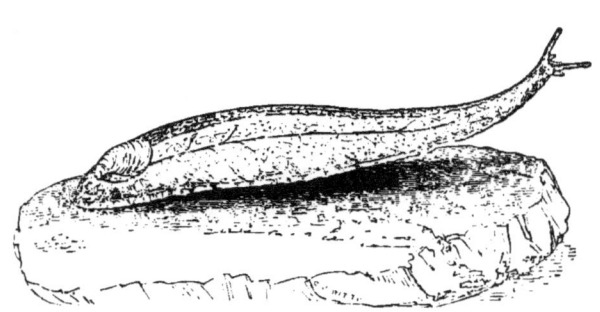

1. **Testacella haliotidea.** *Haliotis-like Testacella.*

Animal; body yellowish, finely veined with linear grooves, speckled with dots and punctures.

Shell; triangularly ear-shaped, depressed, covered by a dark fibrous epidermis.

Testacella, Cuvier (1800), *Leçons d'Anat. Comp.* vol. i. tabl. 5.
Testacella haliotidea, Draparnaud (1801), *Tabl. Moll.* p. 99.
Testacella Europæa, Roissy (1805), *Buff. de Sonn.* vol. v. p. 252. pl. 53. f. 8.
Helix subterranea, Lafon-du-Cujula (1806), Desc. du Lot-et-Garonne, p. 143.
Testacella Galliæ, Oken (1815), *Lehrb. der Naturg.* p. 212. pl. 9. f. 8.
Testacella scutulum, Sowerby (1823), *Genera of Shells*, f. 3.
Testacella bisulcata, Risso (1826), *Hist. Nat. Europ. Mérid.* vol. iv. p. 58.
Testacella Companyonii, Dupuy (1846), *Hist. Moll.* p. 47. pl. 1. f. 3.
Testacella Galloprovincialis, Grateloup (1855), *Limac.* p. 15.
Testacella Canigonensis Limax, Grateloup (1855), *Limac.* p. 15.
Testacella Medii-Templi, Tapping (1856), *Zoologist*, vol. xiv. p. 5105.

Hab.—Canary Islands, Channel Islands, Spain, Algeria, Central and South of France, South of England and Ireland.

The published drawings of *T. haliotidea* vary so much in appearance, that they seem to represent different species, but the truth is, they are made from specimens in different states, and under different conditions. Our figure at the head of the genus is from a drawing of an old and wrinkled individual, whilst that represented opposite is from a young specimen in which the skin is expanded. The animal is always of a light yellowish colour compared with *T. Maugii*, which is of a darker mottled-brown, and the linear veining is much interspersed with dark dots. The shell is small, characteristically triangularly depressed towards the umbo. The principal varieties of the species are those named *T. scutulum*, in which the shell is more than usually attenuated in front, *T. Companyonii* which has the shell large, and *T. bisulcata* in which the shell is so roughly grown as to show two irregular grooves. M. Moquin-Tandon remarks that the animal of the last-named variety has generally a smaller more elongated body of richer colour. Having preserved and studied the character of living specimens sent to him from the environs of Grasse, he came to the conclusion that they were scarcely distinguishable from the original type of the species.

MM. Gassies and Fisher adopt M. Dupuy's *T. Campanyonii* as a distinct species, but they confess to not having seen it for a very long time, and are unable to recall to their recollection its distinctive characters. The grooved form, *T. bisulcata*, Risso, they consider ought to be maintained as a species, not only on account of a difference in the shell, but of the animal and of the form of the egg. So far as the shell is concerned, the grooves appear to be mere variations in the successive additions of growth.

A great deal has been written on the ferocity with which the *Testacellæ* prey upon earthworms, and the manner in which they will follow them in their subterranean windings. M. Gassies, who kept a vivarium of *Testacellæ* of all sizes and ages, gives a lively description of their cunning :—"When a *Testacella* has discovered the prey on which it wishes to make a repast, it moves stealthily to one side of the worm, with an indifference so complete, that one would suppose it had not observed it and disdained it; but suddenly it turns, and while the worm is twisting to the right and to the left, it lifts its head, withdraws its tentacles, dilates enormously its mouth, and throws itself upon its prey, enfixing it by a kind of suction. Contortions of the worm are necessarily the result of the wounds from the palate spines; it wrestles, but in vain. Retained by a multitude of barbs, its movements only serve to engage it more and hasten its passage into the stomach of its voracious enemy."

Originally a native of the Canary Islands, *T. haliotidea* has spread very generally in South-Western Europe, principally in the vicinity of maritime cities.

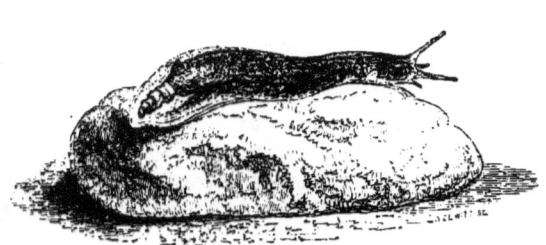

2. **Testacella Maugei.** *Maugé's Testacella.*

Animal; body tawny yellow, darkly mottled with brown, rugosely veined.

Shell; subquadrately oblong, ear-shaped, produced at the umbo, covered by a fibrous epidermis.

Testacella Maugei, Férussac (1819), p. 94, *Hist. Moll.* pl. 8. f. 10, 12.
Testacella Burdigalensis, Gassies (1855), Grateloup, *Limac.* p. 15.
Testacella Oceanica, Grateloup (1855), *Limac.* p. 15.
Testacella Canariensis, Grateloup (1855), *Limac.* p. 15.
Testacella episcia, Bourguignat (1861), *Moll. Alp. Marit.* p. 28.
Hab. Teneriffe, Portugal, Spain, South-West of England, and Ireland.

This species, a native also of the Canary Islands, has not been yet included in the British fauna, nor does M. Moquin-Tandon admit it into that of France. More than fifty years have elapsed since it was transported with plants into the nursery grounds of Messrs. Miller and Sweet, of Clifton, and it has become naturalized in several localities in the West of England. Mr. Norman, writing in 1859 to Mr. Jeffreys, says:—"I have had as many as five dozen sent to me alive at the same time." The animal is of a darker mottled colour than the preceding species, and the shell is of a more oblong compressed form, larger and more produced at the umbo.

Our figure is from a drawing made by the Rev. M. J. Berkeley, as long back as 1829, from a living specimen of rather small dimensions, collected by Mr. Sowerby, in which the shell is held in the embrace of two external lateral lobes; and I have observed the dried remains of a similar condition of the mantle in a specimen collected by Mr. Metcalfe.

With reference to this species in the cocoon, Mr. Jeffreys says, on the authority of some experiments made by Mr. Norman, "While in the encysted state, a thin white membrane, a development of the mantle, is extended from beneath the shell, and stretched over the back and sides of the animal. An admirably designed protective shield is thus formed, which checks evaporation from the surface of the body, and enables the flow of mucus, which is so essential to the life of the animal, still to course along the lateral canals, and thence be distributed through the branching channels over the entire surface of the body. When *T. Maugei* is removed from its cyst and the body moistened with water, the extended membrane is gradually contracted until it is entirely withdrawn beneath the shell." MM. Gassies and Fischer do not believe, apparently, in this separate organization of an enveloping mantle-membrane.

Family II. COLIMACEA.

Respiratory and visceral organs distinct from the main contractile mass of the body, coiled within a spiral shell. Eyes at the extremity of the tentacles.

The above definition applies to all our inoperculated air-breathing mollusks, except the thirteen *Limacinea* just described, and the four *Auriculacea* which follow. They are sixty in number, and are the outlying members of a fauna twenty times as numerous, which has its centre of creation in Southern Europe, and spreads over Asia in an easterly direction to the Himalayas. Of the four principal genera, *Helix, Clausilia, Pupa, Vertigo*, there are certainly eight hundred European to forty British species. The transition between the two families of *Limacinea* and *Colimacea* is shown in Britain by a solitary species, *Vitrina pellucida*, of which there are only six in Europe, but upwards of seventy in the more tropical parts of the eastern hemisphere. It is curiously intermediate in structure between *Limax* and *Zonites*, having the fleshy shield of the former along with the mantle-polished shell of the latter. With *Testacella*, which is an abnormal form of *Limacinea*, the relationship is more with a genus unknown to British soil, *Daudebardia*, in which the respiratory orifice is situated towards the posterior part of the animal, where it is covered by a horny spiral shell, not much unlike that of *Vitrina*. *Daudebardia* is a native of Central and Southern Europe and Syria.

Another genus of *Colimacea* is one of which at least a hundred and forty species have been described from different parts of the world, three of them British, *Succinea;* they are mostly natives of the western hemisphere, where *Vitrina* is unknown, except by a single small species in the United States, descended probably from an exported European individual. *Succinea* has a light, transparent, fusiformly coiled shell, largely inflated, and of a delicate amber colour. The next genus in the series, *Zonites*, has a regular heliciform shell, colourless and polished, arising from the action of a partially reflected lobe of the mantle. We have eight well-defined species in Britain; the foreign species, of which some thirty or forty have been noted, are as yet but imperfectly eliminated from *Helix*.

Of *Helix*, the most numerous and widely distributed of all *Colimacea*, upwards of two thousand species have been described. They inhabit all lands and all elevations within the range of molluscan life. The British *Helices* are twenty-four in number, mostly small. A few are pre-eminent in size and colour, and almost rival tropical species in brilliancy and marking. The fifth genus of the family, *Bulimus*, scarcely differs from *Helix* in the soft parts, but the animal is more arboreal in habit, and has a more elongately convoluted shell, with a more restricted range of habitation. Nine hundred species have been described in all, chiefly natives of the intertropical regions of both hemispheres, and some of them with shells of quite colossal dimensions. We have only three small species in Britain. In the United States there is no *Bulimus* within fifteen degrees of the same latitude. The three genera which follow, *Zua*, *Azeca*, and *Achatina*, represented in Britain by a single species each, are included by M. Moquin-Tandon under *Bulimus*. They are, it appears to me, very distinct. *Zua* has a highly polished, vitrified shell, and a geographical distribution peculiarly its own, the most extended both in space and elevation of all land mollusks. The single known species, for there is no satisfactory evidence of the existence of more than one, ranges throughout Europe and Northern and Central Asia, reaching from Siberia to Cashmere and Thibet, and it appears indigenous in most of the United States. *Azeca*, of which all the described species are probably referable to one, is peculiar in form, contracted and toothed in the aperture. The extra-British distribution of this genus is confined to Central Europe. *Achatina* belongs to a different category. Our minute acicular species is the northernmost member of an intertropical form of which there are two hundred and fifty species, of varying groups, all characterized, however, by an involute truncation of the columella of the shell.

The remaining genera of *Colimacea* are *Clausilia*, *Balea*, *Pupa*, and *Vertigo*, numbering about five hundred and fifty species, of which eighteen are British. Of *Clausilia* we have only four species, although it is an essentially European type. As many as three hundred and twenty species are known, nine-tenths of which are natives of Austria, Hungary, and the islands of the Grecian Archipelago; and the remainder range, in a gracefully developed state, in the more distant kingdoms of the East. The shell of *Clausilia* is always convoluted sinistrally, peculiarly constricted in the neck of the last whorl, and the animal possesses the faculty of closing itself in by means of a spoon-shaped appendage attached by an elastic filament to the columella. *Balea* is a European snail, having a turreted sinistral shell, without any exotic analogue. *Pupa* and *Vertigo*, concluding the series, were formerly united, but they are now shown to be distinct. In *Pupa* the lower tentacles begin to lessen and are reduced to mere protuberances. In *Vertigo* they disappear altogether, and the animal, agreeably with its change of habit in dwelling chiefly on the banks of lakes and rivers, partakes in this respect of the character of the water snails. The species of *Vertigo* are without exception smaller than any *Pupa*. We have only four *Pupæ* in Britain, to a hundred and fifty foreign species, nearly half of which are European. About forty belong to other more tropical Eastern countries, and the remainder to the New World, not less than thirty of them being stationed at the West Indies and the neighbouring mainland. Of *Vertigo* we have nine interesting minute species in Britain. The extra-British species of this genus are about seventy in number, agreeing pretty nearly in their distribution with the *Pupæ*, excepting that there are fewer in proportion in the West Indies, and more in the United States.

The British genera of *Colimacea* are:—

1. **Vitrina.** Animal carrying a lightly whorled, glassy shell, over which a lobe of the mantle is reflected, and in front has a fleshy shield with the respiratory orifice behind it to the right. Shell of three imperforate pellucid whorls.

2. **Succinea.** Animal carrying a thin, submembranaceous shell, head broad and obtuse, lower tentacles short. Shell oblong-

ovate, of three to four transparent amber whorls, the last of which is capaciously inflated.

3. **Zonites.** Animal carrying an umbilicated subdiscoidal shell, occasionally partially enveloped by a reflected lobe of the mantle, lower tentacles short, foot truncated behind. Shell of three to five colourless shining whorls.

4. **Helix.** Animal carrying an orbicularly coiled shell, not enveloped in any part by the mantle, foot lanceolate behind. Shell sometimes solid, brightly coloured, sometimes thin, horny, of four to six whorls.

5. **Bulimus.** Animal carrying a turriculate shell, upper pair of tentacles rather long, foot lanceolate behind. Shell of eight to nine elongately convoluted whorls.

6. **Zua.** Animal carrying an oblong glossy shell, upper pair of tentacles long, with the bulbous extremities rather lengthened. Shell of four to five transparent vitrified whorls obsoletely truncated at the end.

7. **Azeca.** Animal carrying a cylindrical chrysalis-shaped shell, upper tentacles slender. Shell attenuately contracted towards the base, of seven smooth whorls, of which the aperture is ear-shaped and toothed.

8. **Achatina.** Animal carrying a narrow hyaline-white shell, upper tentacles slender, eyeless. Shell of six smooth, acicularly convoluted, imperforate whorls.

9. **Clausilia.** Animal carrying a slenderly acuminated, sinistrally convoluted shell, in which it encloses itself by a clausilium, tentacles short. Shell of nine to twelve attenuately elongated whorls, mostly densely wrinkled, contracted and toothed at the aperture.

10. **Balea.** Animal carrying a thin, sinistrally convoluted shell, tentacles short, upper pair approximating. Shell of seven conically turreted, transparent, horny whorls.

11. **Pupa.** Animal carrying a small horny narrow-whorled shell, upper pair of tentacles short, lower pair very short. Shell of six to nine closely convoluted whorls, more or less toothed in the aperture.

12. **Vertigo.** Animal carrying a minute, tumidly whorled shell, upper tentacles rather short, no lower tentacles. Shell of four to six whorls, sometimes convoluted sinistrally, more or less toothed in the aperture.

Vitrina pellucida. (*Slightly enlarged.*)

Genus I. **VITRINA**, *Draparnaud*.

Animal; body elongated, carrying a lightly-whorled glassy shell, mantle produced in front into a conspicuous fleshy shield, with the respiratory orifice behind it to the right, beyond the respiratory orifice reflected on the lip of the shell is a separated lobed extension of the mantle reaching nearly to the apex, tentacles rather short.

Shell; subglobose, thin, pellucid, glassy, smooth, imperforate, composed of three rapidly increasing whorls.

The passage of affinity between the Slug and the Snail is illustrated in an interesting manner by this little mollusk, our figure of which is from a drawing made from the life, on a slightly enlarged scale, by Mr. Berkeley. The genus *Vitrina*, ranging throughout the eastern hemisphere, is represented by about eighty species, all producing a smooth greenish-olive glassy or horny shell of singular uniformity of character, except in some half-dozen species forming the genus *Peltella* of Webb and Van Beneden, in which the base of the shell is so imperfectly membranaceous as to break away in its attachment to the animal. Six are European, but only one of these, *V. pellucida*, is a native of the British Isles. Africa, south, east, and west, contributes fifteen species distinct from the European; Madeira has three species, of which two are of the *Peltella* section; the Philippine Islands have fifteen, Ceylon five, India eight, Burmah and Siam four, Borneo, Malacca, and Celebes four, Tasmania and Australia seven, and New Caledonia and the neighbouring isles three. In the New World, the transition between the non-spiral and the spiral-shelled mollusk is more abrupt. Beyond a single small species in the United States, so like our own little *V. pellucida* that it is only lately Dr. Gould has ventured to consider it distinct, there is no other record of the genus in its typical form. At Mexico and the Brazils, *Vitrina* is represented by quite a different type, *Simulopsis*, in which the shell is ribbed and more inflated; and, as it has a less vitrified polish, it is probably not enveloped to the same extent by the reflected lobe of the mantle.

What, then, is there in the structure of our solitary example of *Vitrina* which renders it of so much interest? The animal, it will be seen, possesses the shield and respiratory orifice of the slug along with the spirally whorled shell of the snail. The mantle, secreting a light glassy-horny shell of three rapidly enlarging whorls, is produced in front of the aperture in a thickened manner, extending conspicuously towards the head; and behind, on the right side, posterior to the respiratory orifice, it is lobed, so as entirely to cover the shell, and, by its action, keep the shell bright and shining. The action of this retractile lobe, when the animal is in motion, tends to give a highly vitrified polish to the shell; and it has been observed to move even when the animal is at rest. The shell of *Vitrina*, not only of the British but of all the numerous foreign species, is composed of the simplest glassy or horny membranaceous substance, more or less soft and yielding to the touch in the living specimen. After death, when the animal dries up, the shell, in the absence of the living body and its mucous secretions, becomes brittle.

For information of the habits of the genus, it is necessary to refer to the foreign as well as to the British species. Mr. Cuming, who collected *Vitrinæ* in abundance in the Philippine Islands, was particularly struck with their activity. He relates, that on placing them on the palm of his hand they kept jumping up, with sudden leaps, by the muscular action of the foot; and Mr. Benson mentions this habit in reference to an Indian species observed to have the power of springing several inches from the ground. Nilsson, the Swedish naturalist, describes some interesting experiments on incubation. Specimens, as related on his authority by Gray, caught at the end of January were kept in a bell glass, and on the 19th of February eggs were observed to have been deposited on putrescent leaves. The eggs were in little tufts of eight or nine, subpellucid, marked with a central opake white spot. In the beginning of March, the opake spot was not increased in size, but showed signs of slow movement, and on the 21st or 22nd of the month, the animals were excluded. When observed with the microscope, the animal was thought to be boring its way through the egg-shell, forming a hole out of which first the head and then the foot were protruded; when first hatched, both animal and shell were fully formed, the tentacles being retracted into the body.

There is a small genus of seven or eight species inhabiting some parts of Germany, Switzerland, North Italy, Sicily, and Syria, *Daudebardia*, which comes nearer than *Vitrina* to *Testacella*, but is more removed from the *Limax* type. It has no shield, and the pulmonary sac is at the posterior extremity of the body, covered by a little spiral *Vitrina*-like shell.

The only British species of *Vitrina* is—

1. **pellucida.** Shell of three imperforate, pellucid whorls.

1. **Vitrina pellucida.** *Transparent Vitrina.*

Shell; imperforate, depressly globose, thin, transparent, glassy, greenish or yellowish, whorls three to three and a half, rapidly increasing, moderately convex, smooth, shining, spire scarcely raised, suture indicated by a fine thread-line; aperture obliquely ovate, large, lip simple, columella thin, excavately incurved.

Helix pellucida, Müller (1774), *Verm. Hist.* vol. ii. p. 15.
Helix diaphana, Poiret (1801), *Coq de l'Aisne, Prod.* p. 77.
Helicolimax pellucida, Férussac (1801), *Syst. Conch.* p. 30.
Helix limacoides, Alten (1812), *Syst. Abhandl.* p. 85, pl. xi. f. 20.
Vitrina pellucida, Gærtner (1813), *Conch. Wett.* p. 34 (not of Draparnaud).
Helix elliptica, Brown (1815), *Mem. Werner. Soc.* vol. ii. part ii. p. 525. pl. 24. f. 8.
Hyalina pellucida, Studer (1820), *Kurz. Verzeichn.* p. 86.
Limacina pellucida, Hartmann (1821), *Syst. Gasterop.* p. 54.
Vitrina beryllina, C. Pfeiffer (1821), *Deutschl. Moll.* p. 47. pl. 3. f. 1.
Vitrina Mülleri and *Dillwynii*, Jeffreys (1830), *Trans. Linn. Soc.* vol. xvi. p. 506.

Hab. Northern and Central Europe, scarcer towards the South. (Under stones, leaves, or moss.)

The *Vitrina* of the British Isles has been described as three species, all certainly referable to one, the *Helix pellucida* of Müller, which, it may be observed by our synonyms, is not the *Vitrina pellucida* of Draparnaud. The animal varies from grey to ochraceous, but very pale, with a tinge of flesh colour. Two transparent violet-grey lines pass from the tentacles along the neck, and a transparent line between two darker lines may be observed at the side of the foot.

The shell varies from a bright crystalline pale green to a rather muddy yellow, and is composed of from three to three and a half rapidly increasing whorls, smooth and shining, but still showing fine pencil lines of growth. The spire is scarcely raised; the suture is indicated by a delicate linear groove. Authors are somewhat divided in opinion as to the animal being able to retire into its shell. M. Moquin-Tandon separates the six French species into two groups, *Hyalina* and *Helicolimax*, characterizing the animal of the first as not wholly withdrawing itself into the shell, and forming no epiphragm; of the second, as being able wholly to withdraw, especially on the approach of winter, and close itself in by a vitreous epiphragm. *V. pellucida* he includes in the latter section, and yet Mr. Berkeley, who has observed this mollusk with great attention, and published its anatomy in the 'Zoological Journal,' writes me word, "the animal never retires completely into its shell."

Vitrina pellucida is common in all parts of the British Isles. It is a lively little mollusk, keeping the reflected lobe of its mantle in constant motion over the shell. Mr. Thompson, speaking of its habitats in Ireland, remarks, that in suitable localities it may be found under the first stones to be met with in going inland from the seashore to as great an altitude in the mountain glens as there are moss and leaves to shelter it; and it is found as far north as Greenland. The United States species, which was thought at first to be identical with our own, is more globose.

Succinea putris.

Genus II. **SUCCINEA**, *Draparnaud*.

Animal; body oblong, swollen, granulated, bearing a very thin capaciously inflated shell, head broad and obtuse, tentacles stout, somewhat cylindrical, lower pair very short, foot acuminated posteriorily.

Shell; imperforate, oblong-ovate, with the spire more or less exserted, very thin and transparent, yellowish or reddish amber, covered with a slight epidermis; whorls three to three and a half, the last capaciously inflated, aperture large, rounded below, more or less contracted above, lip simple.

Succinea is a herbivorous air-breathing mollusk, restricted in its habitat to the immediate vicinity of water or wet places. It lives chiefly on mud and flags by the side of rivers, lakes, and pools. The animal is generally rather conspicuously mottled with brown, and the head is peculiarly broad and obtuse, with the tentacles short and stout, particularly the lower pair. Its more important organs are scarcely distinguishable from those of *Helix*. Its minute palate teeth resemble those of *Vitrina*. The shell, which is extremely thin and of a transparent yellow or reddish amber-colour, is of an oblong fusiformly ovate shape, with the last whorl sufficiently inflated to cover the large pulmonary sac. Only two British species are admitted by Dr. Gray, and by Forbes and Hanley, *S. putris* and *oblonga*. Dr. Pfeiffer increases them to four, by raising to the rank of species two forms, *S. elegans* and *acuta*, which have been regarded as varieties of *S. putris*. A careful examination of the specimens set apart by Dr. Pfeiffer in the Cumingian collection, as the types of these species, induces me to adopt a middle course by regarding the two as one.

The geographical range of *Succinea* over the globe compared with that of *Vitrina*, is altogether different. In the New World, where *Vitrina* is only known by a single small species in the United States, and, by a distinct representative type, *Simulopsis*, of about a dozen species peculiar to Mexico and the Brazils, *Succinea* is most abundant. Thirty species have been described from the islands of the Polynesian Archipelago, twenty-two from the West Indies, sixteen from Bolivia and Mexico, and about five-and-twenty from the great tract of country commencing with the Southern United States, and passing poleward over Oregon and Greenland to Magellan. From all the more explored lands of the Eastern Hemisphere, in which *Vitrinæ* abound, not half the same number of *Succineæ* have been collected. India, China, and the eastern islands contribute sixteen species, Tasmania and Australia five, Africa, south and west, twelve, and Europe eight, which are probably reducible to six, the French list, according to M. Moquin-Tandon, including five.

The animal is somewhat inactive; it retires habitually into its shell when overtaken by drought, and encloses itself by a slight diaphragm. Dr. Gould, speaking of the *Succinea ovalis* of Massachusetts, remarks that it crawls over the mud and up the stalks of plants; and although it seems to be but little incommoded by water, it cannot endure being entirely submerged, and appears not to have the power of directing its way in the water, though it will generally float.

The British species are:—
1. **putris.** Shell ovately inflated, yellowish.
2. **elegans.** Shell constrictedly fusiform, reddish amber.
3. **oblonga.** Shell oblong-turbinated, but little inflated, yellowish.

1. **Succinea putris.** *Filthy Succinea.*

Shell; oblong-ovate, very thin, transparent, yellowish, spire small, whorls slightly contracted at the suture then convex, densely longitudinally plicately striated, last whorl much the longest, inflated; aperture ovate, columella thinly reflected.

 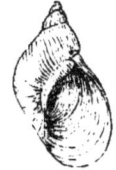

Helix putris, Linnæus (1758), *Syst. Nat.* 10*th edit.* p. 774.
Helix succinea, Müller (1774), *Verm. Hist.* part ii. p. 97.
Turbo trianfractus, Da Costa (1778), *Test. Brit.* p. 72. pl. v. f. 13.
Bulimus succineus, Bruguière (1789), *Enc. Méth.* vol. vi. p. 308.
Helix limosa, Pulteney (1799), *Cat. Dorset.* p. 48.
Succinea amphibia, Draparnaud (1801), *Tabl. Moll.* p. 55.
Amphibulina succinea, Lamarck (1805), *Ann. du Mus.* vol. vi. p. 236.
Lucena putris, Oken (1815), *Lehrb. Nat.* vol. iii. p. 312.
Succinea Mülleri, Leach (1820), *Syn. Moll.* p. 58.
Tapada putris, Studer (1820), *Kurz. Verzeichn.* p. 86.
Amphibina putris, Hartmann (1821), *Neue Alpina*, vol. i. p. 247.
Helix (Cochlohydra) putris, Férussac (1822). *Tabl. Moll.* p. 26. pl. 9. f. 4 to 10. pl. ix. f. 7 to 10.
Succinea putris, Fleming (1828), *Brit. Anim.* p. 267.

Hab. Throughout Europe (on mud and flags).

The typical form of this species is oblong-ovate, with a small spire and only a slight constriction of the sutures; but when limited even to this, and referring the variety *gracilis* to a distinct species, *S. elegans*, there is yet so varied a modification in the coiling of the shell as to induce M. Moquin-Tandon to characterize nine separate varieties. In colour it is uniformly of a yellowish amber. Mr.

Thompson describes having met with individuals of this species adhering to stones in wet spots, at a considerable elevation in the northern mountains of Ireland, but, as may be expected, invariably much dwarfed in size.

2. Succinea elegans. *Elegant Succinea.*

Shell; slenderly ovate, sometimes fusiformly contorted, thin, sometimes rather stouter, reddish amber, spire small, more or less sharply acuminated, whorls constricted at the sutures, then convex, longitudinally plicately striated, last whorl much the longest, rather compressly inflated; aperture oblong, columella very thinly reflected.

Helix angusta, Studer (1798), *Faun. Helv. in Coxe's Travels in Switzerland,* vol. iii. p. 432 (without characters).
Succinea amphibia var., Draparnaud (1805), *Hist. Moll.* p. 58.
Succinea Mülleri var., Leach (1820), *Syn. Moll.* p. 58.
Tapada Succinea, Studer (1820), *Kurz. Verzeichn.* p. 86.
Amphibulina putris var., Hartmann (1821), *Sturm, Deutschl. Faun.,* vol. vi. p. 8. f. 6, 7.
Helix (Cochlohydra) putris var., Férussac (1822), *Tabl. Moll.* p. 26. pl. 11. f. 13.
Succinea elegans, Risso (1826), *Hist. Nat. Europ. Mérid.,* vol. iv. p. 59.
Succinea putris var., Jeffreys (1830), *Trans. Linn. Soc.* vol. xvi. p. 325.
Succinea Pfeifferi, Rossmässler (1835), *Icon. Land und Süssw. Moll.* part 1. p. 92. f. 46.
Succinea Levantina, Deshayes (1836), *Expéd. Scient. Morée, Moll.* p. 170. pl. xix. f. 25 to 27.
Succinea gracilis, Alder (1837), *Mag. Zool. and Bot.* vol. ii. p. 106.
Succinea putris, var. *gracilis,* Macgillivray (1843), *Moll. Aberd.* p. 96.
Succinea Corsica, Shuttleworth (1843), *Moll. Cors.* p. 13.
Succinea acuta, Pfeiffer (1853), *Monog. Helic.* vol. iii. p. 8.
Hab. Throughout Europe.

The shell of *S. elegans,* which varies in being more or less sharply acuminated, is of a narrowly compressed form, constricted at the sutures compared with that of *S. putris,* and the colour inclines more to a rufous amber. The animal is darker, often bluish-black on the neck and sides, and a difference is suspected in the shape of the lingual teeth. Dr. Pfeiffer's *S. acuta* is the slenderly convoluted form of the species represented in our figure as the type. The species was commonly known in collections by Rossmässler's name of *Pfeifferi*, until M. Bourguignat identified it with the original type of Risso's *S. elegans*.

3. Succinea oblonga. *Oblong Succinea.*

Shell; oblong-turbinated, thin, transparent, yellowish, spire exserted, whorls rounded, impressed at the sutures, longitudinally striated, aperture moderate in size, columella thinly reflected.

Helix elongata, Studer (1789), *Faun. Helv. in Coxe's Travels in Switzerland*, vol. iii. p. 432 (without characters).
Succinea oblonga, Draparnaud (1801), *Tabl. Moll.* p. 56.
Amphibulina oblonga, Lamarck (1806), *Ann. du Mus.* vol. vi. p. 306.
Tapada oblonga, Studer (1820), *Kurz. Verzeichn.* p. 86.
Amphibina oblonga, Hartmann (1821), *Neue Alpina*, vol. i. p. 248.
Helix (Cochlohydra) elongata, Férussac (1822), *Tabl. Moll.* p. 26. pl. ix. f. 1, 2.
Succinea arenaria, Bouchard-Chantereaux (1838), *Moll. Pas-de-Calais*, p. 54.
Succinea abbreviata, Morelet (1845), *Moll. du Port.* p. 54. pl. v. f. 4.

Hab. Throughout Europe. Rare and local in Britain. (Near the sea.)

A much smaller species than either of the preceding. The whorls are rounder, and they are impressed, not obliquely constricted, at the sutures; and, increasing more gradually, are less inflated towards the aperture. The mantle of the animal, according to M. Moquin-Tandon, is dotted with dark grey; and the lower tentacles, as is commonly the case with snails affecting watery habitats, are short, reduced almost to tubercles. There is strong reason for believing that *S. arenaria* of Bouchard-Chantereaux is merely a variety of rather solid growth. It is the habit of the species generally to dwell in sandy places, near the sea, burying itself during the winter in an envelope of mucus.

The only localities recorded for this species as British are North Devon, South Wales, South of Ireland, and vicinity of Glasgow, Scotland.

Zonites cellarius.

Genus III. **ZONITES**, *De Montford*.

Animal; rather slenderly elongated, bearing an umbilicated, narrow-whorled subdiscoidal shell, of a semitransparent shining horny substance, over which the mantle is a little reflected, head with the upper pair of tentacles rather long, the lower short, foot obliquely truncated posteriorly, minute palate teeth aculeate.

Shell; conspicuously umbilicated, depressly orbicular, subdiscoidal, semitransparent horny, shining, without spot or marking, spire scarcely exserted, whorls three to five, narrow, rounded; aperture small, lip simple, margins widely removed.

The genus *Zonites* as lately restricted constitutes a very natural group. The animal has a semitransparent shining horny shell, composed of narrow discoidly convoluted whorls, of which the lip is always thin and simple, and has the mantle reflected over it somewhat after the manner of *Vitrina*. The posterior extremity of the foot is more inclined to be truncated than it is in *Helix*, and the minute palate teeth differ in being aculeate. The shell has no epidermis; it is deeply, sometimes largely, umbilicated, and there is no encircling rib within the aperture.

Viewing the genus in a wider geographical range than Britain, its characters, so far as they have been determined by naturalists, are much less perfectly defined. M. Moquin-Tandon refers four distinct forms of *Helicidæ* to *Zonites*, distinguishing them by the sectional names *Conulus*, *Calcarina*, *Aplostoma* and *Verticillus*. The British *Zonites* come into section *Aplostoma*, but he includes with them the large *Helix Algira* and *H. olivetorum* of the European continent, whose shells are not of the same glossy substance though presenting a considerable general resemblance in form and colour. Mr. Shuttleworth refers the *Epistylia* group of *Helices* inhabiting Jamaica to this genus; and Messrs. Adams refer some forty or fifty foreign *Helices* to the genus.

The British *Zonites* are eight in number, three of moderate size and five much smaller; the shells vary in being more or less glossy with no perceptible epidermis, and the variation in size of the umbilicus, according as the whorls are more or less closely coiled, is an obvious specific character. All the species inhabit wet and sheltered situations, chiefly damp cellars, in preference to dry and exposed places; and as the animal has less need of enclosing the aperture of its shell with any calcareous covering, the diaphragm is reduced to a few filaments. According to Mr. Jeffreys, they greedily devour animal food whether fresh or putrid. It is certain they are often found in places where there is no sort of vegetation beyond funguses. The only British *Zonites* not observed on the Continent is *Z. excavatus;* it may have been passed over as a largely umbilicated variety of *Z. nitidus*, but there are good reasons for keeping it distinct. Two species, *Z. cellarius* and *radiatulus*, are included in the fauna of the United States, having been transported accidentally from Europe in casks or other packages, and become naturalized there.

The British species of *Zonites* are :—

1. **cellarius.** Shell comparatively large, rather narrowly umbilicated, pale glossy, spire convexly flattened.
2. **alliaria.** Shell smaller, narrowly umbilicated, transparent horny, inclined to glassy, spire slightly convex.
3. **nitidulus.** Shell rather larger than *alliaria*, but not quite so large as *cellarius*, more openly umbilicated, rufous above, dull opake yellowish beneath.
4. **purus.** Shell very small, rather largely umbilicated, greenish glassy.
5. **radiatulus.** Shell very small, moderately umbilicated, greenish glassy, obviously radiately striated.
6. **nitidus.** Shell much larger than *purus* and *radiatulus*, smaller than *alliaria*, rather largely umbilicated, brownish fulvous, shining.
7. **excavatus.** Shell rather larger than *nitidus*, often smaller, conspicuously excavately umbilicated, fulvous horny.

8. **crystallinus.** Shell very small, rather smaller than *purus*, narrowly umbilicated, bright crystalline opal horny.

---

1. Zonites cellarius. *Cellar Snail.*

Shell; rather narrowly deeply umbilicated, depressly orbicular, greenish-yellow, pale, very glossy, spire convexly flattened, sutures linearly channelled, whorls six, narrow, increasing slowly, longitudinally obscurely plicately striated; aperture obliquely lunar.

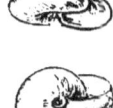

Helix cellaria, Müller (1774), *Verm. Hist.* vol. ii. p. 28.
Helix lucida, Montagu (1803), *Test. Brit.* p. 425. (not of Draparnaud nor Studer).
Helix nitens, Maton and Rackett (1807), *Trans. Linn. Soc.* vol. viii. p. 198. pl. v. f. 7. (not of Gmelin nor Sheppard).
Vortex cellaria, Oken (1815), *Lehrb. Nat.* vol. iii. p. 314.
Helix glaphyra, Say (1817), *Nich. Encycl.* vol. iv. pl. i. f. 3.
Zonites lucidus, Leach (1820), *Syn. Moll.* p. 75.
Oxychilus cellarius, Fitzinger (1833), *Syst. Verzeichn.* p. 100.
Helicella cellaria, Beck (1837), *Ind. Moll.* p. 6.
Polita cellaria, Held (1837), *Isis*, p. 916.
Zonites cellarius, Gray (1840), *Turt. Man.* p. 170.
Helix (Lucilla) cellaria, Lowe (1854), *Pro. Zool. Soc.* part 22. 177.
Zonites (Aplostoma) cellarius, Moquin-Tandon (1855), *Hist. Moll.* vol. ii. p. 78. pl. 9. f. 12.

Hab. Throughout Europe. Madeira. Northern, Eastern, and Middle States of America. (Chiefly in cellars and drains, under loose bricks or among stones.)

This is the largest species of the genus. *Z. nitidulus* comes very near to it in size, but *Z. cellarius* may be distinguished by its more depressly discoidal form and glossy substance, uniform pallid transparency, and smaller contracted umbilicus. It is very generally distributed throughout the British Isles, having a preference for damp places in drains and cellars. Mr. Lowe collected *Z. cellarius* in Madeira, and it has become widely naturalized in the United States, from being transported with casks or other packages. It was first noticed in America by Mr. Say, who collected it in 1817, in gardens, in the vicinity of Philadelphia, and described it as a new species, under the name of *Helix glaphyra*.

The animal is mostly of a lead-blue colour, varying to dingy yellow.

---

2. Zonites alliaria. *Garlic Zonites.*

Shell; narrowly deeply umbilicated, convexly orbicular, transparent horny, inclined to glassy, whorls five, slopingly convex, obscurely plicately striated; aperture obliquely lunar.

Helix alliaria, Miller (1822), *Ann. Phil. New Series*, vol. viii. p. 379.
Helix glabra, Studer (1822), *Féruss. Tabl. Syst.* p. 45.
Helix nitens, Sheppard (1825), *Trans. Linn. Soc.* vol. xiv. p. 160.
Helix fœtida, Stark (1828), *Elem. Nat. Hist.* vol. ii. p. 59.
Helix alliacea, Jeffreys (1830), *Trans. Linn. Soc.* vol. xvi. p. 341.
Helicella alliaria and *glabra*, Beck (1837), *Ind. Moll.* p. 6, 7.
Polita glabra, Held (1837), *Isis*, p. 619.
Zonites alliarius, Gray (1840), *Turt. Man.* p. 168.
Zonites (Aplostoma) alliarius and *glaber*, Moquin-Tandon (1855), *Hist. Moll.* pp. 80, 83, pl. 9. f. 3 to 11.

Hab. Throughout Europe (in gardens, among leaves and under stones).

The names that have been given to this species well express its distinctive characters; it is particularly smooth and shining, and the mucous secretions of the animal give out a fetid odour of garlic. Mr. Jeffreys remarks that this smell is very strong and pungent, especially when the animal is irritated. "I have perceived it," he says, "at a distance of several feet from the spot. Having found living specimens under stones in a bed of wild garlic, I thought at first that they might have fed upon this herb and have thus acquired the peculiar odour; but I afterwards observed that this scent was quite as powerful in specimens collected on an open down where there was no garlic." Dr. Johnston and Mr. Norman have both, however, borne testimony that the scent varies in intensity and is sometimes little perceptible, even after considerable irritation of the animal.

Compared with *Z. cellarius*, the shell is smaller, and composed of a whorl less, with the spire rather more convexly raised, but the whorls are broader in proportion, and it results from their more contracted coiling that the shell is less widely umbilicated.

The animal varies in both species, from blue-black to violet; a yellow variety noticed in *Z. cellarius* has not been observed in *Z. alliaria*.

3. **Zonites nitidulus.** *Dull Zonites.*

Shell; rather openly umbilicated, depressly convex, dull horny, rufous above, opake yellowish-white beneath, whorls five, rather impressed at the sutures, then narrowly rounded; aperture rotundately lunar.

Helix nitidula, Draparnaud (1805), *Hist. Moll.* p. 117.
Oxychilus nitidulus, Fitzinger (1833), *Syst. Verzeichn.* p. 100.
Helicella nitidula (1837), *Beck. Ind. Moll.* p. 6.
Polita nitidula, Held (1837), *Isis*, p. 916.
Zonites nitidulus, Gray (1840), *Turt. Man.* p. 172.
Helix Helmii, Gilbertson (1840), *Gray, Turt. Man.* p. 173.
Zonites (Aplostoma) nitidulus, Moquin-Tandon (1855), *Hist. Moll.* vol. ii. p. 83. pl. ix. f. 12, 13.

Hab. Throughout Europe (among moss and under stones in sheltered places).

Larger than the preceding species, but hardly so large as *Z. cellarius*, *Z. nitidulus* has not the glossy shell of either. The upper surface of the shell is characterized by a dull reddish-fawn colour, while around the umbilicus, which is more perspectively open, it is of a dull opake yellowish-white. Our figure of this species is drawn from an unusually large specimen. Mr. Jeffreys remarks that the animal is shy, and delights in dark places.

4. **Zonites purus.** *Transparent Zonites.*

Shell; largely umbilicated, depressly orbicular, greenish glassy, spire convex, sutures rather impressed, whorls three and a half to four, slopingly rounded, narrow, faintly striated; aperture obliquely lunar.

Helix nitidula var., Draparnaud (1805), *Hist. Moll.* p. 117. pl. viii. f. 21, 22.
Helix nitidiosa, Férussac (1822), *Tabl. Syst.* p. 45 (without characters).
Helix pura, Alder (1830), *Trans. Nat. Hist. Soc. Northumb.* vol. i. p. 37.

Helix lenticula, Held (1837), *Isis*, p. 304.
Polita nitidiosa, Held (1837), *Isis*, p. 916.
Helicella nitidiosa, Beck (1837), *Ind. Moll.* p. 6.
Zonites purus, Gray (1840), *Turt. Man.* p. 171. pl. iv. f. 43.
Zonites (Aplostoma) purus, Moquin-Tandon (1855), *Hist. Moll.* vol. ii. p. 87. pl. 9. f. 22 to 25.

Hab. Northern and Central Europe. Throughout Britain (in woods).

A small transparent greenish glassy shell, of which the animal is described by its discoverer, Mr. Alder, as being white with two black cervical lines, with the mantle also white, speckled with black. The shell is rather broadly umbilicated, composed of from three and a half to four whorls, impressed at the sutures, and slopingly rounded towards the periphery. *Z. radiatulus*, which, in my monograph of *Helix* in Conch. Iconica, I quoted erroneously as a variety of this species, is sculptured with fine radiating striæ, and has a more contracted umbilicus.

Until within the last few years *Z. purus* was only found in the northern parts of France and England; it has now been collected pretty generally in Britain, and on the Continent, from Siberia to Switzerland.

---

5. **Zonites radiatulus.**   *Finely rayed Zonites.*

Shell; moderately umbilicated, rather depressly orbicular, greenish glassy, spire convex, sutures a little impressed, whorls three and a half to four, rounded, radiately finely striated, the last whorl somewhat produced; aperture broadly obliquely lunar.

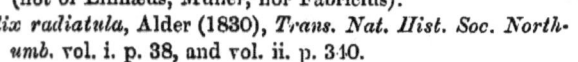

Helix nitidula var., Draparnaud (1805), *Hist. Moll.* p. 117. pl. viii. f. 21, 22.
Helix striatula, Gray (1821), *Lond. Med. Repos.* vol. xv. p. 239 (not of Linnæus, Müller, nor Fabricius).
Helix radiatula, Alder (1830), *Trans. Nat. Hist. Soc. Northumb.* vol. i. p. 38, and vol. ii. p. 340.
Zonites radiatulus, Gray (1840), *Turt. Man.* p. 173. pl. xii. f. 137.
Helix pura, var., Pfeiffer (1848), *Mon. Helic.* vol. i. p. 96.
Zonites (Aplostoma) striatula, Moquin-Tandon (1855), *Hist. Moll.* vol. ii. p. 86. pl. ix. f. 19 to 21.

Hab. Central Europe. Throughout Britain. United States. (Among moss and under stones).

This well-marked species has been confounded by Pfeiffer and other conchologists, myself among the number, with *Z. purus*, with which it has a close resemblance, but is constantly distinct. The shell, besides being regularly radiately striated, acquires a different form from the last whorl being more outwardly produced at the periphery; and yet the umbilicus is smaller. It is of the same transparent greenish-glassy substance as *Z. purus;* and a careful application of the lens is necessary to bring out its specific characters in obvious relief. The animal, according to Mr. Alder, is just the reverse in colour, black in place of white. It appears to be very generally distributed throughout Central Europe and the British Isles, and it has been transported and become naturalized in the United States. Mr. Thompson, one of the most diligent of our observers, describes having procured it in Ireland, in moist spots in the wildest and bleakest localities, as well as in woods.

---

6. **Zonites nitidus.** *Shining Zonites.*

Shell; rather largely umbilicated, moderately depressed, brownish fulvous, shining, spire convex, sutures impressly channelled, whorls four to four and a half, narrow, slopingly rounded, finely radiately striated; aperture obliquely lunar.

Helix nitida, Müller (1774), *Term. Hist.* vol. ii. p. 32 (not of Gmelin nor Draparnaud).
Helix succinea, Studer (1789), *Faun. Helv. in Coxe's Travels in Switzerland*, vol. iii. p. 429 (not of Müller).
Helix lucida, Draparnaud (1805), *Hist. Moll.* p. 103.
Helicella nitida, Risso (1826), *Hist. Nat. Europ. Mérid.* vol. iv. p. 72.
Oxychilus lucidus, Fitzinger (1833), *Syst. Verzeichn.* p. 100.
Polita lucida, Held (1837), *Isis*, p. 916.
Helicella succinea, Beck (1837), *Ind. Moll.* p. 7.
Zonites lucidus, Gray (1840), *Turt. Man.* p. 174. pl. iv. f. 38.
Zonites (Aplostoma) nitidus, Moquin-Tandon (1855), vol. ii. p. 72. pl. vii. f. 11 to 15.

Hab. Central and Southern Europe. Iskardo, Thibet. United States. Britain (under stones in shady places, in pine beds and orchid houses).

The most obvious characteristic of this species is its brownish fulvous colour and shining aspect. It is a much larger shell than *Z. purus* or *radiatulus*, rather more openly umbilicated; and the sutures of the spire are impressed into a narrow channel. The radiating longitudinal striæ, though delicate, are very apparent, but they are not so defined over the periphery as in *Z. radiatulus*. Mr. Benson is erroneously quoted by Gray, Forbes and Hanley, and Moquin-Tandon, as having given the generic name of *Tanychlamys* to this species. It was for Dr. Gray's genus *Nanina* that Mr. Benson proposed that name, referring at the same time to an Indian species, which he merely described in general terms, as resembling *Z. nitidus*, but of larger size. *Z. nitidus* is well known to gardeners, from its habit of infesting pine beds and orchid houses.

---

7. **Zonites excavatus.** *Excavated Zonites.*

Shell; largely excavately umbilicated, convexly depressed, fulvous horny, spire rather exserted, suture moderately impressed, whorls four to four and a half, slopingly convex, enlarging slowly, aperture small, striated throughout; aperture obliquely lunar-rounded.

Helix vitrina, Férussac (1821), *Tabl. Syst.* p. 41 (without characters).
Helix excarata, Bean (1830), *Alder, Trans. Nat. Hist. Soc. Northumb.* vol. i. p. 38.
Helix viridula, Menke (1830), *Syn. Moll.* p. 20.
Helix nitida var., Jeffreys (1830), *Trans. Linn. Soc.* vol. xvi. pp. 339, 511.
Helix petronella, Charpentier (1837), *Moll. de la Suisse.*
Zonites excavatus, Gray (1840), *Turt. Man.* p. 175. pl. xiii. f. 138.

Hab. Northern and Central Europe. Throughout Britain, but local (under decayed timber).

To a casual observer, this shell has very much the appearance of *Z. nitidus*, but it will be seen on comparison to be composed of narrower whorls, increasing more slowly in width, and coiling round a broader axis. The umbilicus thus acquires a perspectively excavated *Rotella*-like appearance. The whorls drop a little more in the coiling, which renders the spire more exserted, while the sutural line is less strongly impressed. Dr. Gray states that the animal is lead-coloured but lighter, and frequents a different situation, being found under decayed wood and timber that has lain some time on the ground.

Mr. Jeffreys has lately recorded his opinion that *Helix viridula*, Menke, and *Zonites excavatus*, Bean, described in the same year, are one and the same species, having identified British specimens of the latter with Continental specimens of the former collected by himself at a height of about seven thousand feet, on the Gorner glacier in Switzerland. The circumstance is interesting, from the fact of this species having been for many years regarded as peculiar to Britain; as a solitary instance of a British land snail, not inhabiting the Continent. It now appears beyond all doubt that the British Isles have no land mollusk of their own of any kind; wherever the origin of a species has been attributed to Britain, its history has been surrounded with doubt and ultimately shown to be incorrect.

8 Zonites crystallinus. *Crystalline Zonites.*

Shell; narrowly deeply umbilicated, subglobosely discoid, shining opal horny, spire rather depressed, suture linearly channelled, whorls from four to four and a half, rounded, finely striated, striæ minutely plicated at the suture; aperture obliquely lunar-rounded.

Helix crystallina, Müller (1774), *Verm. Hist.* vol. ii. p. 23.
Helix eburnea, Hartmann (1821), *Neue Alp.* vol. i. p. 231.
Helix vitrea, Brown (1827), *Edin. Journ.* vol. i. pl. i. f. 12 to 14.
Discus crystallinus, Fitzinger (1833), *Syst. Verzeichn.* p. 99.
Helicella crystallina, Beck (1837), *Ind. Moll.* p. 7.
Polita crystallina, Held (1837), *Isis*, p. 916.
Helix hydatina, Rossmässler (1838), *Icon. Land und Süss. Moll.* p. 36. f. 529.
Zonites crystallinus, Gray (1840), *Turt. Man.* p. 176. pl. iv. f. 42.
Zonites (Aplostoma) crystallinus, Moquin-Tandon (1855), vol. ii. p. 89. pl. ix. f. 26 to 29.

Hab. Europe. Throughout Britain (under stones and among moss).

The animal of this universally distributed species, the smallest of the genus, is milk-white, and the shell always retains a shining greenish crystalline or opal-like appearance. The whorls, which are from four to four and a half in number, are convoluted, rather closely and almost discoidly upon one another, the umbilicus being contracted almost to a puncture. They are extremely regular, and the delicate striæ with which they are sculptured are minutely puckered in the suture. *Zonites crystallinus* occurs among moss and dead leaves, under stones and upon decaying wood, both in dry and wet situations, but more frequently in the latter. M. Bouchard-Chantereaux remarks, that it is particularly common among plants in very wet places upon the banks of rivers.

Helix aspersa.

Genus IV. **HELIX**, *Linnæus*.

Animal; sometimes stout, sometimes slender, bearing an orbicularly coiled shell, varying considerably in general character; mantle not reflected over the shell, head with the upper pair of tentacles rather long, the lower short, foot lanceolate posteriorly, never truncate, minute palate teeth serrated.

Shell; mostly umbilicated, globose, varying to depressly orbicular, sometimes solid, gaily coloured, more frequently thin and horny, spire only moderately exserted, whorls varying from about four to six, mostly striated in the direction parallel with the lines of growth, aperture more or less lunar-rounded, lip in the solid species reflected, in the thin horny species simple.

Compared with *Zonites*, the animal of *Helix* differs in being lanceolate and never truncate at the tail extremity, and in not reflecting the edge of the mantle over the shell. The polished surface imparted by this action to the shell of *Zonites*, is never seen in *Helix*, so that the genera are distinguished from each other by good conchological characters, resulting from a corresponding difference in the animal. In neither genus has the animal any operculum. It encloses itself within its shell by a secretion of calcareous mucus, which, becoming dried, and more or less hardened according to the species, answers the purpose of an operculum. There is, however, no structural analogy between them; the hardened mucus, or epiphragm as it is technically called, is broken away and renewed, as often as the animal desires to crawl forth and again enclose itself.

One-third of our land mollusks belong to this genus. Eleven species occur throughout our islands, three nearly throughout, two in the central and south parts, seven in the south only, and one in the north only. As a general rule, the *Helices* increase in number of species and individuals towards the south. All the twenty-four species inhabit some or other of our southern counties, excepting the minute *H. lamellata*, which has not been collected in England south of Scarborough. Four species, *H. aspersa, arbustorum, nemoralis,* and *pomatia* are pre-eminent in size, and in the colours and development of their shell they are little surpassed in brilliancy, considering the difference of latitude, by their analogues of the intertropical islands of Eastern Asia. The last-mentioned, *H. pomatia*, is, however, local in Britain, and can hardly be said to be indigenous. The next most conspicuous of our snails are *H. Cantiana, Carthusiana, Pisana, virgata, fasciolata,* and *ericetorum*, natives chiefly of the south and eastern maritime counties of the chalk formation. Then we have a single lens-shaped species, *H. lapicida,* belonging to a type which has its maximum of beauty also in the Eastern Asiatic Islands. *H. obvoluta* is a solitary form of doubtful British parentage, colonized within a very limited area in the county of Hampshire. The remaining half of the British *Helices* are small species with thin horny shells, covered more or less with a fibrous, often hairy, epidermis, dwelling in darker and more concealed places of habitation, under stones or logs of wood, among damp moss or decaying leaves.

The range of our *Helices* in other parts of the world extends over the chief part of Europe, and in some instances to a large portion of Western Asia, between Siberia and Thibet. *H. hispida, sericea, fulva, pulchella* and *pygmæa* appear in the list of *Helices* described by Gerstfeldt from the district of the Amoor, and *H. pulchella* was collected by Dr. Thomson on the north side of the western Himalayas. Southwards, our British *Helices* range in four instances to North Africa, *H. aspersa, virgata, fasciolata,* and *rufescens*, neither of which are found in Northern or North-central Europe. Six species, *H. aspersa, Pisana, fulva, pulchella, aculeata,* and *pygmæa*, appear in the Azores, and three, *H. Pisana, rufescens,* and *pulchella*, in Madeira and the Canary Islands; but in localities like Madeira, which have a *Helix* fauna of their own, the presence of European types is probably due to transportation. *H. aspersa* has become naturalized by transportation, both in the United States and in Brazil. Varieties of *H. nemoralis* are also located in the States. The only British *Helix* which appears to be indigenous to the United States is that which has the widest distribution in the Eastern Hemisphere, the little *H. pulchella*,—a very interesting feature in the phenomena of distribution. There are, however, two other species, *H. fulva* and *rupestris*, which have very near representatives, in the United States, in *H. chersina* and *saxicola*.

Distribution of British Helices.

Species.	Britain.	Europe.	Other Parts of the World.
aspersa.	Throughout.	Central and South.	N. Africa, Azores, United States, Brazil.
arbustorum.	Throughout.	Throughout.	
nemoralis.	Throughout.	Throughout.	United States.
hispida.	Throughout.	Central and South.	Siberia.
fusca.	Throughout.	West France.	
fulva.	Throughout.	Throughout.	Siberia, Azores.
aculeata.	Throughout.	Throughout.	Azores.
pulchella.	Throughout.	Throughout.	Siberia, Thibet, Madeira, Azores, United States.
rotundata.	Throughout.	Throughout.	Azores.
rupestris.	Throughout.	Central and South.	
pygmæa.	Throughout.	Throughout.	Siberia, Azores.
virgata.	Nearly throughout.	Central and South.	N. Africa.
fasciolata.	Nearly throughout.	Central and South.	N. Africa.
ericetorum.	Nearly throughout.	Central and South.	
lapicida.	Central and South.	North and Central.	
rufescens.	Central and South.	Central and South.	N. Africa, Madeira.
Cantiana.	Chiefly South-east.	South-west.	
Carthusiana.	Chiefly South-east.	Central and South.	
Pisana.	South and West.	South.	Canaries, Azores, N. Africa.
sericea.	Chiefly South-west.	Throughout.	Siberia, Caucasus.
revelata.	South and West-south.	West France.	
pomatia.	South and Central.	Central.	
obvoluta.	South (doubtful).	Central.	
lamellata.	North.	North.	

We come now to the inquiry, How is the *Helix* fauna represented in other parts of the world? Of all mollusks, whether terrestrial or marine, the *Helices* are the most universally distributed over the globe. Dr. Pfeiffer catalogues upwards of two thousand species. About seventeen hundred is the number whose specific value has been tested by comparison, and of about fifteen hundred of these the habitats are known. It will be found that the Western Hemisphere contributes three hundred and sixty species, representing five natural provinces of distribution, the North American, West Indian, Central American, South-west American, and Brazilian. The Eastern Hemisphere contributes more than three times as many, which, divided into five natural Provinces, may be classed under the heads, Caucasian, Malayan, African, Australian, and Polynesian. The Eastern assemblages of types, owing to the more irregular configuration of the land, are of very unequal proportions. The Malayan province, for example, which includes the brilliant Molucca and Philippine species, is restricted to the Eastern Asiatic islands, lying between Corea and Japan and the Straits of Malacca. Islands like the Philippine group, and the West Indies and Madeira, having a *Helix* fauna of their own, abound in species. Allowing for the multiplication of names, it may still be reckoned that the Philippine Islands have two hundred species, the West Indies a hundred and fifty, and Madeira a hundred, all emanating from their own respective areas of creation. The British Islands have no *Helix* fauna of their own. Our four-and-twenty *Helices* are outlying members of the great Caucasian province, which includes five hundred and sixty species, spread over Europe and Asia Minor to the Himalayas, and comprising Western China and Hindoostan. The English Channel is probably some impediment to the western march of the Caucasian snails. The nearest continental land to Britain has twice as many species of *Helix*; and they are equally numerous in the parts of Germany that come within the same isothermal latitude.

The British *Helices* are:—

1. **aspersa.** Shell large, imperforate, solid, of from four to four and a half whorls, blotched and banded with brown, lip expandedly reflected.
2. **pomatia.** Shell very large, ventricose, with a half-covered umbilicus, of from four to four and a half whorls, fawn-coloured, faintly banded, lip thinly dilated.

3. **arbustorum.** Shell rather large, with a nearly covered umbilicus, solid, of five and a half whorls, brown, freckled with fawn-yellow and rust, lip expandedly reflected.
4. **nemoralis.** Shell rather large, imperforate, solid, of five to five and a half whorls, mostly yellowish, variously banded with chocolate brown, lip expanded and a little reflected.
5. **Cantiana.** Shell moderate, rather narrowly umbilicated, thin, semitransparent, of five and a half to six whorls, rufous, lip expanded.
6. **Carthusiana.** Shell smaller, minutely umbilicated, semitransparent horny, of six whorls, lip expanded, ribbed within.
7. **Pisana.** Shell rather large, very narrowly umbilicated, of five whorls, thin, opake rust-white encircled with black interrupted pencillings; aperture rose-tinted, lip simple.
8. **virgata.** Shell moderate, slightly umbilicated, of five and a half whorls, rather thin, dead white variously banded with rust-brown, lip simple, ribbed within.
9. **fasciolata.** Shell smaller, rather largely umbilicated, of five and a half obliquely wrinkled whorls, dead white variously banded with rust-brown, lip simple, ribbed within.
10. **ericetorum.** Shell rather large, depressed, broadly umbilicated, of five and a half whorls, buff-tinted, variously banded; aperture nearly circular.
11. **obvoluta.** Shell moderate, depressed, largely umbilicated, of six whorls, opake brown, obscurely downy; aperture triangularly contorted, obtusely toothed within.
12. **lapicida.** Shell rather large, lens-shaped, broadly umbilicated, of five whorls, keeled at the periphery, fawn-brown, lip expandedly reflected.
13. **rufescens.** Shell moderate, deeply umbilicated, thin, of six whorls; yellowish or rufous, lip thinly expanded, white-ribbed within.
14. **hispida.** Shell small, rather openly umbilicated, of five to six whorls, yellowish horny, downy with short hairs, lip thinly expanded.
15. **sericea.** Shell small, minutely umbilicated, of five to six whorls, whitish horny, with short, rather distant, silky hairs, lip thinly reflected.
16. **revelata.** Shell small, minutely umbilicated, of four whorls, wrinkly membranaceous, greenish, lip scarcely reflected.

FAMILY COLIMACEA.

17. **fusca.** Shell small, scarcely umbilicated, of five whorls, membranaceous, glossy fuscous olive, plicately wrinkled, lip thin.
18. **fulva.** Shell smaller, minutely umbilicated, trochiform, of six whorls, very thin, glossy fulvous, lip thin.
19. **lamellata.** Shell very small, minutely umbilicated, subglobose, horny, of six whorls, fuscous horny, crossed by membranaceous lamellæ, lip thin.
20. **aculeata.** Shell very small, moderately umbilicated, horny, of four whorls, crossed by fibrous lamellæ elongated into a corona of lashes.
21. **pulchella.** Shell minute, largely umbilicated, subhyaline, of three to four whorls, trumpet-shaped at the aperture.
22. **rotundata.** Shell small, largely umbilicated, depressed fulvous horny, red-spotted, of six whorls, closely rib-striated, lip simple.
23. **rupestris.** Shell minute, largely umbilicated, brown horny, of four to five whorls, lip simple.
24. **pygmæa.** Shell more minute, largely umbilicated, brown horny, of three to four whorls, lip simple.

1. Helix aspersa. *Besprinkled Helix.*

Shell: imperforate, obliquely conoid-globose, thin, greyish yellow, blotched and banded with dark brown, sometimes of a delicate straw-colour without spot or band, spire short, whorls four to four and a half, slopingly convex, swollen, densely concentrically plicately striated, striæ rendered more and more obscure towards the aperture; by a copious shagreen of very irregular opake wrinkles, last whorl suddenly descending at maturity; aperture somewhat squarely ovate, lip expanded, a little reflected, white, margins inclined to approximate, columellar margin appressly dilated.

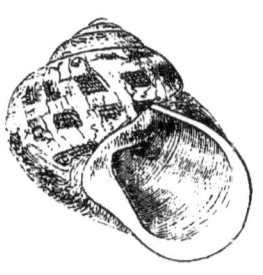

Helix aspersa, Müller (1774), *Verm. Hist.* p. 59.
Helix hortensis, Pennant (1777), *Brit. Zool.* vol. iv. p. 136. pl. lxxxiv. f. 129 (not of Müller).
Cochlea vulgaris, Da Costa (1778), *Test. Brit.* p. 72. pl. iv. f. 1.

Helix lucorum, Razoumowsky (1789), *Hist. Nat. Jor.* vol. i. p. 274 (not of Linnæus).

Helix variegata, Gmelin (1788), *Syst. Nat.* p. 3650.

Pomatia adspersa, Beck (1837), *Ind. Moll.* p. 44.

Cœnatoria aspersa, Held (1837), *Isis*, p. 911.

Helix (Cryptomphalus) aspersa, Moquin-Tandon (1855), *Hist. Moll.* vol. ii. p. 174. pl. xiii. f. 14 to 32.

Hab. Throughout Central and Southern Europe. Algeria. The Azores, United States. Brazil. (In gardens and woods, in crevices of old walls, etc.)

With the exception of *H. pomatia*, which can hardly be said to be indigenous to Britain, *H. aspersa* is the largest of our land mollusks. It occurs abundantly with little variation in all parts of our islands, and it is equally abundant in Central and Southern Europe. In Northern and some parts of North-Central Europe it is unknown. It was not known to Linnæus. In its more extended range the species presents a greater variety of typical forms, passing from the obliquely deflected brown mottled type with which we are familiar in this country, to a more conically convoluted banded type approximating to *H. Mazzullii*, which MM. Deshayes and Moquin-Tandon regard as a variety of it.

On examining the shell of *H. aspersa* with the lens, it will be seen that the whorls are crowded in the ordinary manner with concentric striæ of growth, somewhat rudely puckered at the sutures, but towards the last whorl they are rendered obscure by a copious shagreen of very irregular opake wrinkles, the interspaces between which have a malleated appearance. The predominant colouring of the shell is to be largely blotched in a banded manner with dark brown; a very characteristic variety occurs of a delicate unspotted straw colour. M. Moquin-Tandon has some curious observations on the occurrence of *Helix aspersa* in varieties in the South of France. After describing and naming sixteen varieties, among which *H. Mazzullii* stands as var. ρ *crispata*, he goes on to remark, "On the 15th of August, 1852, I gathered in the Garden of Plants of Toulouse, 817 specimens. There were 729 more or less characteristic types, some of which formed a passage to var. *obscurata*, 51 *zonata*, 8 *grisea*, and 29 *marmorata*.

Helix aspersa has been frequently eaten, or used in the preparation of a mucilaginous broth, but *H. pomatia* is the edible snail *par excellence*.

2. **Helix pomatia.** *Apple Helix.*

Shell; with a half-covered umbilicus, globose, inflated, fawn-coloured, encircled with two to three faint reddish-brown bands; spire moderately exserted, whorls four to four and a half, ventricosely rounded, longitudinally rugosely striated, spirally faintly impressly lineated; aperture lunar oval, lip thinly dilated, broadly reflected over the umbilicus.

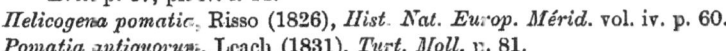

Helix pomatia, Linnæus (1758), *Syst. Nat.* 10th edit. p. 771.
Cochlea pomatia, Da Costa (1778), *Test. Brit.* p. 67, pl. iv. f. 14.
Helicogena pomatia, Risso (1826), *Hist. Nat. Europ. Mérid.* vol. iv. p. 60.
Pomatia antiquorum, Leach (1831), *Turt. Moll.* p. 81.
Pomatia pomatia, Beck (1837), *Ind. Moll.* p. 43.
Cœnatoria pomatia, Held (1837), *Isis*, p. 911.
Helix (Helicogena) pomatia, Moquin-Tandon (1855), *Hist. Moll.* vol. ii. p. 179. pl. xiv. f. 1 to 9.
Hab. Central Europe, but local. Southern counties of England.

The Apple, Edible, or Vine Snail, as it has been variously called, belongs to a characteristic Continental type, of which *H. lucorum, Taurica Gussoniana, ligata, lutescens, grisea,* may be quoted as examples. It is said to have been introduced into England about the middle of the sixteenth century, either as a foreign delicacy or as a cure for consumption. The species, if not indigenous, has become fully naturalized in our southern counties, but it is not generally common. Few Englishmen venture to partake of *H. pomatia*, though Ben Jonson and Lister have extolled it as a dainty dish. Some of our more epicurean neighbours on the Continent, however, still follow the ancient Romans in keeping preserves of Apple Snails, and fattening them in sties for the table. Few of the Parisian dealers in comestibles, in the present day, are without a bowl of *H. pomatia* among the *pièces de résistance*, temptingly displayed in their shop-windows; and they may be as commonly seen in Germany, Switzerland, and Italy. To its efficacy in cases of consumption I am able to testify on personal knowledge. Mr. Barlow, of the firm of John Dickinson and Co., paper-makers, informs me that he has a brother who was in the last stage of consumption, when their father resolved to try the experiment of a diet of Apple Snails. Specimens had been transported from Italy some thirty years before by an English nobleman, for the same purpose, to the range of hills in the neighbourhood of Reigate and Box Hill, in Surrey; and as they had bred abundantly, Mr. Barlow was induced to take a house in that locality. The expressed mucilaginous juice of the snail was administered to the patient, without his knowledge, in every conceivable form. It was taken in jellies and conserves, in gravies and with *entremets* of meats. In the course of a twelvemonth, the invalid was entirely cured, and went to the Crimea, and is living at this moment a strong hearty man.

The shell is of a light globose form, very limited in its variation of colour, which is a light fawn, encircled with three rather distant bands of darker hue. The axis of the whorls shows a distinct umbilicus, but it is almost concealed from view by the broadly reflected dilation of the columellar margin at its junction with the body-whorl. The close-set longitudinal striæ of the whorls, puckered at the suture, and indicating accessions of growth, are not obscured, as in *H. aspersa*, by a shagreen of wrinkles; the interstices between them are decussated with regularity by delicately impressed spiral lines. The animal is described as being neither so lively nor so irritable as *H. aspersa*, but of great strength, carrying its shell vigorously upright. Owing to its large size it has been a frequent object of anatomical study, the most elaborate memoir being that of M. Gaspard, of which Professor Bell published an abstract in the 'Zoological Journal.'

Helix pomatia, as already noticed, is partial in its distribution in England, not occurring further north than Gloucestershire or Wilts. It is also partial on the Continent. It is extremely abundant in Germany; while walking in the neighbourhood of Stolzenfels on one occasion, after a shower, I could scarcely step for some distance without fear of treading on one. It is also common in the north of France, while in the south the species is unknown. It has not bred in the United States; and attempts made to breed it in the north of England and Ireland have proved unsuccessful. The diaphragm secreted by this species during the time of hybernation, is unusually solid and calcareous, and it is supposed by some writers that Linnæus derived his name *pomatia*, not from *pomum* an apple, but from πῶμα a lid. The author of the 'Systema Naturæ' seldom, however, went to the Greek lexicon for his derivatives.

FAMILY COLIMACEA.

3. **Helix arbustorum.** *Tree Helix.*

Shell; with a nearly covered umbilicus, depressly globose, rather solid, livid-brown, freckled in a zigzag manner with fawn-yellow and rust, mostly encircled at the periphery, with a dark linear band, spire convex, moderately exserted, whorls five and a half, densely obliquely arcuately striated, striæ puckered at the sutures, interstices between them malleated with crowded minute spiral impressed lines, last whorl a little descending in front; aperture lunar-rounded, rather contracted, lip sharply expandedly reflected, white, broadly dilated over the umbilicus at its junction with the body whorl.

Helix arbustorum, Linnæus (1758), *Syst. Nat.* 10th edit. p. 771.
Helix Gothica, Linnæus (1758), *Syst. Nat.* 10th edit. p. 770.
Cochlea unifasciata, Da Costa (1778), *Test. Brit.* p. 75. pl. xvii. f. 6.
Helix turgidula, Wood (1828), *Ind. Test. Suppl.* pl. vii. f. 6.
Arianta arbustorum, Leach (1831), *Brit. Moll.* p. 86.
Helix Canigonensis, Boubée (1833), *Bull. Hist. Nat.* p. 36.
Helix Xartartii, Farines (1834), *Ann. Sc. Nat.* vol. ii. p. 122.
Helix Wittmanni, Zawadsky (1837), *Rossmässler, Icon. Land und Süssw. Moll.* part 5. f. 279 D.
Helix alpestris, Ziegler (1834), *Rossmässler, Icon. Land und Süssw. Moll.* part 5. f. 279 D.
Cingulifera arbustorum, Held (1837), *Isis*, p. 911.
Helix planospira, Gras (1840), *Desc. Moll. de l'Isère*, p. 36. pl. iii. f. 11.
Helix (Arianta) arbustorum, Moquin-Tandon (1855), *Hist. Moll.* vol. ii. p. 123. pl. xi. f. 1 to 4.

Hab. Throughout Europe (in damp places in woods and gardens, or under rocks).

Helix arbustorum is a granular livid leaden-blue or greenish snail, carrying a solid, globularly convoluted, dark brown shell, freckled in an irregular zigzag manner with fawn-yellow, often broken up into dots; and the last whorl is mostly encircled at the periphery, with a sharply defined, linear dark band. The lip is always white, rather sharply expandedly reflected, and at its junction with the body whorl it is dilated over the umbilicus, so as nearly to conceal it from view. There is not much variation in the species. The most striking variety is one of a yellowish horn colour, in which the freckled reticulations are opake white.

The favourite habitat of *H. arbustorum* is in damp places in shady woods and gardens, or under boulders of granite on mountain sides. In a more stunted form it extends its range to a considerable elevation. On the Alps and in the Jura it approaches nearly to the snow-line. The animal is eaten, says M. Moquin-Tandon, but is not much esteemed.

4. Helix nemoralis. *Helix of the woods.*

Shell; imperforated, subglobose, rather solid, yellowish, mostly banded with chocolate-brown, spire moderately convex, whorls five to five and a half, convex, densely arcuately striated and corrugately malleated, last whorl suddenly descending in front; aperture obliquely subquadrately lunar, lip expanded, a little reflected, sometimes chocolate-brown, sometimes white, columella margin straightly drawn out.

Helix nemoralis, Linnæus (1758), *Syst. Nat.* 10th edit. p. 773.
Helix hortensis, Müller (1774), *Verm. Hist.* part 2. p. 57 (not of Pennant).
Cochlea fasciata, Da Costa (1778), *Test. Brit.* p. 76. pl. v. f. 1 to 5.
Helix hybrida and *fusca,* Poiret (1801), *Coq. fluv. et terr. de l'Aisne,* p. 71.
Helix turturum, Stewart (1817), *Elem. Nat. Hist.* vol. ii. p. 413.
Helix mutabilis pars, Hartmann (1821), *Neue Alpina,* vol. i. p. 242.
Helix cincta and *quinquefasciata,* Sheppard (1825), *Trans. Linn. Soc.* vol. xiv. p. 163.
Helicogena nemoralis, Risso (1826), *Hist. Nat. Europ. Mérid.* vol. iv. p. 60.
Helicogena libellula, Risso (1826), *Hist. Nat. Europ. Mérid.* vol. iv. p. 62. pl. iii. f. 21*.
Tachea nemoralis and *hortensis,* Leach (1831), *Turt. Man.* pp. 84, 85.
Cepæa nemoralis and *hortensis,* Held (1837), *Isis,* p. 910.
Helicogena hortensis and *hybrida,* Beck (1837), p. 39.
Helix subglobosa, Binney (1837), *Journ. Nat. Hist. Soc. Boston, U.S.,* vol. i. p. 485. f. 7.
Helix lucifuga, Ziegler (1840), *Hartmann, Erd. und Süssw. Gast.* vol. i. p. 191. pl. lxx.
Helix (Tachea), nemoralis and *hortensis,* Moquin-Tandon (1855), *Hist. Moll.* vol. ii. p. 162-171, pl. xiii. f. 1 to 2.

Hab. Throughout Europe. United States. (In gardens, woods, and fields, also on marine sand-hills, and among chalk cliffs and quarries.)

Of all our land-shells, this is the most gaily painted and tropical-looking. In the New World there is no species of *Helix* so brightly coloured within many degrees of the same isothermal latitude. The animal is mostly of an olivaceous yellow, the shell mostly of a pale yellow often inclining to red, and it is generally banded with chocolate brown. The name *nemoralis* has been restricted to varieties with dark-stained lips, *hortensis* to those with white lips, and *hybrida* to intermediate varieties; but the varieties are endless. M. Moquin-Tandon has given names to seventy-seven varieties of the dark-lipped specimens, and to forty-six of the white-lipped, and he adds some amusing details of the contents of a basket of them bought in the market of Toulouse. Out of 1468 specimens he found 684 marked with distinct bands, 39 in which the bands were more or less broken up, and 745 without bands. C. Pfeiffer indicates 67 varieties of the species; and one author, M. Albin Gras, has gone so far as to enumerate 198 varieties of the dark-lipped specimens alone. In France, where this snail is sold as an article of food, collectors have ready opportunities of indulging their fancy for arranging and naming varieties. Some specimens, however, change the character of their painting during growth. In the early whorls the bands will be separate, while at maturity they become united; and the describer is fairly baffled in his attempt to name them.

H. nemoralis inhabits Europe throughout, but it is less common in the south. It has been transported to the United States, and keeps to the eastern parts near the sea, especially the lower extremity of Cape Cod and Cape Ann. Mr. Binney remarks that in the neighbouring islands, " each island is inhabited by a variety peculiar to itself, showing that the variety which happened to be introduced there, has propagated itself, without a tendency to run into other variations." " Thus, on one islet," he adds, " we have the yellowish-green unicoloured variety, once described as *H. subglobosa*, and on another, within a very short distance, we find a banded variety, and none others."

The reticulated wrinkling of the surface of the shell of *H. nemoralis*, is a character that comes with age. As the growth of the whorls matures, the arcuated striæ become obscure, and give place to the sculpture expressed by the term 'corrugately malleated.'

F

5. Helix Cantiana. *Kentish Helix.*

Shell; rather narrowly umbilicated, globosely depressed, subdiscoid, thin, semitransparent, rufous, dull, livid white towards the apex, spire but little raised, suture rather impressed, whorls five and a half to six, rounded, densely rudely arcuately plicately striated; aperture lunar-circular, lip sharp, a little expanded, edged within, columellar margin shortly dilately reflected, but not covering the umbilicus.

Helix Carthusiana, Draparnaud (1801), *Tabl. Moll.* p. 86 (not of Müller).
Helix Cantiana, Montagu (1803), *Test. Brit.* p. 422 and Supp. p. 145. pl. xxiii. f. 1.
Helix pallida, Donovan (1803), *Brit. Shells*, vol. v. pl. clvii. f. 2.
Theba Carthusiana, Risso (1826), *Hist. Nat. Europ. Mérid.* vol. iv. p. 74.
Teba Cantiana, Leach (1831), *Turt. Man.* p. 94.
Fruticicola Carthusiana, Held (1837), *Isis*, p. 914.
Bradybæna Cantiana and *Brunonensis*, Beck (1837), *Ind. Moll.* p. 19.
Helix Carthusianella, pars, Morelet (1845), *Desc. Moll. du Port.* p. 62.
Helix Galloprovincialis, Dupuy (1848), vol. ii. p. 204. pl. xvi. f. 9 to 12.
Helix (Zenobia) Cantiana, Moquin-Tandon (1855), *Hist. Moll.* vol. ii. p. 201. pl. xvii. f. 9 to 13.

Hab. Central and South France. Italy. Portugal. Chiefly south-eastern counties of England. (Among hedges, mostly in chalk districts.)

This and the following species should be studied together. They have a certain general resemblance, and are frequently confounded by authors, but they are in reality very distinct. *H. Cantiana*, which Draparnaud originally took for Müller's *H. Carthusiana*, and, to mend the matter, named the true species *H. Carthusianella*, is much the larger shell, and of more irregular rudely striated growth. The striæ where they emerge from the sutures are even clumsily puckered; and colour comes to our aid in characterizing this species. The lower half of the shell is always tinged with a rufous foxy rust colour, while towards the apex the shell is whitish and rather opake. In *H. Carthusiana*, as we shall presently see, the shell is smaller, and, though more transparent, it is yet firmer, and more finely and regularly striated. The aperture, as in *H. Cantiana*, has an internal rib, but it is more developed and more conspicuously milk-white, edged with red.

The geographical range of *H. Cantiana* is confined chiefly in this country to the south-eastern counties, including Essex. Specimens have been taken in the north and west of England, but it is hardly indigenous in those parts. Mr. Guise informs me that he lately met with *H. Cantiana* in abundance near the sea, between Swansea and Oystermouth, in South Wales, under circumstances which left no doubt in his mind that it had been transported thither. In Ireland it is unknown. In France it inhabits chiefly the central and southern departments. In Portugal it is recorded by M. Morelet as *H. Carthusianella*, larger opal-white variety, met with by him only in the environs of Oporto.

In my monograph of *Helix* in Conch. Iconica (pl. clxii. fig. 1079), I figured *H. limbata*, Drap., as a British species, on the strength of the discovery of some specimens by Mr. Sowerby, in the neighbourhood of Hampstead. Its claim to a place in our fauna has not however been confirmed by subsequent collectors. It is a native of Central and South France, and comes as near to us as the environs of Dieppe.

6. **Helix Carthusiana.** *Carthusian Helix.*

Shell; minutely umbilicated, globosely depressed, whitish horny, semitransparent, spire but little raised, whorls six, rounded, rather produced at the periphery, minutely arcuately striated, shining; aperture broadly lunar, ribbed within, rib milk-white, lip sharp, a little expanded, more or less stained with red, columellar margin dilately reflected, but not sufficiently to hide the umbilicus.

Helix Carthusiana, Müller (1775), *Verm. Hist.* part ii. p. 75.
Helix nitida, Chemnitz (1786), *Conch. Cab.* vol. ix part 2. p. 103. pl. cxxvii. f. 1130, 1131.
Helix arenaria, Olivi (1792), *Zool. Adriat.* p. 178.
Helix Carthusianella, Draparnaud (1801), *Tabl. Moll.* p. 86.
Helix (Zenobia) bimarginata, Gray (1821), *Lond. Med. Repos.* p. 239.
Theba Carthusianella and *Charpentieri*, Risso (1826), *Hist. Nat. Europ. Mérid.* vol. iv. p. 75.
Helix Gibsii, Montagu (1827), *fide* Leach, Brown, *Illus. Conch. Brit.* pl. xl. f. 49, 51.
Helix rufilabris, Jeffreys (1833), *Trans. Linn. Soc.* vol. xvi. p. 509.
Monacha Carthusianella, Fitzinger (1833), *Syst. Verz.* p. 95.

Bradybæna Carthusiana, Beck (1837), *Ind. Moll.* p. 19.
Fruticicola Carthusianella, Held (1837), *Isis*, p. 914.

Hab. Central and Southern Europe. South-eastern counties of England (on the chalk downs).

The true *H. Carthusiana*, as we have it in the south-eastern counties of England upon the chalk downs, is a smaller and more symmetrically developed shell than *H. Cantiana*. On the Continent the two forms come nearer to each other, and they have been frequently recorded as varieties of the same species under Draparnaud's inadmissible name of *Carthusianella*. The shell is of a firm growth characterized by a livid transparency, minutely striated throughout, and there is a conspicuous milk-white rib within the aperture, between which and the lip the shell is coloured red.

H. Carthusiana is less common than *H. Cantiana*, and more restricted in its range of habitation.

---

7. Helix Pisana. *Pisa Helix.*

Shell; very narrowly umbilicated, subglobose, rather thin; rusty or fawn-white, encircled with interrupted pencillings of black, sometimes chestnut-filleted, spire rather obtuse, apex red-brown, whorls five, arcuately rudely striated, minutely decussated with very fine spiral striæ, convex; aperture lunar-circular, lip simple, columellar margin rather broadly dilately reflected, sometimes, together with the body whorl, tinged with rose.

Helix Pisana, Müller (1774), *Verm. Hist.* part ii. p. 60.
Helix zonaria, Pennant (1777), *Brit. Zool.* p. 137. pl. lxxxv. f. 133.
Helix petholata, Olivi (1792), *Zool. Adriat.* p. 178.
Helix rhodostoma, Draparnaud (1801), *Tabl. Moll.* p. 74.
Helix cingenda, Montagu (1803), *Test. Brit.* p. 418. Supp. pl. xxiv. f. 4.
Helix strigata, Dillwyn (1817), *Recent Shells*, vol. ii. p. 911.
Theba Pisana, Risso (1826), *Hist. Nat. Europ. Mérid.* vol. iv. p. 73.
Helix albella, Fleming (1828), *Brit. Anim.* p. 260.
Carocolla maculata, Menke (1828), *Syn. Meth. Moll.* p. 25.
Teba cingenda, Leach (1831), *Turt. Man.* p. 92.
Xerophila Pisana, Held (1837), *Isis*, p. 913.

FAMILY COLIMACEA.

Euparypha rhodostoma, Hartmann (1840), *Erd. und Süssw. Gast.* part 1. p. 204. pl. lxxix, lxxx.
Helix (Heliomane) Pisana, Moquin-Tandon (1855), *Hist. Moll.* vol. ii. p. 259. pl. xix. f. 9 to 20.
Hab. Southern Europe. North Africa. Canary Isles and Azores. West of England, south-east of Ireland and Jersey. (In fields and on sand-hills near the sea.)

Helix Pisana would seem to be rather an intruder in the British fauna. It has only been found in Cornwall and South Wales, in the south-east of Ireland and in Jersey. Its native home is in Italy and the south of France, Spain, Portugal, and North Africa. In Spain it is the common snail of the country. In the north of France and in Germany, the species is unknown. The animal is yellowish-grey, purplish towards the head, and the shell is cream-coloured buff, banded with interrupted pencillings of black or brown and sometimes tinged about the aperture and body-whorl with rose. Mr. Jeffreys has remarked that the more sheltered the place it inhabits, the less rosy the shell is; and Dr. Gray deduces from this observation, that the animal thrives better in sunny places, as being most like the warmer climate which is its more natural temperature.

---

8. Helix virgata. *Striped Helix.*

Shell; moderately umbilicated, subglobosely turbinated, rather thin, dead white or buff-tinted, variously encircled with rust-brown bands and lines, spire somewhat raised, horny at the apex, whorls five and a half to six, slopingly rounded, sometimes obtusely keeled at the periphery, closely rather roughly arcuately striated; aperture lunar-rounded, lip simple, strongly internally ribbed, tinged with flesh-pink, columellar margin dilated partly round the circumference of the umbilicus.

Cochlea virgata, Da Costa (1778), *Brit. Conch.* p. 79, pl. iv. f. 7.
Helix virgata, Pulteney (1799), *Hist. Dorset*, p. 47.
Helix zonaria, Donovan (1800), *Brit. Shells*, vol. ii. p. 65.
Helix variabilis, Draparnaud (1801), *Tabl. Moll.* p. 73.
Helix subalbida, Poiret (1801), *Coq. de l'Aisne*, p. 83.
Helix elegans, Brown (1817), *Wern. Trans.* vol. vi. p. 524. pl. xxiv. f. 9.

Helix disjuncta, Turton (1819), *Conch. Dict.* p. 61. f. 63.
Helicella variabilis, Risso (1826), *Hist. Nat. Europ. Mérid.* vol. iv. p. 71.
Helix monilifera, Menke (1830), *Syn. Moll.* 2nd edit. p. 22.
Teba virgata, Leach (1831), *Turt. Man.* p. 93.
Helix Terverii, Michaud (1831), *Comp. Moll. Drap.* p. 26. pl. xiv. f. 20, 21.
Xerophila variabilis and *Terverii*, Held (1837), *Isis*, p. 913.
Thea virgata and *Terverii*, Beck (1837), *Ind. Moll.* p. 12.
Helix (Heliomane) variabilis, Moquin-Tandon (1855), *Hist. Moll.* vol. ii. p. 262. pl. xix. f. 21 to 26.

Hab. Central and Southern Europe. North Africa. Nearly throughout Britain. (Chiefly in sandy and chalky districts near the sea.)

This and the two following species are the snails seen in such myriads on our chalk downs after a shower of rain, crawling up the blades of grass; and it is said to be owing in some measure to the relish of the sheep for these dainties, that our Southdown mutton has attained its celebrity. *H. virgata* has a subglobose buff-tinted shell, very prettily painted with rust-brown bands, of various widths, and lines, and the aperture is often tinged with a blush of flesh-pink, just within the lip, where there is an internal rib. The animal is rather coarsely wrinkled, pale at the sides, and livid-purple about the neck and head. It is much less abundant in our inland counties. On the Continent it follows the usual range of southern species to North Africa.

9. **Helix fasciolata.** *Finely-banded Helix.*

Shell; rather largely umbilicated, depressly globose, dead-white or buff-tinted, variously encircled with occasionally interrupted rust-brown bands and lines, spire but little exserted, horny at the apex, whorls five and a half, rather narrow, slopingly rounded, faintly obtusely keeled at the periphery, obliquely wrinkled throughout with close-set rib-like striæ; aperture small, lunar-rounded, with an internal milk-white rib, lip simple, ribbed within, sometimes tinged with flesh-pink.

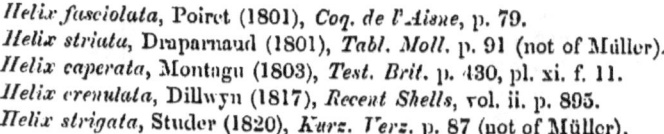

Helix fasciolata, Poiret (1801), *Coq. de l'Aisne*, p. 79.
Helix striata, Draparnaud (1801), *Tabl. Moll.* p. 91 (not of Müller).
Helix caperata, Montagu (1803), *Test. Brit.* p. 430, pl. xi. f. 11.
Helix crenulata, Dillwyn (1817), *Recent Shells*, vol. ii. p. 895.
Helix strigata, Studer (1820), *Kurz. Verz.* p. 87 (not of Müller).

Teba caperata, Leach (1831), *Turt. Man.* p. 97.
Xerophila striata, Held (1837), *Isis*, p. 913.
Helix (Helicella) fasciolata, Moquin-Tandon (1855), *Hist. Moll.* vol. ii. p. 239. pl. xviii. f. 7 to 10.

Hab. Central and Southern Europe. Algeria. Throughout Britain, but local. (Under stones and in sand-banks, principally near the sea.)

To English collectors, this pretty wrinkle-striated *Helix* is best known by the name of *caperata*. Montagu was the first to give a figure of it; but M. Moquin-Tandon has satisfactorily shown that Poiret's *H. fasciolata* can be no other than this species, and his name takes precedence. It is a smaller shell than *H. virgata*, more depressly convoluted, and the spiral bands are more frequently interrupted. The whorls are narrower, and they form a larger umbilicus; the principal specific character is, however, the close sculpture of oblique rib-like striæ, which is the same in all, whatever may be the variety of painting. The animal is of a silvery grey, streaked and stained about the neck and head with brown. Its range of habitation, like the preceding species, extends from Central Europe to North Africa, but it is less abundant.

10. Helix ericetorum. *Heath Helix.*

Shell; broadly perspectively umbilicated, depressly orbicular, somewhat discoid, buff-tinted white variously linearly banded with pale brown, spire scarcely exserted, whorls five and a half, narrow, rounded, obliquely closely plicately striated, last whorl a little descending in front; aperture small, nearly circular, sometimes faintly ridged within, lip simple, margins approximating.

Cochlea ericetorum, Lister (1678), *Hist. Anim. Ang.* p. 126.
Helix Itala, Linnæus (1758), *Syst. Nat.* p. 1245.
Helix ericetorum and *striata*, Müller (1774), *Verm. Hist.* part ii. p. 33. 38.
Helix albella, Pennant (1777), *Brit. Zool.* ed. iv. vol. iv. p. 132. pl. lxxxv. f. 122 (not of Linnæus nor Draparnaud).
Helix erica, Da Costa (1778), *Test. Brit.* p. 53. pl. iv. f. 8.
Helix nivea, Gmelin (1788), *Syst. Nat.* p. 3639.
Helix neglecta, Hartmann (1821), *Syst. Gast.* p. 51 (not of Draparnaud).

Helicella ericetorum, Risso (1826), *Hist. Nat. Europ. Mérid.* vol. iv. p. 7.
Zonites ericetorum, Leach (1831), *Turt. Man.* p. 101.
Oxychilus ericetorum, Fitzinger (1833), *Syst. Verz.* p. 100.
Theba ericetorum, Beck (1837), *Ind. Moll.* p. 13.
Xerophila ericetorum, Held (1837), *Isis*, p. 913.
Helix arenosa, Ziegler (1838), *Rossm. Icon.* vol. vii. p. 34. f. 119.
Helix (Helicella) ericetorum, Moquin-Tandon (1855), *Hist. Moll.* vol. ii. p. 252. pl. xviii. f. 30 to 33.

Hab. Central and Southern Europe. Nearly throughout Britain. (Principally on heaths and downs near the sea.)

The well-known Heath Snail may be recognized by the depressed subdiscoidal form and large umbilicus of the shell, the whorls being narrow and widely tubularly convoluted. The painting appears in bands and pencillings of semitransparent brown above the periphery upon a buff-tinted ground. *H. ericetorum* is as abundant and has much the same habit as *H. virgata*. It is more essentially a southern species affecting a calcareous soil. After a shower of rain, it may be seen by myriads attached to blades of grass on the downs and roadside pastures near the sea; but plentiful as they are in Europe, neither species has been transplanted to the United States, nor have they any representative species in that country. Out of from sixty to seventy *Helices* indigenous to the United States, there are only two with banded or variegated shells, *H. alternata* and *Cumberlandiana*, and these are of quite a distinct pattern.

M. Moquin-Tandon gives some interesting particulars of the anatomy and habits of *H. ericetorum*. It is, he says, a shy, inactive mollusk, withdrawing itself on the slightest touch into its shell, which is carried inclined while crawling. The epiphragm is more or less painted, very thin and transparent, iridescent and membranous, but it has a small calcareous spot upon it of the size of the respiratory orifice. Lister stated, and Mr. Jeffreys corroborates the statement, that continued rain kills *H. ericetorum* in numbers.

Linnæus named this species *Helix Itala*. The original type, in the collection of the Linnean Society of London, has the figures 598, referring to the number of the species in the 'Systema Naturæ,' still remaining upon it in Linnæus's handwriting. But it was known in this country nearly a century before by the name of *ericetorum*, which has ever since been universally adopted.

11. Helix obvoluta. *Obvolute Helix.*

Shell; openly umbilicated, flatly discoid, brown opake, obscurely downy, spire concavely immersed, whorls six, increasing slowly, narrowly appressed, densely finely striated, impressed at the suture, last whorl a little descending in front, constricted behind the lip; aperture small, obtusely triangular, lip reflected, lilac-tinged within, basal and right margins obtusely one-toothed within.

Helix obvoluta, Müller (1774), *Verm. Hist.* part ii. p. 27.
Helix holoserica, Gmelin (1788), *Syst. Nat.* p. 3641 (not of Studer).
Helix bilabiata, Olivi (1792), *Zool. Adriat.* p. 177.
Helix trigonophora, Lamarck (1792), *Journ. Hist. Nat.* vol. ii. p. 349. pl. xlii. f. 2, *a*, *b*.
Planorbis obvolutus, Poiret (1801), *Coq. de l'Aisne,* p. 89.
Helicodonta obvoluta, Risso (1826), *Hist. Nat. Europ. Mérid.* vol. iv. p. 65.
Trigonostoma obvolutum, Fitzinger (1833), *Syst. Verz.* p. 98.
Vortex obvoluta, Beck (1837), *Ind. Moll.* p. 29.
Gonostoma obvoluta, Held (1837), *Isis,* p. 915.
Helix (Trigonostoma) obvoluta, Moquin-Tandon (1855), *Hist. Moll.* vol. ii. p. 114. pl. x. f. 26 to 30.

Hab. Central Europe. Rare in Britain (among moss at the roots of trees at Ditcham Wood, Hunts, and, on the northern side of the chalk escarpment of the South Downs).

This very distinct form of *Helix* is rare in Britain, and doubts are still entertained as to whether it is indigenous. It is unknown in Scotland or Ireland, and has only been found in England at Ditcham and Stoner Hill, in Hampshire, and, according to Forbes and Hanley, for some distance on the northern side of the chalk escarpment of the South Downs. Throughout France, especially in the northern parts, *H. obvoluta* is comparatively abundant. It is essentially a Continental type, being represented in a more angularly compressed form by *H. angigyra,* and in a similar form with the labial teeth more prominently developed by *H. holoserica.* The animal is described by Forbes and Hanley as being dusky, with the head and neck nearly black. The shell is a closely convoluted discoid with the spire concavely immersed, dark umber-brown in colour, obscurely covered with a fine down. The aperture is curiously developed into a compressed angle with indications of teeth on the inner margin.

12. Helix lapicida. *Stone-cutter Helix.*

Shell; openly deeply umbilicated, lenticularly depressed, fawn brown, variegated, spire obtusely convex, whorls five, slopingly convex, sharply keeled at the periphery, minutely granosely shagreened throughout, last whorl suddenly deflected at the aperture, constricted behind the lip; aperture horizontal, transversely oval, lip continuous, expandedly reflected, white.

Helix lapicida, Linnæus (1758), *Syst. Nat.* 10th edit. p. 768.
Helix acuta, Da Costa (1778), *Test. Brit.* p. 55. pl. iv. f. 9.
Helix affinis, Gmelin (1788), *Syst. Nat.* p. 3621.
Vortex lapicida, Oken (1815), *Lehrb. Nat.* vol. iii. p. 314.
Carocolla lapicida, Lamarck (1822), *Anim. sans vert.* vol. vi. part 2. p. 99.
Helicigona lapicida, Risso (1826), *Hist. Nat. Europ. Mérid.* vol. iv. p. 66.
Chilotrema lapicida, Leach (1831), *Turt. Man.* p. 106.
Latomus lapicida, Fitzinger (1833), *Syst. Verz.* p. 97.
Lenticula lapicida, Held (1837), *Isis*, p. 913.
Helix (Vortex) lapicida, Moquin-Tandon (1855), *Hist. Moll.* vol. ii. p. 137. pl. xi. f. 22 to 27.

Hab. Northern and Central Europe. Central and Southern England. (In limestone and chalky districts, chiefly among rocks.)

In this species we have another isolated type, *Carocolla* of Schumacher and Lamarck, in which the shell is compressed into the shape of a lens, with the whorls attenuated and keeled at the periphery. There are not many of the type in Europe; it is an intertropical form, and has its maximum development in the *H. Listeri* and *parmula* of the Philippine Islands. *H. lapicida* has rather a curious range of geographical distribution. It is not found south of the Pyrenees, but extends northward from that latitude as far as Sweden; yet it is not found in Scotland and the northern counties of England, nor has it been collected in Wales or Ireland. From Yorkshire and Lincolnshire to Portland Island and North Devon, *H. lapicida* appears in occasional plenty. The animal is a pallid rufous brown, grey towards the posterior end, covered throughout with promiscuous dark tubercles. The shell has a shagreen surface of fine grains. The last whorl being suddenly deflected, the aperture is very obliquely horizontal, and the margin is white, conspicuously expandedly reflected. This colourless expansion of the lip arises from the mantle of the animal being produced around the aperture in concentric overlapping segments.

The name "Stone-cutter" was probably given to this species by Linnæus, from the circumstance of the animal having been observed to absorb indentations on the surface of limestone rocks, after the manner of *H. aspersa* and others.

13. Helix rufescens.   *Rufous Helix.*

Shell; moderately deeply umbilicated, subglobosely depressed, thin, reddish or pale yellow, spire obtusely convex, whorls six, slopingly convex, densely minutely striated, faintly keeled and pale-banded at the periphery; aperture obliquely lunar-circular, lip thin, a little expanded, a milk-white rib at a short distance in the interior.

Helix rufescens, Pennant (1777), *Brit. Zool.* ed. iv. vol. iv. p. 134. pl. lxxxv. f. 127.
Helix circinata, cælata, and *montana*, Studer (1820), *Kurz. Verz.* p. 86.
Helix corrugata and *clandestina*, Hartmann (1821), *Neue Alp.* vol. i. p. 256.
Helix striolata, C. Pfeiffer (1828), *Deutsch. Moll.* vol. iii. p. 28. pl. vi. f. 8.
Teba rufescens, Leach (1831), *Turt. Man.* p. 96.
Bradybæna rufescens, cælata, and *circinata*, Beck (1837), *Ind. Moll.* p. 20.
Fruticicola circinata, Held (1837), *Isis*, p. 914.
Helix rufina, Parreys (1841), *Pfeiff. Symb.* part 1. p. 39.
Helix (Zenobia) rufescens, Moquin-Tandon (1855), *Hist. Moll.* vol. ii. p. 206. pl. xvi. f. 18, 19.

Hab. Central and Southern Europe. Central and Southern counties of England and Ireland. North Africa. Madeira. (In gardens and hedges.)

Helix rufescens has a shell of delicate texture roughly striated, with traces, especially in the suture, of a fine pilous epidermis. As the name implies, it is often red, a kind of rust-red, but the prevailing colour is a dull pallid yellow, with an indistinct band round the periphery. The animal is more variable, passing from yellow to ash-brown, and even black, and it is sometimes striped. This species is not uncommon in gardens and hedges in the central and southern parts of England and Ireland, and ranges to Algeria and Madeira. It is not found in the north of France, and Mr. Jeffreys gives Westmoreland as its northern limit in Britain.

14. Helix hispida. *Hairy Helix*.

Shell; rather openly umbilicated, orbicularly depressed, rufous or yellowish horny, faintly banded at the periphery, semitransparent, downy with short hairs; spire convex, impressed at the sutures, whorls five to six, rather narrow, sometimes obtusely produced at the periphery, densely obliquely striated, aperture rather small, broadly lunar, lip thinly expanded, white-ribbed at some distance within, basal margin a little drawn out.

Helix hispida, Linnæus (1758), *Syst. Nat.* 10th edit. f. 771.
Trochulus hispidus, Chemnitz (1766), *Conch. Cab.* vol. ix. part ii. p. 52. pl. cxxii. f. 1057, 1058.
Helix plebeium, Draparnaud (1805), *Hist. Moll.* p. 105. pl. vii. f. 5.
Helix rudis and *rufescens*, Studer (1820), *Kurz. Verz.* p. 86, 87.
Helicella hispida, Risso (1826), *Hist. Nat. Europ. Mérid.* vol. iv. p. 72.
Helix lurida, Ziegler (1828), *Pfeiff. Deutsch. Moll.* vol. ii. p. 33. pl. vi. f. 14, 15.
Helix concinna and *plebeia*, Jeffreys (1833), *Trans. Linn. Soc.* vol. xvi. p. 510.
Bradybæna hispida and *plebeia*, Beck (1837), *Ind. Moll.* p. 20.
Fruticicola hispida, Held (1837), *Isis*, p. 914.
Helix depilata, Alder (1837), *Mag. Zool. and Bot.* vol. ii. p. 107 (not of Draparnaud).
Helix (Zenobia) concinna, hispida, and *plebeia,* Moquin-Tandon (1855) *Hist. Moll.* vol. ii. p. 221, 224, 225. pl. 17, f. 8, 9, 14 to 18.

Hab. Central and Southern Europe. Common throughout Britain. Siberia. (Under stones, fallen trees, decaying leaves, etc.)

This species somewhat resembles *H. rufescens* in form, but the shell is much smaller and more hairy. There are four well-marked varieties. In a typical state the epidermis is covered with a down of fine hairs; in the variety of which Mr. Jeffreys has made a species, *H. concinna*, the whorls are rather produced at the periphery, and the hairs are fewer and more deciduous; in a variety named by Mr. Alder *H. depilata*, the shell is bald; lastly there is a small dwarfed form, called the mountain variety. The spire is rather more depressed in some species than in others, and the whorls in such specimens are less closely convoluted, leaving a more open umbilicus. The animal is mostly of a mottled grey colour, lighter towards the edge of the foot.

H. hispida is common throughout Central and Southern Europe, and is quoted by Gerstfeldt in his list of Siberian shells collected at the mouth of the Ussuri, a tributary of the Amoor.

15. Helix sericea. *Silky Helix.*

Shell; minutely deeply umbilicated, subglobose, whitish horny, subhyaline, covered with short bristly silky hairs, spire conoidly convex, whorls five to six, rather broad, obliquely closely striated; aperture broadly lunar, lip thinly reflected, basal margin partially dilated round the umbilicus.

Helix sericea, Draparnaud (1801), *Tabl. Moll.* p. 85.
Helix albula, Studer (1820), *Kurz. Verz.* p. 87.
Helix granulata, Alder (1830), *Trans. Nat. Hist. Soc. Northumb.* vol. i. p. 39.
Helix globularis, Jeffrey (1833), *Trans. Linn. Soc.* vol. xvi. p. 507.
Monacha sericea, Fitzinger (1833), *Syst. Verz.* p. 95.
Fruticicola sericea, Held (1837), *Isis*, p. 914.
Helix piligera, Ziegler (1839), *Anton, Verz. Conch.* p. 36.
Helix (Zenobia) sericea, Moquin-Tandon (1855), *Hist. Moll.* p. 219. pl. xvii. f. 6, 7.

Hab. Throughout Europe, but local and rare. England, principally in the western and southern counties. Irkutsk, Siberia. Caucasus. (Vicinity of damp mossy banks, and under stones.)

Compared with *H. hispida*, the shell of this species may be at once recognized by its globose form, more pallid subhyaline substance, and minute umbilicus. The hair with which it is covered is not a fine down, but bristly and silky, rather distantly rooted on the surface, so firmly so in some specimens as to render them slightly granular to the touch. Moquin-Tandon describes the animal as being timid, irritable, slow, carrying the shell a little inclined when crawling.

The species is not common, and the records of its geographical range are comparatively few. It appears to be rather scattered in England, principally in the southern and western counties, local and rare. Its presence in Ireland is rather doubtful, although Mr. Thompson gives Lagan, near Belfast. Gerstfeldt gives Irkutsk as its Siberian habitat, and Krynicki the Caucasus, but the species was formerly, and is still, frequently confounded with *H. hispidus*.

16. Helix revelata. *Discovered Helix.*

Shell; minutely umbilicated, subglobose, very thin, membranaceous, greenish umber, spire convex, whorls four, impressed at the suture, then rounded, obliquely plicately wrinkled in a crimped manner, hairy, the hairs being comparatively distant, short and rather rigid; aperture lunar-circular, lip thin, scarcely reflected, basal margin a little dilated round the umbilicus.

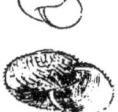

Helix revelata, Férussac (1821), *Tab. Syst.* p. 44. no. 273.
Helicella revelata, Beck (1837), *Ind. Moll.* p. 7.
Helix subviridis, Bellamy (1841), *Brit. Assoc. Rep.*
Hab. France. Channel Islands. South-west of England. (Among grass and at the roots of shrubs.)

A delicate greenish-umber membranaceous shell, having somewhat the appearance of a globose exotic *Vitrina*. It is, however, obliquely plicately wrinkled in a crimped manner, and beset throughout with fine rather distant bristly hairs, erectly rooted in the epidermis, as in *H. sericea*. The lip is scarcely reflected except at its junction with the body-whorl, where it is a little dilated round the small umbilicus. *H. revelata* was not observed in Britain until several years after its discovery in France, and English conchologists hesitated to regard it as anything more than the young of *H. sericea*. It was detected first in Britain in Guernsey, by Professor Forbes, then in Cornwall by Mr. A. E. Benson, son of the well-known Indian conchologist, and it has been since collected in Devonshire and in the Scilly Isles. Mr. Jeffreys, says that "in winter and dry weather, it buries itself rather deep in the earth, and must be looked for by pulling up tufts of grass and large stones which are sunk in the ground, as well as by searching among the roots and furze-bushes."

Dupuy, Moquin-Tandon, and Jeffreys have referred to *H. revelata* the *H. occidentalis*, Recluz (*H. Ponentina*, Morelet), which is quite another species, a firm opake shell with a broadly reflected lip admirably figured by Morelet (*Moll. du Port.* pl. vi. f. 4), and of which there are well authenticated specimens named by Pfeiffer, in the collection of Mr. Cuming. Mr. Jeffreys' description of the animal of *H. revelata*, which partakes very much of the details of Moquin-Tandon's description of *H. occidentalis*, must be received with caution.

17. Helix fusca. *Fuscous Helix.*

Shell; scarcely umbilicated, depressed, very thin, membranaceous, glossy fuscous-olive, spire convex, sutures rather impressed, whorls five, arcuately plicately wrinkled, minutely hairy, slopingly convex, last whorl obsoletely angled at the periphery; aperture lunar, lip simple, basal margin shortly dilated round the umbilicus.

Helix fusca, Montagu (1803), *Test. Brit.* p. 424. pl. xiii. f. 1.
Helix (Zenobia) corrugata, Gray (1820), *Lond. Med. Repos.* vol. xv. p. 239.
Helix subrufescens, Miller (1829), *Ann. Phil.* New Series, vol. iii. p. 379.
Zonites fuscus, Macgillivray (1843), *Moll. Aberd.* p. 93.
Vitrina membranacea and *margaritacea*, Brown (1845), *Illus. Conch. Brit.* pl. xl. f. 3 to 5 and 54 to 56.
Helix (Zenobia) fusca, Moquin-Tandon (1855), *Hist. Moll.* vol. ii. p. 212. pl. 15. f. 33 to 36.

Hab. Throughout Britain. Western maritime parts of France. (In woods and bushy places, under leaves or upon brambles.)

The shell of *H. fusca* is larger and more depressed than that of *H. revelata*, but of the same membranaceous substance, wanting the greenish cast of colour. The umbilicus is even smaller, often scarcely distinguishable, and the hair is so fine and deciduous that the surface is frequently supposed to be without hair. M. Moquin-Tandon describes the animal as being irritable, especially in front, keeping the upper tentacles always in motion, and crawling with some rapidity. "Individuals of this species," he adds, "love to congregate, and reciprocally polish each others' shell with the foot."

A peculiar interest attaches to this species. It is found in all parts of Britain south of Aberdeen, and has been known from the commencement of the present century. But *H. fusca* was not observed on the Continent until 1838, and then only in the part nearest to Britain. The first record of its appearance out of Britain is by M. Bouchard-Chantereaux, who, under the name of *H. revelata*, for which species he mistook it, describes it as being common in the neighbourhood of Boulogne. The Abbé Dupuy gives also Mont-de-Marsan as a habitat; and M. Grateloup has collected it at Dax, between Bordeaux and the Pyrenees. Is it then an instance of an originally British type, spreading in a direction contrary to that of all the other species of the genus, in which the theoretical law of migration points to a north-westerly course?

18. Helix fulva. *Fulvous Helix.*

Shell; minutely umbilicated, globosely trochiform, very thin, glossy, fulvous, spire obtusely conoid, whorls six, convex, narrow, last whorl flatly convex at the base, very finely striated; aperture depressed, narrowly lunar, lip simple, thinly dilated over the umbilicus.

Helix fulva, Müller (1774), *Verm. Hist.* part 2. p. 56.
Trochus terrestris, Da Costa (1778), *Test. Brit.* p. 36 (not of Chemnitz).
Helix trochiformis, Montagu (1803), *Test. Brit.* p. 427, pl. xi. f. 9.
Helix nitidula, Alton (1812), *Syst. Abhandl.* p. 53. pl. liv. f. 8 (not of Draparnaud).
Helix trochulus, Dillwyn (1817), *Desc. Rec. Shells,* vol. ii. p. 916 (not of Hartmann).
Theba fulva, Leach (1831), *Turton, Man.* p. 99.
Helix Mortoni, Jeffreys (1833), *Trans. Linn. Soc.* vol. xvi. p. 332.
Conulus fulvus, Fitzinger (1833), *Syst. Verz.* p. 94.
Polita fulva, Held (1837), *Isis,* p. 916.
Petasia trochiformis, Beck (1837), *Ind. Moll.* p. 21.
Zonites (Conulus) fulvus, Moquin-Tandon (1855), *Hist. Moll.* vol. ii. p. 67. pl. viii. f. 1 to 4.

Hab. Throughout Europe. Siberia. Azores. (Under stones, decaying leaves and moss, in damp and shady places.)

Authors are somewhat divided in opinion, as to whether this little species should be referred to *Helix* or *Zonites.* It has a decidedly heliciform shell, composed of six narrow whorls convoluted into a globose trochus-shape, so closely as only to leave the smallest possible umbilicus; while the animal is described as closely resembling that of *Zonites nitidus.* It is found in all parts of Britain, and is very generally distributed throughout Europe, passing to Siberia in the north and to the Azores in the south.

H. fulva is represented by a very similar and equally widely spread species in the United States, *H. chersina,* Say, which some writers incline to think is the same.

FAMILY COLIMACEA.

19. Helix lamellata. *Lamellated Helix.*

Shell; minutely deeply umbilicated, conoidly globose, olive horny, subfuscous, spire obtusely raised, whorls six, narrow, rounded, crossed obliquely with membranaceous lamellæ, last whorl swollen, excavated around the umbilicus; aperture lunar, lip simple.

Helix lamellata, Jeffreys (1830), *Trans. Linn. Soc.* vol. xvi. p. 338.
Helix Scarburgensis, Bean (1830), Alder, *Trans. Nat. Hist. Soc. Northumb.* vol. i. p. 39.
Helix seminulum, Rossmässler (1838), *Icon. Moll.* f. 533.
Hab. North Britain. North Germany. Sweden. (In woods, among dead leaves.)

A very interesting small species, for the discovery of which we are indebted to Mr. Jeffreys. The shell is as round as a pea, globosely swollen, peculiarly impressly excavated around the umbilicus, which is minute and deep, and obliquely ribbed throughout with membranaceous lamellæ. It is essentially a northern species. It is widely spread in Scotland and Ireland; in England it has not been found south of Scarborough. The only Continental habitats yet recorded are North Germany and Sweden.

---

20. Helix aculeata. *Prickly Helix.*

Shell; moderately umbilicated, globosely pyramidal, fuscous horny, spire raised, sutures impressed, whorls four to four and a half, rounded at the upper part, obtusely angled at the periphery, very finely striated, crossed obliquely throughout by fibrous epidermic lamellæ which at the angle of the periphery are elongated into lashes; aperture rounded, lip slightly expanded, margins approximating.

Helix aculeata, Müller (1774), *Verm. Hist.* p. 81.
Helix spinulosa, Lightfoot (1786), *Phil. Trans. Roy. Soc.* vol. lxxvi. p. 166. pl. ii. f. 2.
Teba spinulosa, Leach (1831), *Turt. Man.* p. 100.
Fruticicola aculeata, Held (1837), *Isis*. p. 914.
Helix Granatelli, Bivon (1839), *Occh. Giorn. Palerm. Mag.* n. 9. f. 2.

Helix (Fruticicola) aculeata, Moquin-Tandon (1855), *Hist. Moll.* vol. ii. p. 189. pl. xv. f. 5 to 9.

Hab. Throughout Europe. Azores. (Under dead leaves, moss, and stones, in damp places.)

Mollusks affecting damp places of habitation secrete in general a copious supply of mucus, and produce sombre-coloured shells, which they invest with a more than ordinarily fibrous epidermis. This is especially seen in the *Melaniadæ*, inhabiting the marshes and sluggish streams of Pernambuco, Borneo, and the Philippine Islands; and in *Melania setosa* of the last-named locality, the shell is coronated with a row of flexible epidermic bristles. The same law may be observed among our *Helices*. Species like *H. virgata* and *ericetorum* which inhabit the chalk districts secrete light variegated shells, combining a larger proportion of lime; while, in the shells of *H. hispida* and *sericea*, dwelling under stones and among wet moss, the shell is horny, and the constituent elements partake more largely of animal matter. In our elegant little *H. aculeata*, the epidermis is somewhat of the character of that of the intertropical *Melania setosa*; it is deposited on the shell in obliquely longitudinal lamellæ, and these are prolonged beyond the angle of the whorls into a corona of lashes.

H. aculeata occurs in all parts of the British Isles, and throughout Europe generally, reaching to the Azores.

---

21. **Helix pulchella.** *Pretty Helix.*

Shell; largely umbilicated, depressed, whitish, subhyaline, smooth, covered with a plicately wrinkled epidermis, spire small, raised at the apex, whorls three to four, rounded, the last trumpet-like towards the aperture; aperture round, lip callously expandedly reflected with the margins almost continuous.

Helix pulchella and *costata*, Müller (1774), *Verm. Hist.* part 2. p. 30, 31.
Helix paludosa, Da Costa (1778), *Test. Brit.* p. 59.
Turbo helicinus, Lightfoot (1786), *Phil. Trans. Roy. Soc.* vol. lxxvi. p. 167.
Helix crenella, Montagu (1803), *Test. Brit.* p. 441. pl. xiii. f. 3.
Helix minuta, Say (1818), *Nich. Ency. Phil.* 2nd edit. vol. iv.
Turbo paludosus, Turton (1819), *Conch. Dict.* p. 228.

Lucena pulchella, Hartmann (1821), *Syst. Gast.* p. 54.
Vallonia rosalia, Risso (1826), *Hist. Nat. Europ. Mérid.* vol. iv. p. 102. pl. iii. f. 30.
Amplexus paludosus and *crenellus*, Brown (1827), *Ill. Conch.* pl. xli. f. 70 to 78.
Chilostoma pulchella, Fitzinger (1833), *Syst. Verz.* p. 98.
Circinnaria pulchella, Beck (1837), *Ind. Moll.* p. 23.
Corneola pulchella, Held (1837), *Isis*, p. 912.
Helix (Lucena) pulchella, Moquin-Tandon (1855), *Hist. Moll.* vol. ii. p. 140. pl. xi. f. 28 to 34.

Hab. Throughout Europe. Madeira. Azores. Thibet. United States. (Under stones or wood, or on walls, either in wet or dry places.)

Of all *Helices* this minute species is the widest distributed over the globe. It is generally diffused throughout Europe, including the British Isles; it is recorded by Gerstfeldt from Siberia; it is found at Madeira and the Azores; it was collected by Dr. Thomson, in 1848, in Thibet; and it inhabits the United States from Maine to South Carolina, and from Vermont to the Missouri. The animal and shell are alike delicate and colourless, save with a faint tinge of yellow; the shell is of an elegant *Cyclostoma*-like form, the aperture expanding at maturity like the mouth of a trumpet, with a thickened reflected lip, of which the margins are almost continuous. The epidermis forms rib-like folds, so characteristic in appearance that they have been mistaken for calcareous ribs, while denuded specimens have been described as another species.

---

22. Helix rotundata. *Rounded Helix.*

Shell; very largely perspectively umbilicated, thinly depressed, almost discoid, horny yellow or fulvous fawn, mostly stained or tessellately spotted with fuscous red, spire more or less convex, sometimes rather flat, sutures impressed, whorls six, narrow, rounded, densely arcuately rib-striated throughout, obtusely produced at the periphery; aperture depressly lunar, lip simple.

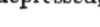

Helix rotundata, Müller (1774), *Verm. Hist.* part 2. p. 29.
Helix radiata, Da Costa (1778), *Test. Brit.* p. 57. pl. iv. f. 15, 16.
Helix Turtoni, Fleming (1828), *Brit. Anim.* p. 269.

Zonites radiatus, Leach (1831), *Turt. Man.* p. 102.
Discus rotundatus, Fitzinger (1833), *Syst. Verz.* p. 99.
Euryomphala rotundata, Beck (1837), *Ind. Moll.* p. 9.
Patula rotundata, Held (1837), *Isis*, p. 916.
Zonites rotundatus, Gray (1840), *Turt. Man.* p. 165. pl. r. f. 44.
Helix (Delomphalus) rotundatus, Moquin-Tandon (1855), *Hist. Moll.* vol. ii. p. 107. pl. x. f. 9 to 12.

Hab. Throughout Europe. Azores. (Under stones and ruins.)

This and the following species have shells composed of narrow subtubular whorls convoluted round an axis which leaves a wide umbilicus open to the interior of the apex. *H. rotundata* is common throughout the British Isles and in all parts of Europe, and it is very closely represented in the United States by a species in which the whorls are still more openly convoluted, *H. perspectiva*, Say. *H. Gueriniana*, Lowe, a Madeiran species, and *H. engonata*, Shuttleworth, a native of Teneriffe, are also of the same typical form, with the shell more angled at the periphery. The whorls of *H. rotundata* are obtusely produced, not to say angled, at the periphery, and they are ribbed throughout with conspicuously raised striæ. The spire is faintly tessellated with brown, but sometimes colourless and translucid, in which state it has been mistaken for a *Zonites*.

---

23. **Helix rupestris.** *Rock Helix.*

Shell; very largely perspectively umbilicated, depressly orbicular, brown horny, spire convex, sutures impressed, whorls four to five, rounded, smooth or minutely striated; aperture small, lunar-rounded, lip simple.

Helix rupestris, Studer (1789), *Faun. Helvet. in Coxe, Trav. Switz.* vol. iii. p. 430 (without characters).
Helix rupestris, Draparnaud (1801), *Tabl. Moll.* p. 71.
Helix pusilla, Vallot (1801), *Exerc. d'Hist.* p. 5.
Helix umbilicata, Montagu (1803), *Test. Brit.* p. 434. pl. xiii. f. 2.
Helix saxatilis, Hartmann (1821), *Syst. Gast.* p. 52.
Helicella rupestris, Risso (1826), *Hist. Nat. Europ. Mérid.* vol. iv. p. 69.
Zonites rupestris, Leach (1831), *Turt. Man.* p. 103.
Turbo Myrmecidis, Scacchi (1833), *Osserv. Zool.* vol. i. p. 11.
Pyramidula rupestris, Fitzinger (1833), *Syst. Verz.* p. 95.
Euryomphala umbilicata, Beck (1837), *Ind. Moll.* p. 9.
Patula rupestris, Held (1837, *Isis*, p. 916.

Delomphalus rupestris and *saxatilis*, Hartmann (1840), *Syst. Gast.* vol. i.
    p. 120 and 122. pl. xxxvii. f. 1 to 6.
Helix spirula, Villa (1841), *Disp. Syst. Conch.* p. 53.
Helix aliena, Ziegler (1841), *Pfeiff. Symb.* part 1. p. 39.
Helix (Hygromane) rupestris, Moquin-Tandon (1855), *Hist. Moll.* vol. ii.
    p. 192. pl. xv. f. 10 to 13.

Hab. Central and Southern Europe. Madeira. Throughout Britain. (Chiefly among rocks in mountainous places.)

One of the most conspicuous characteristics of the shell of this species, and from which Montagu derived his name, is its largely excavated perspective umbilicus. *H. rotundata* has an almost equally large umbilicus, but the shell is much larger, red-spotted, and densely rib-striated throughout. Here it is smooth, and the species is altogether smaller. "The animal," says Moquin-Tandon, "is sluggish and irritable, especially when exposed to the light, secreting a very abundant mucus." It has been noticed from the earliest observers to dwell chiefly among rocks in mountainous districts in all parts of Britain and in Central and Southern Europe. It has a characteristic analogue, very similar in character and of similar habit, in *H. saxicola* of Cuba and Texas.

Dr. Gray complains that English conchologists have done an injustice to their countryman Montagu, in not having more generally adopted his name of *umbilicata* for this species. He claims priority for it on the ground that Montagu's name was published in 1803, whilst the name of *rupestris* appeared for the first time in the posthumous Hist. Moll. of Draparnaud, published by his widow in 1805. The name *rupestris* was given to this *Helix* by Studer, in his appendix to 'Coxe's Travels in Switzerland in 1789,' and it was adopted by Draparnaud in the 'Tableau' published during his lifetime in 1801, with the following very accurate description, not only of the shell, but of the animal and its habits:—" Coquille brune, torse ; spire élevée, ouverture ronde, ombilic évasé. Haut. 2 mill., larg. $2\frac{1}{3}$, diam. $2\frac{1}{4}$. Habite France méridionale sur les rochers élevés (4 tours). Animal noirâtre, plus pâle en dessous. Tentacules supérieurs courts, gros et très-obtus; inférieurs à peine visibles à la loupe, et semblables à de petits tubercules. Il redresse sa coquille, et la porte très-élevée lorsqu'il marche." The book is very scarce. My own library copy, formerly in the library of Professor Brongniart, and generously presented to me by M. Bourguignat, is the only one that I have seen.

24. Helix pygmæa. *Pigmy Helix.*

Shell; minute, largely umbilicated, orbicularly convex, brownish horny, spire but little exserted, whorls three to four, convex, narrow, smooth or finely striated; aperture lunar, lip simple.

Helix minuta, Studer (1789), *Faun. Helvet. in Coxe, Trav. Switz.* vol. iii. p. 428 (without characters).
Helix pygmæa, Draparnaud (1801), *Tabl. Moll.* p. 93.
Helix Kirbii, Sheppard 1823), *Trans. Linn. Soc.* vol. xiv. p. 162.
Discus pygmæus, Fitzinger (1833), *Syst. Verz.* p. 99.
Euryomphala pygmæa, Beck (1837), *Ind. Moll.* p. 9.
Patula pygmæa, Held (1837), *Isis*, p. 916.
Zonites pygmæus, Gray (1840), *Turt. Man.* p. 167.
Helix (Delomphalus) pygmæa, Moquin-Tandon (1855), *Hist. Moll.* vol. ii. p. 103. pl. x. f. 2 to 6.

Hab. Throughout Europe. Siberia. Azores. (Under stones, and among decaying leaves in moist woods, or among grass.)

Conchologists are pretty well agreed upon this being a species distinct from the preceding, but it has sometimes been taken for the young of it. The shell of *H. pygmæa* is much smaller than that of *H. rupestris*, and composed of fewer whorls. The umbilicus is rather variable, always large, however; and the whorls being rather more depressly convoluted, it is not so deeply perspective. The animal is a dingy black, or grey freckled with black, extremely timid and irritable, according to Moquin-Tandon, avoiding the light of day and enclosing itself within its shell on the slightest touch. "When this mollusk is contracted," he adds, "the upper tentacles may be seen to be directed towards the umbilicus. The eyes are as large as those of *Pupa*."

Helix pygmæa is common in all parts of the British Isles, but being very minute, the smallest of our *Helices*, may easily escape notice. Dr. Turton's plan of collecting them, Mr. Jeffreys says, was to get a quantity of dead and rather moist leaves and spread them on a sheet of paper to dry, when the refuse yielded a good harvest. Others collect them by brushing wet grass with an entomologist's gauze net.

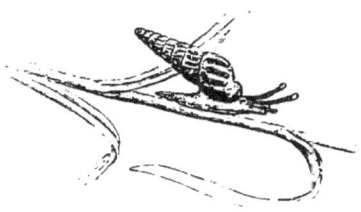

Bulimus acutus.

Genus V. **BULIMUS**, *Scopoli*.

Animal; slender, bearing a turriculate shell of from eight to nine whorls, mantle not reflected over the shell, head with the upper pair of tentacles rather long, the lower short, foot lanceolate, posteriorly minute, palate, teeth serrated.

Shell; minutely umbilicated, turriculate or conically cylindrical, rather thin, variegated or plain horny, whorls eight to nine, moderately convex, striated in the direction parallel with the lines of growth, aperture small, lip mostly dilated over the umbilicus.

The value of the genus *Bulimus* rests very much on its geographical limits. The importance hitherto attached to the number of whorls of the shell and to their elongately drawn out mode of convolution is now confirmed by observations made on the habits and distribution of the species. The British species are too few to furnish materials for generalization. Comparing the foreign *Bulimi* with *Helices*, it will be found that they are more arboreal in tropical countries, and less numerous and smaller in the temperate and sub-temperate zones. At the Philippine Islands, in New Granada, at Natal, Brazil, and Bolivia, *Bulimus* exists of very large size, much surpassing any *Helix* in dimensions. In Europe, *Bulimus* is greatly surpassed by *Helix* in size, and the number of species is quite insignificant in comparison. The number of *Bulimi* throughout the globe compared with the *Helices* (including *Zonites*) is about one to two. The latest census by Pfeiffer gives *Bulimus* 1100, *Helix* 2050. In France, the proportion is one in eight; in Britain, only one in ten. In the United States there is no *Bulimus* at all north of Tennessee, which is fifteen degrees south of Britain geographically, and nearly five isothermally. The chief habitats of the genus are the Central American, Chilian, and Bolivian provinces in the new world, and the Malayan in the old. Another feature in the distribution of the *Bulimi* which adds to their generic importance is that whilst the proportion in number of *Helices* in the two hemispheres is nearly three to one, of *Bulimi* it is as two to three. The denser and more widely spread forest vegetation of South and Central America appears to be more favourable to the production and habits of *Bulimus* than of *Helix;* in the rainy districts of Venezuela the shell is of quite a solid growth, with a richly variegated epidermis.

In Britain we have only three small ground species; one, *B. acutus,* inhabiting chiefly the south and western chalk downs near the sea; a second, *B. montanus,* local in comparatively few places in the south and midland counties; and a third, *B. obscurus,* of smaller size, which is very general throughout.

1. **acutus.** Shell conically turreted, of eight to ten whorls, light horny or brown, marbled with opake-white streaks, black-banded below.
2. **montanus.** Shell conically cylindrical, of seven whorls, olive horny, minutely shagreened.
3. **obscurus.** Shell smaller, turriculate cylindrical, fuscous horny, glossy.

---

1. Bulimus acutus. *Sharp Bulimus.*

Shell; conically turreted, minutely umbilicated, thin, light horny, or brown, marbled with opake cream-coloured streaks and blotches, last whorl mostly encircled round the lower part by a fuscous black band; whorls eight to nine, striated in a wrinkled manner, finely plicately crenulated at the sutures; aperture ovate, lip simple, broadly reflected over the umbilicus.

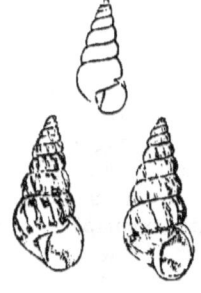

Helix acuta, Müller (1774), *Verm. Hist.* part ii. p. 100 (not of Lamarck).
Turbo fasciatus, Pennant (1777), *Brit. Zool.* ed. iv. vol. iv. p. 131. pl. lxxxii. f. 119.
Helix cretacea, Chemnitz (1786), *Conch. Cab.* vol. ix. part 2. p. 190. pl. cxxxvi. f. 1263.

FAMILY COLIMACEA.

Bulimus acutus, Bruguière (1789), *Enc. Méth.* vol. vi. part 1. p. 323.
Turbo Turricula Maroccana, Chemnitz (1795), *Conch. Cab.* vol. ii. p. 280. pl. ccix. f. 2063.
Helix bifasciata, Pulteney (1799), *Cat. Dorset.* p. 40.
Bulimus variabilis, Hartmann (1815), *Sturm, Faun.* vol. vi. n. 12.
Bulimus articulatus, Lamarck (1819), *Anim. sans vert.* vol. vi. part 2. p. 124.
Cochlicella meridionalis, Risso (1826), *Hist. Nat. Europ. Mérid.* vol. iv. p. 78. pl. iii. f. 26.
Lymnæa fasciata, Fleming (1830), *Edin. Ency.* vol. vii. part 1. p. 78.
Elisma fasciata, Leach (1831), *Brit. Moll.* p. 109.
Bulimus ventricosus, Turton (1831), *Man.* p. 84. f. 69.
Bulimus turritella, Andrz. (1832), *Kryn. Bull. Mosc.* vol. vi. p. 415.
Bulimus elongatus, Cristofori and Jan (1832), *Cat. Conch.* n. 1772.
Cochlicellus acutus and *Maroccanus*, Beck (1837), *Ind. Moll.* p. 63.
Bulimus litoralis, Brumati (1838), *Conch. Montfalc.* p. 34. f. 9.
Helix (Cochlicella) acuta, Moquin-Tandon (1855), *Hist. Moll.* vol. ii. p. 280. pl. xx. f. 27 to 32.

Hab. South-western Europe. Chiefly south and west of England and Ireland. (Abundant near the sea, especially on chalky or sandy soils.)

To an attentive observer of shell-structure viewed in relation with the physical conditions with which the animal producing it is surrounded, *Bulimus acutus* may be readily seen to be an inhabitant of a dry soil, rather bare of vegetation, in the vicinity of the sea. The shells of mollusks affecting this habitat are generally opake-white, more or less variegated and banded, as in *Helix virgata* and *ericetorum*, or horny, obliquely streaked with opake-white. This type of structure is especially developed in the shells of the numerous small *Bulimi* which inhabit the sandy plains and elevated grounds of Chili and Peru; and our *Bulimus acutus*, of similar habit, has a shell of much the same character. It is of a delicate horny substance, composed of from eight to nine whorls slenderly acuminately convoluted, obliquely streaked throughout with opake-white, on a yellowish, sometimes a brown ground; and round the base of the whorl there is almost invariably a fuscous-black band, which is concealed from view in all but the last whorl by the overlapping of one whorl upon the other. The animal, M. Mcquin-Tandon says, is that of *Helix*, the structure of the jaw leaving no doubt on this subject; it is, therefore, a form intermediate between the two genera. It is, he adds, an active but very irritable creature, withdrawing itself into the shell on the slightest touch.

Bulimus acutus is found abundantly on our south-western downs attached to blades of grass, as represented in our vignette, and it is supposed to share largely with *Helix virgata* and *ericetorum* in the honour of fattening our famous Southdown mutton.

2. Bulimus montanus. *Mountain Bulimus.*

Shell; conically cylindrical, minutely compressly umbilicated, dark olive-horny, whorls seven, obliquely plicately striated, minutely shagreened throughout; aperture somewhat squarely ovate, lip thinly reflected, whitish, broadly dilated next the umbilicus.

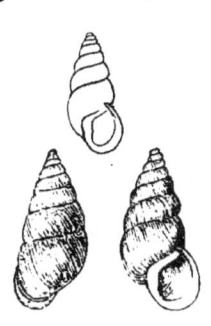

Helix sylvestris, Studer (1789), *Faun. Helv. Coxe, Trav. Switz.* vol. iii. p. 43 (without characters).
Bulimus montanus, Draparnaud (1801), *Tabl. Moll.* p. 65.
Helix Lackhamensis, Montagu (1803), *Test. Brit.* p. 394. pl. ii. f. 3.
Helix buccinata, Alten (1812), *Syst. Abhand.* p. 100. pl. xii. f. 22.
Lymnæa Lackhamensis, Fleming (1814), *Edin. Encyc.* vol. vii. part 1. p. 78.
Bulimus obscurus var., Hartmann (1821), *Syst. Gast.* p. 50.
Helix montana, Férussac (1822), *Tabl. Syst.* p. 60.
Bulimus Lackhamensis, Fleming (1828), *Brit. Anim.* p. 265.
Bulimus Montacuti, Jeffreys (1830), *Trans. Linn. Soc.* vol. xvi. p. 345.
Ena montana, Leach (1831), *Brit. Moll.* p. 112.
Buliminus Lackhamensis, Beck (1837), *Ind. Moll.* p. 71.
Merdigera montana, Held (1837), *Isis*, p. 917.
Bulimus (Ena) montanus, Moquin-Tandon (1855), *Hist. Moll.* vol. ii. p. 289. pl. xxi. f. 1 to 4.

Hab. Central Europe. South and west-central counties of England. (In wooded districts, on trees and among decaying leaves.)

This and the next species live, as their dull fuscous horny shells, without pattern or marking of any kind, indicate, in moist woody districts, on trees or among decaying leaves. *B. montanus*, which is the larger, has an obtuse conically cylindrical shell of seven whorls, obliquely striated and minutely shagreened throughout. The animal is described as being like that of *B. obscurus*, but paler. It inhabits the south and west-central counties of England, and is very local. On the Continent it appears to have the same kind of distribution, in rather more elevated situations.

3. **Bulimus obscurus.**   *Concealed Bulimus.*

Shell; acuminately cylindrical, slightly turriculated, minutely umbilicated, fuscous horny, semitransparent; whorls six and a half to seven, very finely obliquely striated; aperture small, lip somewhat squarely ovate, lip callously expanded, shining white, dilated next the umbilicus, margins inclined to approximate.

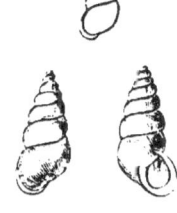

Helix obscura, Müller (1774), *Term. Hist.* part 2. p. 103.
Turbo rupium, Da Costa (1778), *Test. Brit.* p. 90.
Bulimus hordeaceus, Bruguière (1789), *Enc. Méth. Vers*, p. 331.
Helix stagnorum, Pulteney (1799), *Cat. Dorset.* p. 49. pl. xix. f. 27.
Bulimus obscurus, Draparnaud (1801), *Tabl. Moll.* p. 65 (not of Poiret).
Lymnæa obscura, Fleming (1814), *Edin. Encyc.* vol. vii. part 1. p. 78.
Ena obscura, Leach (1831), *Brit. Moll.* p. 113.
Buliminus obscurus, Beck (1837), *Ind. Moll.* p. 71.
Merdigera obscura, Held (1837), *Isis*, p. 917.
Bulimus Astierianus, Dupuy (1849), *Hist. Moll.* vol. iii. p. 320. pl. xv. f. 7.
Bulimus (Ena) obscurus, Moquin-Tandon (1855), *Hist. Moll.* vol. ii. p. 291. pl. xxi. f. 5 to 10.
Bulimus Humberti, Bourguignat (1857), *Rev. et Mag. de Zool.* No. 1. *Amén. Malac.* vol. i. p. 28. pl. ii. f. 5 to 7.

Hab. Throughout Europe. Siberia. (In old walls and among ruins, under stones, in woods, on trees, and among moss.)

A species of general occurrence throughout Britain and the Continent, reaching across Siberia to the district of the Amoor. The shell is much smaller than that of *B. montanus*, and is composed of about half a whorl less, convoluted in a more turriculate cylindrical manner. The surface, not being shagreened, is more transparent and glossy, and the reflected margin of the aperture is more callously expanded.

The animal is grey or dark mottled, somewhat roughly tubercled for its size, and discharges a rather copious supply of mucus over its shell, which is frequently encrusted from this cause by dirt and such other substances as may become agglutinated to it. "It is slow and sluggish," says M. Moquin-Tandon, "in its habits, irritable, delighting in moisture, and carrying its shell horizontally when crawling."

Zua subcylindrica. (*Very much enlarged.*)

Genus VI. ZUA, *Leach*.

Animal; body oblong, obtusely rounded in front, converging to an angular point behind, dingy grey, but little tubercled, carrying a rather narrow, oblong, glossy, transparent shell, upper pair of tentacles much the longer, with the bulbous extremities rather lengthened.

Shell; imperforate, oblong-cylindrical, of five to five and a half, smooth, glossy, transparent whorls; aperture small, lip simple, callous, right margin obsoletely truncated at the end.

This little snail has been separated as a genus chiefly on account of the shining vitrified surface and marginal subtruncation of the shell, which somewhat partakes in these respects of the characters of the little glassy *Achatinæ* of the tropics. The animal is the same as that of *Helix*, excepting that the bulbous extremities of the tentacles are rather more lengthened. Risso and Jeffreys first separated it from *Bulimus*, but included it with some others in rather an incongruous medley. Dr. Leach was the first to distinguish it as a genus by itself. M. Moquin-Tandon retains *Zua* in his genus *Bulimus*, adopting Risso's appellation of *Cochlicopa* in a subgeneric sense. Forbes and Hanley consider it entitled to be kept apart generically from *Bulimus*.

Zua has the widest distribution, in space and elevation, of all our land mollusks. It occurs in countless numbers, both in wet and dry places, throughout the whole of the Caucasian province of distribution—enclosing Europe, North Africa, and Western Asia—as well as throughout the chief portion of the United States.

1. **subcylindrica.** Shell imperforated, oblong, glossy, of five to five and a half whorls, columellar lip obsoletely truncated at the end.

1. Zua subcylindrica. *Subcylindrical Zua.*

Shell; cylindrically oblong, imperforate, very transparent, smooth, shining, horny, spire obtuse, whorls five to five and a half, convex, the last rather produced; aperture rather small, lip simple, opakely callous, left margin obsoletely truncated at the end.

Helix subcylindrica, Linnæus (1767), *Syst. Nat.* 12th edit. p. 1248 (not of Montagu).
Helix lubrica, Müller (1774), *Verm. Hist.* part 2. p. 104.
Turbo glaber, Da Costa (1778), *Test. Brit.* p. 87. pl. v. f. 18.
Bulimus lubricus, Bruguière (1789), *Enc. Méth. Vers.* vol. i. p. 311.
Lymnæa lubrica, Fleming (1814), *Edin. Encyc.* vol. vii. part 1. p. 78.
Cochlicopa lubrica, Risso (1826), *Hist. Nat. Europ. Mérid.* vol. iv. p. 80.
Cionella lubrica, Jeffreys (1830), *Trans. Linn. Soc.* vol. xvi. p. 347.
Achatina lubrica, Menke (1830), *Syn. Moll.* p. 29.
Zua lubrica, Leach (1831), *Brit. Moll.* p. 114.
Columna lubricus, Cristofori and Jan (1832), *Cat.* n. 6.
Styloides lubricus, Fitzinger (1833), *Syst. Verz.* p. 105.
Achatina subcylindrica, Deshayes (1839), *Anton, Verz. Conch.* p. 44.
Zua Boissii, Dupuy (1850), *Hist. Moll.* vol. iv. p. 332. pl. xv. f. 9.
Bulimus (Cochlicopa) subcylindricus, Moquin-Tandon (1855), *Hist. Moll.* vol. ii. p. 304. pl. xxii. f. 15 to 19.
Ferussacia subcylindrica, Bourguignat (1856), *Aménu. Malac.* vol. i. p. 209.

Hab. Throughout Europe. Northern and Central Asia. North Africa. Madeira. United States. (Under stones, logs, and leaves, both in wet and dry places.)

Zua subcylindrica has of all our land mollusks the most extended distribution in height and in space, being found both in wet and dry places, on low ground as well as at a considerable elevation on the mountain side. It occurs throughout Europe, in Central and Western Asia, from the Amoor to Cashmere and Thibet, and in North America it is found in the north-western State of Ohio, in the Middle United States, and in all the States of New England. "On visiting Oak Island, Chelsea," says Dr. Gould in his 'Report of the Invertebrata of Massachusetts,' "I found the surface of the ground covered with these shells in incalculable numbers. Hundreds might be taken up clinging to a single fallen leaf; as the moisture evaporated they all disappeared beneath the leaves."

The shell, which is not umbilicated, is composed of five whorls, closely convoluted into a rather narrow, subcylindrical, oblong form, transparent in substance, and with a shining, glossy surface. The aperture is small, and the lip is peculiarly callous, the base of the right margin being marked by a subtruncate, faintly angular indentation. The species is more generally known to collectors by its name of *lubrica*, but modern writers have restored to it the Linnean name *subcylindrica*. Doubts have been entertained of this being the Linnean *Helix subcylindrica*, because that species is described in the 'Systema Naturæ' as inhabiting water. The animal can, however, exist some time under water, and is not unfrequently found amid the floods and overflows of rivers.

Azeca tridens. (*Enlarged.*)

Genus VII. **AZECA**, *Leach*.

Animal; body oblong, attenuately rounded in front, converging to a point behind, dingy speckled grey, moderately tubercled, carrying a cylindrical chrysalis-shaped shell, upper pair of tentacles slender where exserted.

Shell; cylindrical, attenuately contracted at the base, shining horny, smooth, whorls seven, convex; aperture small, contractedly ear-shaped, toothed within, lip forming an opake continuous rim.

Azeca is by no means, as some have ventured to assert, the same generic form as *Zua*, with no other difference than a toothed aperture to its shell. It is a mollusk of different distribution and habit, and the shell has a totally distinct typical structure. On reaching maturity, instead of being expanded, it is attenuately contracted, and the aperture has a compressly distorted ear-like form. The interior is furnished with three prominent teeth within, and sometimes with one or two smaller teeth.

We have only one species of *Azeca* in Britain. It occurs but sparingly in England, and has not been collected in Scotland or

Ireland. Germany appears to be its northern limit on the Continent. Bourgnignat refers to this genus as many as twelve additional species, from Sicily, Algeria, Madeira, and the Canary Islands, described by himself under *Azeca* and by Cantraine, Lowe, Webb and Berthelot, Roth and Morelet, under *Bulimus, Achatina, Glandina, Ferussacia,* and *Tornatellina.* The genus has not been detected in the United States.

1. **tridens.** Shell cylindrically ovate, shining, of seven whorls, contracted at the base, with a small three-toothed ear-shaped aperture.

1. **Azeca tridens.** *Three-toothed Azeca.*

Shell; imperforate, cylindrically ovate, transparent, smooth, shining, horny, spire, rather obtuse, whorls seven, compressly convex, the last short, attenuately contracted towards the aperture, which is small and obliquely ear-shaped, furnished interiorly with three prominent teeth, and sometimes with two or more smaller teeth, lip slightly sinuous at the upper part, extending round the aperture in a continuous callous rim.

Turbo tridens, Pulteney (1799), *Cat. Dorset.* p. 46. pl. xix. f. 12.
Carychium Menkeanum, C. Pfeiffer (1828), *Deutsch. Moll.* vol. i. p. 70. pl. iii. f. 42.
Helix Goodalli, Férussac (1822), *Tabl. Syst.* p. 75 (not of Miller).
Pupa tridens, Gray (1820), *Ann. Phil.* vol. ix. p. 413 (not of Draparnaud).
Pupa Menkeana, C. Pfeiffer (1828), *Deutsch. Moll.* vol. iii. p. 67. pl. vii. f. 7, 8.
Carychium politum, Jeffreys (1830), *Trans. Linn. Soc.* vol. xvi. p. 365.
Azeca tridens, Leach (1831), *Brit. Moll.* p. 122.
Pupa Goodallii, Michaud (1831), *Comp.* p. 67. vol. xv. f. 39, 40.
Azeca Mateni, Turton (1831), *Brit. Shells,* p. 68. f. 52.
Achatina Goodallii, Rossmässler (1839), *Icon.* vol. ix. p. 33. f. 654.
Achatina tridens, Pfeiffer (1846), *Zeitschr. Malak.* p. 162.
Azeca Nouletiana, Dupuy (1849), *Hist. Moll.* p. 358. pl. xv. f. 12.
Bulimus (Azeca) Menkeanus, Moquin-Tandon (1855), *Hist. Moll.* vol. ii. f. 302. pl. xxii. f. 7 to 14.
Cochlicopa tridens, Jeffreys (1862), *Brit. Conch.* vol. i. p. 290.
Hab. Central Europe. England. (Widely but sparingly distributed in wooded districts.)

The shell of *Azeca tridens* is of a peculiar chrysalis form, transparent and shining, looking very much as if it were membranaceous and flexible during the life of the animal. The aperture is small, sloping and attenuated towards the base, ear-shaped, and a little distorted. Within are three prominent teeth, with sometimes one or two smaller teeth, including a small threadlike plait winding into the interior.

It is only sparingly diffused in England, between the northern and south-western counties, and has not been observed in Scotland or Ireland. France and Germany appear to be the only habitats recorded on the Continent.

Achatina acicula. (*Considerably enlarged.*)

Genus VIII. ACHATINA, *Lamarck.*

Animal; body slender, attenuated to a point behind, white, carrying a narrowly convoluted colourless hyaline shell, upper pair of tentacles slender, lower pair very short, eyeless.

Shell; imperforate, elongated, narrow, hyaline, colourless, whorls smooth, margined at the sutures; aperture small, columella involute, truncated at the end.

The brilliant tropical genus *Achatina*, including the largest of all land mollusks, is represented at its northernmost limit, in Britain, by a single small species, which, with comparatively few exceptions, is the smallest of all. In the sultry woods of West Africa, *Achatina* produces a shell as large as an ostrich's egg. The shell of our solitary British species scarcely measures the fifth of an inch in length. It belongs, however, rather to the *Glandina* section of the genus, inhabiting chiefly Central America and the West Indies, but well represented, though more sparingly, in Southern Europe and Algeria, Madeira, Ceylon, and Hindostan. The nearest allied foreign species to *A. acicula*, so near indeed as to be mistaken for it, is a minute Indian species, *A. Balanus*, inhabiting the banks of the Jumna. The only other European *Achatina* is one of which the shell is an inch and three-quarters in length, *A. Algira*, a native of the provinces of Austria and Italy, the Morea, some of the islands of the Mediterranean, and Algeria.

A. acicula lives buried under stones or at a depth of several inches in the soil, and both animal and shell are colourless. The animal is also supposed to be without eyes.

1. **acicula.** Shell cylindrically subulate, diaphanous, of six whorls, columella thinly involute, truncated.

---

1. **Achatina acicula.** *Needle Achatina.*

Shell; imperforate, cylindrically subulate, diaphanous white, smooth, transparent, spire obtuse at the apex, whorls six, flatly convex, margined at the suture; aperture narrowly ovate, columella arched, involute, truncated at the end.

Buccinum acicula, Müller (1774), *Verm. Hist.* part 2. p. 150.
Bulimus acicula, Bruguière (1789), *Enc. Méth. Vers,* vol. i. p. 311.
Helix acicula, Studer (1789), *Faun. Helv. in Coxe, Trav. Switz.* vol. iii. p. 431.
Buccinum terrestre, Montagu (1803), *Test. Brit.* p. 248. pl. viii. f. 3.
Achatina acicula, Lamarck (1822), *Anim. sans vert.* vol. vi. part 2. p. 133.
Acicula eburnea, Risso (1826), *Hist. Nat. Europ. Mérid.* vol. iv. p. 81.
Cionella acicula, Jeffreys (1830), *Trans. Linn. Soc.* vol. xvi. part 2. p. 347.
Styloides acicula, Fitzinger (1833). *Syst. Verz.* p. 105.
Achatina acuta, Aleron (1837), *Moll. Pyr. Bull. Soc. Phil. Perpig.* vol. iii. p. 92.
Acicula acicula, Beck (1837), *Ind. Moll.* p. 79.
Polyphemus acicula, Villa (1841), *Disp. Conch.* p. 20.
Cæcilioides acicula, Beck (1846), *Amtl. Ber. Vers. Kiel.* p. 122.
Glandina acicula, Albus (1851), *Malak. Mäder.* p. 59. pl. xv. f. 17, 18.
Bulimus (Acicula) acicula, Moquin-Tandon (1855), *Hist. Moll.* vol. ii. p. 309. pl. xxii. f. 32 to 34.
Cæcilianella acicula, anglica, and *Liesvillei,* Bourguignat (1856), *Mag. Zool.* p. 382.

Hab. Throughout Europe. Algeria. Madeira. (Buried in loose earth, among roots, or under stones.)

This minute species is a mollusk of very secluded habits, living among roots or under stones, buried, not unfrequently, several inches in the ground. Its shell is perfectly colourless, and of almost glassy tenuity, but distinctly characterized by an arched, involute columella, truncated at the end. It ranges throughout Europe from Sweden to the Mediterranean; and it is more than probable that M. Bourguignat's *Cæcilianellæ raphidia, tumulorum, Brandellii, subsaxana, nanodea, Syriaca,* and *nyctelia* from Greece, Syria, Algeria, and Madeira, are merely local varieties of it. In Britain it is diffused sparingly throughout England and Ireland, and the Channel Islands, but not in Scotland or the Isle of Man.

Clausilia laminata. (*Moderately enlarged.*)

Genus IX. CLAUSILIA, *Draparnaud.*

Animal; sometimes rather broad, sometimes narrow, carrying a sinistrally convoluted, cylindrically acuminated shell, foot rather obtuse at the posterior end, head with the upper pair of tentacles stout, clavate, the lower ones small.

Shell; sinistral, imperforate, fusiformly tapering, of from nine to twelve whorls, somewhat papillary at the apex, whorls sometimes smooth, generally densely wrinkled, last whorl constricted, with the aperture small, surrounded by a continuous lip, furnished with teeth and internal thread-like plaits, enclosing a lamellar valve or clausilium.

Clausilia is a mollusk of much interest, considered in reference to its shell, its habits, and its geographical distribution. In the soft parts it scarcely differs from *Achatina* or *Bulimus*. The shell is of a narrow cylindrical form, varying little in shape and dimensions through a very considerable number of species, always convoluted sinistrally; and the aperture is more or less furnished at maturity with teeth and internal thread-like plaits. Its chief peculiarity of habit, denoted by the name of the genus, is a faculty which the animal possesses of closing itself in the shell by means of a calcareous appendage, a spoon-shaped lamina, conforming to the contour of the aperture, attached by an elastic filament to the columella. When the animal crawls forth, the clausilium is pushed aside against the columella; when retiring into its shell, it closes on the retreating animal by the aid of the elastic filament.

The geographical distribution of the genus is altogether peculiar. Nine-tenths of the *Clausiliæ* are of the true Caucasian type, having their centre of creation developed to a most prolific extent within a comparatively limited area in the south-eastern parts of Europe, including the islands of the Grecian Archipelago, and in Asia Minor. Upwards of three hundred species have been described from this locality, the greater portion of them being inhabitants of Austria and Hungary; but their progress westward is curiously limited. Not more than a dozen well marked species inhabit France, and only four of these range into Britain, one alone reaching Ireland. In addition to the European and Western Asiatic species, a few *Clausiliæ* differing very little in typical character appear in wide-spread localities; Madeira has six of its own, Java five, Burmah four, India five, China ten, Japan five, Borneo one, the Philippine Islands one; a species is recorded from Cairo, one from Sennaar, and an interesting and beautiful species has been collected within the present year by M. Mouhot in Cambojia. There are no *Clausiliæ* in the United States, nor in any part of the Western Hemisphere, excepting five in the West Indies and Central America. A species has been recorded from Peru, but some doubt attaches to this statement. Of the British species, one, *C. perversa*, is universal, the others are confined to the central and southern counties of England. They are:—

1. **laminata.** Shell moderately large, comparatively smooth, semitransparent, rufous brown or straw-colour.
2. **biplicata.** Shell rather larger, densely ridge-wrinkled throughout, dingy olive-brown.

3. **Rolphii.** Shell smaller, rather swollen, finely wrinkled throughout, light fulvous brown, additional small teeth in aperture.

4. **perversa.** Shell small, narrowly acuminated, finely wrinkled throughout, dark chocolate-brown.

---

1. **Clausilia laminata.** *Laminated Clausilia.*

Shell; cylindrically fusiform, transparent horny, rufous brown or straw-colour, shining, whorls eleven to twelve, depressly convex, smooth, obscurely malleated, faintly obliquely striated; aperture fusiformly ovate, lip continuous, callously expanded, columella strongly two-plaited, with four fine internal thread-like plaits within.

Helix bidens, Müller (1774), *Verm. Hist.* part 2. p. 116.
Turbo bidens, Pennant (1777), *Brit. Zool.* ed. iv. vol. iv. p. 131 (not of Linnæus).
Bulimus bidens, Bruguière (1792), *Enc. Méth. Vers,* p. 2. f. 352.
Pupa bidens, Draparnaud (1801), *Tabl. Moll.* p. 61.
Turbo laminatus, Montagu (1803), *Test. Brit.* p. 359. pl. ii. f. 4.
Clausilia bidens, Draparnaud (1805), *Hist. Moll.* p. 68. pl. iv. f. 5 to 7.
Odostomia laminata, Fleming (1814), *Edin. Encyc.* vol. vii. part i. p. 77.
Clausilia (Marpessa) bidens, Gray (1821), *Lond. Nat. Med. Repos.* vol. xv. p. 239.
Clausilia ampla, Hartmann (1821), *Syst. Gast.* p. 50.
Helix derugata, Férussac (1822), *Tabl. Syst.* p. 67.
Clausilia derugata, Jeffreys (1830), *Trans. Linn. Soc.* vol. xvi. part ii. p. 354.
Clausilia lucida, Menke (1830), *Syn. Moll.* ed. ii. p. 129.
Clausilia lamellata, Leach (1831), *Brit. Moll.* p. 118.
Clausilia laminata, Turton (1831), *Brit. Moll.* p. 70.
Clausilia (Marpessa) laminata, Moquin-Tandon (1855), *Hist. Moll.* vol. ii. p. 318. pl. xxiii. f. 2 to 9.

Hab. Nearly throughout Europe. In England chiefly towards the south. Rare in Ireland. Not in Scotland. (Mostly in beech woods, among decayed leaves about the trunks of trees.)

Of the four species of *Clausilia* inhabiting Britain, this is the only one of which the shell is nearly smooth. Beneath the lens it is abundantly striated in the same order in which the shell of the other three species is plicately wrinkled, but to the naked eye the surface is smooth and glossy, rufous brown, and sometimes of a delicate pellucid straw-colour. The soft parts, as seen in our vignette of the living animal, are mottled with reddish-brown at the side, the upper portion of the body being of a dingy yellowish-grey.

Clausilia laminata does not appear in the Scottish lists of mollusks, and it is a rarity in Ireland. In England it is of general occurrence, but infrequent and local, increasing towards the south. It ranges throughout the Continent from Sweden southwards, especially in the central parts. Some confusion has attended the naming of the species in consequence of its having been taken for the Linnean *Turbo bidens*, which is a Continental species, *C. papillaris*, Draparnaud, not found in Britain.

2. Clausilia biplicata. *Two-plaited Clausilia.*

Shell; elongately fusiform, dingy olive-brown, marked at intervals next the sutures with short whitish hair lines, whorls twelve, flatly convex, densely ridge-wrinkled throughout; aperture pyriformly ovate, lip continuous, callously expanded, columella strongly two-plaited, with two or more fine straggling thread-plaits within.

Helix perversa pars, Müller (1774), *Verm. Hist.* part 2. p. 118.
Turbo biplicatus, Montagu (1803), *Test. Brit.* p. 361. pl. xi. f. 5.
Clausilia plicata, Gærtner (1813), *Syst. Wett.* p. 22 (not of Draparnaud).
Odostomia biplicata, Fleming (1814), *Edin. Encyc.* vol. vii. part ii. p. 77.
Clausilia biplicata and *ventricosa*, C. Pfeiffer (1821), *Deutsch. Moll.* part i. p. 61 and 63. pl. iii. f. 27, 29.
Clausilia Montagui, Gray (1825), *Ann. Phil.* p. 413.
Clausilia similis, Charpentier (1835), *Rossm. Icon.* vol. i. p. 77. f. 30.
Clausilia vivipara, Held (1837), *Isis*, p. 309.

Clausilia cordata, vulnerata, infulæformis, radicans, rostrata, and *quadrata*, Forster (1841), *Nov. Act. Leop.* vol. xix. p. 269. pl. lviii. f. 1 to 6.
Clausilia (Iphigena) biplicata, Moquin-Tandon (1855), *Hist. Moll.* vol. ii. p. 337. pl. xxiv. f. 11, 12.

Hab. Central Europe. Central and southern counties of England. Not in Scotland or Ireland. (In woods and hedges, among the roots of shrubs.)

This is the largest of the British *Clausiliæ*, the shell being more acuminately convoluted and composed of a whorl more than the preceding species. It may be recognized by its brown, rather irregular, roughly wrinkled sculpture, marked at intervals next the suture by faint patches of whitish hair lines. The localities of its habitat are few in number, the chief being the banks of the Thames in the neighbourhood of London.

3. Clausilia Rolphii. *Rolph's Clausilia.*

Shell; oblong-fusiform, semitransparent, fulvous brown, whorls nine to ten, moderately convex, densely striately wrinkled throughout; aperture obliquely subquadrately pyriform, lip thinly callous, columella strongly two-plaited, with sometimes two or three fine intermediate marginal plaits.

Clausilia (Iphigenia) Rolphii, Gray (1821), *Med. Repos.* vol. xv. p. 239.
Clausilia Mortilletii, Dumont (1853), *Bull. Sav. Soc.* (*fide* Moq.-Tand.).
Clausilia plicatula, Forbes and Hanley (1853), *Brit. Moll.* vol. iv. p. 120. pl. cxxix. f. 3 (not of Draparnaud).
Clausilia (Iphigena) Rolphii, Moquin-Tandon (1855), vol. ii. p. 343. pl. xxiv. f. 32 to 35.

Hab. Central Europe. Midland and southern counties of England. (Among dead leaves, beneath the bark of trees, etc.)

Clausilia Rolphii is scarce and not very generally known to collectors. The only recorded habitats of the species in England are Kent, Sussex, Hampshire, and Gloucestershire; it has not been found in Scotland, Ireland, or the Channel Isles. The shell is smaller than that of *C. biplicata*, rather more ventricose in proportion, and of lighter colour, more finely and delicately wrinkled.

Forbes and Hanley considered it identical with a widely diffused Continental species, *C. plicatula*, Draparnaud, but that has a smaller and more slender shell; and the aperture in *C. Rolphii* is distinctly characterized by the presence of three, and sometimes four, fine marginal plaits between the teeth.

4. Clausilia perversa.   *Reversed Clausilia.*

Shell; slenderly acuminately fusiform, dark shining chocolate-brown, subtransparent, marked at intervals near the sutures with short indistinct whitish hair lines, whorls ten to eleven, flatly convex, densely striately wrinkled throughout; aperture obliquely pyriformly ovate, lip callously expanded, columella widely two-plaited, with two or more fine internal thread-like plaits within.

Helix perversa, Müller (1774), *Verm. Hist.* part 2. p. 118 (not of Linnæus, Chemnitz, nor Férussac).
Turbo perversus, Pennant (1777), *Brit. Zool.* p. 130.
Bulimus perversus, Bruguière (1792), *Enc. Méth.* vol. ii. p. 351.
Pupa rugosa, Draparnaud (1801), *Tabl. Moll.* p 63.
Clausilia rugosa and *dubia*, Draparnaud (1805), *Hist. Moll.* p. 70 and 73. pl. iv. f. 10, 19, 20.
Turbo nigricans, Maton and Rackett (1807), *Trans. Linn. Soc.* vol. viii. p. 130.
Odostomia nigricans, Fleming (1814), *Edin. Encyc.* vol. vii. part i. p. 77.
Clausilia roscida and *cruciata*, Studer (1820), *Kurz. Verz.* p. 20.
Clausilia obtusa, C. Pfeiffer (1821), *Deutsch. Moll.* part i. p. 65. pl. iii. f. 33, 34.
Clausilia Everettii, Miller (1822), *Ann. Phil. new series*, vol. iii. p. 377.
Clausilia parvula, Turton (1826), *Zool. Journ.* vol. ii. p. 566.
Clausilia nigricans, Jeffreys (1830), *Trans. Linn. Soc.* vol. xvi. p. 351.
Stomodonta rugosa, Mermet (1843), *Moll. Pyr.-Occid.* p. 147.
Clausilia abietina and *Reboudii*, Dupuy (1851), *Hist. Moll.* vol. v. p. 356 and 358. pl. xvii. f. 5. and pl. xviii. f. 3, 4.
Clausilia (Iphigena) perversa and *nigricans*, Moquin-Tandon (1855), *Hist. Moll.* vol. ii. p. 332 and 334. pl. xxiv. f. 17 to 27.

Hab. Throughout Europe.   (In the crevices of walls, rocks, and trees, and under stones.)

There is no part of the British Isles in which this *Clausilia* may not be found. It is the smallest of our four species, and the shell is the most slenderly convoluted and darkest in colour, a rich chocolate-brown, with traces of the whitish hair lines noticed in *C. biplicata* less disposed in patches. The animal of *C. perversa* is of a dark mottled-grey colour, small and attenuated.

Agreeably with the suggestion of Deshayes (Anim. sans Vert. vol. viii. p. 201, note), I restore to this species the name given to it by Müller, and very generally adopted by the earlier British authors. It is true that Linnæus had a *Helix perversa*, but it is a large Malayan *Bulimus* (*B. citrinus*, Bruguière), not a *Clausilia*. Chemnitz's *H. perversa* is a Tahiti *Partula*, and Férussac's *H. perversa* is the British *Balea*. Draparnaud separated Müller's *Clausilia perversa*, into two species, giving to them the names *C. rugosa* and *dubia*, and Moquin-Tandon has followed Draparnaud, substituting the name *nigricans* for *dubia*, assuming, on the authority apparently of Dr. Gray, that it was so named by Pulteney in 1799. This is a mistake. There is no *Clausilia nigricans* in Dr. Pulteney's original work. The species under consideration is there correctly named *C. perversa;* and in the edition of 1813, *C. perversa* is incorrectly illustrated. The true *C. perversa* is there figured with the name *C. nigricans* improperly given to it by Maton and Rackett.

Balea perversa. (*Enlarged.*)

Genus X. **BALEA**, *Prideaux.*

Animal; lanceolate, rounded in front, attenuated behind, blackish, passing into grey, minutely speckled with dark tubercles, carrying a slight sinistrally convoluted turreted shell, tentacles short, upper pair approximating.

Shell; sinistral, minutely umbilicated, conically turreted, of seven whorls, transparent horny, finely striated, aperture rather small, lip thin, thinly reflected.

We have in Britain, plentifully diffused from North Scotland to the Channel Islands, plentifully diffused also throughout Europe from Sweden to the islands of the Mediterranean and Atlantic, a little moss-buried snail, carrying a horny sinistrally convoluted turreted shell of only seven whorls, which belongs neither to *Bulimus*, *Clausilia*, nor *Pupa*. It is not like the shell of *Bulimus*, for it is of very delicate transparent horny substance coiled sinistrally; it is not like that of *Clausilia*, which is of a different texture, more cylindrically acuminated, and has the special provision of a clausilium, with teeth and winding plates in the mature aperture, which *Balea* has not; it is not like *Pupa*, which, including *Vertigo*, is of a peculiar shortly cylindrical form, mostly furnished in the mature aperture with teeth.

In attempting to monograph this genus, Dr. Pfeiffer has included species, such as the large Brazilian *Pupa elatior* of Spix, which have no relation whatever with it, and has abandoned them; and in a subsequent edition of his monograph, he included two Sandwich Island snails described by Dr. Newcomb as *Balea*, which he has had to abandon. In the third and latest edition of Pfeiffer's monograph of *Balea*, a Cuba shell described by Gundlach (*B. Funcki*), and a shell described by myself thirteen years ago, from the Andes of Caxamarca, Peru (*Bulimus Clausilioides*), were included, but neither of these can be regarded as representatives in the Western Hemisphere of the European *Balea*. It is more than probable that the genus will have to be restricted to the European species, which, in addition to *B. perversa*, distributed throughout, include *B. Sarsi*, described as inhabiting Norway and Sweden, and *B. Tristensis* and *ventricosa* inhabiting the island of Tristan d'Acunha. Even these may prove to be varieties.

The animal of *Balea* scarcely differs from that of the genera already cited.

1. **perversa.** Shell small, turreted, transparent horny, of six to seven finely striated whorls convoluted sinistrally, with a simple small aperture having a thinly reflected lip.

1. Balea perversa. *Reversed Balea.*

Shell; sinistral, minutely umbilicated, conically turreted, fulvous-olive, very thin, horny, semitransparent, glossy, crossed irregularly with opake, whitish lines, whorls seven, rather flatly convex, beneath the lens minutely wrinkle-striated, rather impressed at the sutures, last whorl sometimes furnished with a small callosity or lamella; aperture pyriformly subquadrate, lip thin, simple, slightly reflected, sinuous above, dilated over the umbilicus.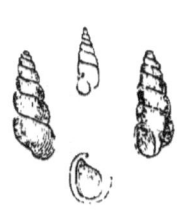

Turbo perversus, Linnæus (1758), *Syst. Nat.* 10th edit. p. 767.
Bulimus perversus, Poiret (1801), *Coq. de l'Aisne, Prod.* p. 57.
Pupa fragilis, Draparnaud (1801), *Tabl. Moll.* p. 64.
Clausia parvula, Gærtner (1813), *Conch. Wetter.* p. 22.
Odostomia perversa, Fleming (1814), *Edin. Encyc.* vol. vi. p. 76.
Clausilia fragilis, Studer (1820), *Kurz. Verz.* p. 89.
Helix perversa, Férussac (1822), *Tabl. Syst.* p. 66 (not of Linnæus).
Balea fragilis, Prideaux (1824), *Zool. Journ.* vol. i. p. 61. pl. vi.
Balea perversa, Fleming (1828), *Brit. Anim.* p. 271.
Fusulus fragilis, Fitzinger (1833), *Syst. Verz.* p. 105.
Clausilia perversa, Charpentier (1837), *Moll. Suiss.* p. 17.
Pupa perversa, Potiez and Michaud (1838), *Gal. Moll. Douai*, p. 166.
Eruca fragilis, Swainson (1840), *Treat. Malac.* p. 334.
Clausilia uniplicata, Calcara (1840), *Effem. Sicil.* p. 82.
Stomodonta fragilis, Mermet (1843), *Moll. Pyr.-Occid.* p. 48.
Pupa (Balea) perversa, Moquin-Tandon (1855), *Hist. Moll.* vol. ii. p. 349. pl. xxv. f. 6 to 14.

Hab. Throughout Europe. Azores. Madeira. (Among moss and lichens in crevices of walls, rocks, or trees.)

The shell of *Balea perversa* may be readily distinguished from those of *Bulimus, Clausilia*, or *Pupa*, by its simple sinistrally coiled form of seven whorls, and glossy fulvous horny substance, scarcely striated, but characteristically marked in the direction of the lines of growth with irregularly developed whitish opake lines. There is no contraction of the last whorl, and no indication of teeth or winding plaits in the aperture. The only appearance of internal sculpture is the occasional presence of a small callosity or lamella on the body-whorl. The lip is thinly, very thinly, reflected, and it is just sufficiently sinuated to impart a squarish contour to the aperture.

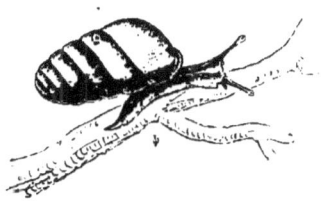

Pupa muscorum. (*Much enlarged.*)

Genus XI. **PUPA**, *Lamarck*.

Animal; body rather short, slenderly acuminated towards the tail, carrying a horny narrow-whorled shell, mostly brownish or slate-grey, dark-streaked on the neck and back, pale towards the sides; upper pair of tentacles rather short, lower pair very short.

Shell; cylindrical, narrowly umbilicated, composed of from six to nine semitransparent horny glossy whorls, mostly smooth, rather obtuse at the apex, rounded at the base; aperture somewhat triangularly ovate, generally more or less toothed within, the teeth having a parietal form, winding in thread-like ridges into the interior.

Land snails are supplied with two pairs of tentacles; water snails with only one pair. In the genera on the confines of these two primary divisions, *Pupa* and *Vertigo* on the one hand, *Carychium* and *Auricula* on the other, an intermediate state of these organs exists. The lower tentacles lessen until they are reduced to rudimentary protuberances; finally they are represented by mere specks, and disappear without a trace of any kind.

In *Pupa*, the lower tentacles are always present, but they vary considerably in their development between short symmetrical tentacles and blunt protuberances. *Pupa* lives chiefly in damp places among moss or under stones. *Vertigo* is almost amphibious in habit. It lives, with rare exception, at the roots of grass in wet places, and the lower tentacles are either represented in the most rudimentary form or are wanting.

We have only four *Pupæ* in Britain, outlying forms of a generic type which, like *Clausilia*, has its centre of creation in Southern

Europe, where the species are most abundant and the shells larger and more opake. The British species have light transparent horny shells. *P. secale*, the largest, is found nearly throughout England, but chiefly among the chalky districts of the south and west parts. It does not inhabit Scotland, nor is it found on the Continent north of Germany. The aperture of *P. secale* is contracted by not fewer than seven internal parietal teeth. Our most common species are two of much smaller dimensions, without internal parietal teeth, *P. muscorum* and *cylindracea*. The first ranges on the Continent from Lapland to Sicily, the second has a rather more southern range, extending to Algeria. Both inhabit the British Islands throughout, *P. cylindracea* being the commoner of the two. The remaining species, *P. Anglica*,—with two or more internal teeth, the smallest of the genus, but larger than any *Vertigo*,—is more scarce. Abroad it has only been observed in Portugal and the south of France, and in Algeria.

The *Pupæ* are chiefly members of the great Caucasian province of the Eastern Hemisphere. After eliminating about seventy from the described list as belonging to *Vertigo*, about one hundred and fifty remain, nearly half of which are South European. Of the remaining seventy-five, rather more than half are distributed in other parts of the Eastern Hemisphere, as follows:—Madeira, fourteen; Mauritius and Madagascar, twelve; South Africa, five; West Africa and Canary Islands, five; India, four; Ceylon, two; Australia, one; New Zealand, one. The remainder belonging to the Western Hemisphere, range as follows:—West Indies, the next principal station to Europe, thirty; South United States, three; Bolivia, two.

The British species are:—

1. **secale.** Shell rather large, elongately cylindrical, of from eight to nine densely ridge-striated whorls; aperture seven-toothed.
2. **muscorum.** Shell small, oblong-cylindrical, of from six to seven glossy semitransparent whorls; aperture sometimes toothless, sometimes one-toothed.
3. **cylindracea.** Shell very small, shortly cylindrical, of six transparent rather swollen whorls; aperture one-toothed.
4. **Anglica.** Shell rather minute, of six semitransparent somewhat narrow whorls; aperture subtriangular five-toothed.

1. Pupa secale. *Rye-grain Pupa.*

Shell; elongately cylindrical, compressly umbilicated, fulvous brown, glossy, whorls eight to nine, convex, obliquely densely ridge-striated, aperture squarely ovate, slightly sinuated above, furnished with seven conspicuously developed parietal teeth winding into the interior.

Pupa secale, Draparnaud (1801), *Tabl. Moll.* p. 59.
Turbo juniperi, Montagu (1803), *Test. Brit.* p. 340. pl. xii. f. 12.
Odostomia juniperi, Fleming (1814), *Edin. Encyc.* vol. vii. part i. p. 76.
Torquilla secale, Studer (1820), *Kurz. Verz.* p. 89.
Chondrus secale, Hartmann (1821), *Syst. Gast.* p. 50.
Helix (Cochlodonta) secale, Férussac (1822), *Tabl. Syst.* p. 64.
Jaminia secale, Risso (1826), *Hist. Nat. Europ. Mérid.* p. 89.
Abida secale, Leach (1831), *Brit. Moll.* p. 165.
Vertigo secale, Turton (1831), *Brit. Moll.* p. 101.
Granaria secale, Held (1837), *Isis,* p. 918.
Pupa juniperi, Gray (1840), *Turt. Man.* p. 197.
Stomodonta secale, Mermet (1843), *Moll. Pyr.-Occid.* p. 51.
Pupa (Torquilla) secale, Moquin-Tandon (1855), *Hist. Moll.* vol. ii. p. 366. pl. xxvi. f. 26 to 29.

Hab. Central and Southern Europe. Nearly throughout England, but chiefly in the south and west. Not in Scotland or Ireland. (Under stones and in crevices of rocks and trees, or among wet moss, chiefly in chalky districts.)

In this, the largest of our *Pupæ,* the shell is composed of from eight to nine whorls, two whorls more than that of any other British species, and it is more elongated. *Pupa secale* is moreover the only British species in which the surface of the shell is densely obliquely ridge-striated throughout, after the manner, though in a more delicate form, of the *Clausiliæ*. It is the best species for examining the structure of the aperture. The projections usually called teeth are, it will be seen, parietal ridges winding into the interior, and may be seen through the shell from the outside. "They are," Dr. Gray well observes, "foldings of the substance of the shell, caused by some withdrawing of the mantle of the animal in the part immediately in connection with them. They are produced by a sudden contraction of the part which forms a mould for the newly deposited portion of the shell."

The animal of *Pupa secale* is of a dingy grey colour, rather slender, with the tentacles short and thick. It is found in England as far north as Westmoreland, but inhabits chiefly the south and west parts in chalky districts. Germany is its northern limit on the Continent; southwards it becomes scarce towards Spain and Portugal.

2. Pupa muscorum. *Moss-dwelling Pupa.*

Shell; oblong-cylindrical, deeply umbilicated, rufous horny, semi-transparent, glossy, whorls six to seven, narrow, increasing slowly, moderately convex, very delicately striated, rather constricted at the sutures; aperture small, ovately rounded, sometimes toothless, sometimes furnished with a parietal tooth on the body whorl, lip thinly callously reflected.

Turbo muscorum, Linnæus (1758), *Syst. Nat.* 10th edit. p. 767 (not of Montagu).
Helix muscorum, Müller (1774), *Verm. Hist.* part 2. p. 105.
Bulimus muscorum, pars, Bruguière (1789), *Enc. Méth. Vers*, vol. i. p. 334.
Pupa marginata, Draparnaud (1801), *Tabl. Moll.* p. 58.
Turbo chrysalis, Turton (1819), *Conch. Dict.* p. 220.
Pupa muscorum, C. Pfeiffer (1821), *Deutsch. Moll.* vol. i. p. 57. pl. iii. f. 17.
Turbo marginatus, Sheppard (1823), *Trans. Linn. Soc.* vol. xiv. p. 152 (not of Brown).
Jaminia marginata, Risso (1826), *Hist. Nat. Europ. Mérid.* vol. iv. p. 89.
Alæa marginata, Jeffreys (1830), *Trans. Linn. Soc.* vol. xvi. part 2. p. 357.
Pupilla marginata, Leach (1831), *Turt. Man.* p. 127.
Pupilla muscorum, Beck (1837), *Ind. Moll.* p. 84.
Torquatella muscorum, Held (1837), *Isis*, p. 919.
Pupa bigranata, Rossmässler (1839), *Icon.* vol. ix. p. 25. f. 645.
Stomodonta marginata, Mermet (1843), *Moll. Pyr.-Occid.* p. 53.
Pupa (Odostomia) muscorum, Moquin-Tandon (1855), *Hist. Moll.* vol. ii. p. 392. pl. xxviii. f. 5 to 15.

Hab. Throughout Europe, from Iceland to Sicily. Siberia. (Under stones, and among dead leaves and moss.)

In this *Pupa* the shell has only an occasional tooth in the aperture on the body-whorl, and even then it is scarcely more than a superficial callosity. More frequently, so far as my own experience in collecting goes, it is without any callosity, as in the specimen figured. Shells which show that the calcifying functions of the mollusk have been exercised more vigorously than usual in the secretion of superficial sculpture are invariably smaller than specimens of the same species in which any kind of decorative sculpture is avoided. Granulated varieties of *Cones*, for example, are always smaller than smooth varieties of the same species. It is the same with *Pupa muscorum*. When there is no tooth or callosity, the whorls are of a lighter substance and more tumid growth. In either case they are narrow, increasing but slowly on one another, and forming an unusually straightly cylindrical chrysalis-like shell.

The animal of *Pupa muscorum*, as might be expected in a narrow-whorled shell, is rather slender. In colour it is dark leaden blue, obscurely lineated and speckled. It is generally distributed throughout the British Isles, more particularly in chalky districts; and it is equally universal on the Continent, reaching in the north to Iceland, Lapland, and Siberia, and in the south to Sicily.

3. Pupa cylindracea. *Cylindrical Pupa.*

Shell; shortly cylindrical, rather conspicuously umbilicated, olive-brown, transparent horny, whorls six, rather tumidly convex, scarcely striated, last whorl rather large in proportion to the others; aperture small, triangularly ovate, with a single parietal tooth on the body-whorl, tooth sometimes nearly obsolete, lip broad, whitish, callously, flatly expanded, sometimes swelling a little inwardly.

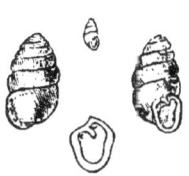

Turbo cylindraceus, Da Costa (1778), *Test. Brit.* p. 89. pl. v. f. 16.
Bulimus muscorum, pars, Bruguière (1789), *Enc. Méth. Vers*, vol. i. p. 334.
Pupa umbilicata, Draparnaud (1801), *Tabl. Moll.* p. 58.
Bulimus unidentatus, Vallot (1801), *Exerc. d'Hist. Nat.* p. 6.
Turbo muscorum, Montagu (1803), *Test. Brit.* p. 335, Supp. pl. xxii. f. 3 (not of Linnæus).
Odostomia muscorum, Fleming (1814), *Edin. Encyc.* vol. vii. pl. i. p. 76.
Helix (Cochlodonta) umbilicata, Férussac (1822), *Tabl. Syst.* p. 59.
Jaminia muscorum, Risso (1826), *Hist. Nat. Europ. Mérid.* vol. iv. p. 88.
Pupilla Draparnaudii, Leach (1831), *Turt. Man.* p. 126.
Pupilla umbilicata, Beck (1837), *Ind. Moll.* p. 84.
Pupa Sempronii, Charpentier (1837), *Moll. Suiss.* p. 15. pl. xi. f. 4.
Eruca umbilicata, Swainson (1840), *Treat. Malac.* p. 334.
Stomodonta umbilicata, Mermet (1843), *Moll. Pyr.-Occid.* p. 53.

Pupa cylindracea, Moquin-Tandon (1849), *Act. Soc. Linn. Bord.* vol. xv.
Pupa (Odostomia) cylindracea, Moquin-Tandon (1855), *Hist. Moll.* vol. ii. p. 390. pl. xxvii. f. 42 and 43, and pl. xxviii. f. 1 to 4.
Hab. Throughout Europe. Algeria. (Under stones and about hedges, among moss, etc.)

Pupa cylindracea is even more generally and more plentifully distributed throughout Europe than *P. muscorum*, and is often taken for it. It abounds in all localities in the British Isles, among moss on walls or under stones, and especially under bark, and in the crevices of old trees. The shell is more shortly cylindrical than that of *P. muscorum*, and has hardly so many whorls. The aperture is rather triangularly compressed, and the tooth, though not unfrequently a little obsolete, is rarely absent. Generally it is rather conspicuous. The lip is also conspicuously flatly expanded and opake. The animal presents scarcely any appreciable variation, being of the same dark leaden blue colour, speckled and lineated at the sides, and white towards the sole. *P. cylindracea* has a wide distribution on the Continent, but it is not clear whether it extends so far north as Iceland or Lapland. It does not appear in the lists of Siberian land shells.

4. **Pupa Anglica.** *English Pupa.*

Shell; ovately cylindrical, rather broadly umbilicated, thin, fulvous horny, semitransparent, whorls six, rather narrow, obsoletely finely striated; aperture compressly triangularly ovate, sinuated above, rounded below, with five parietal teeth, of which two are more internal and obscure, lip rather callous, whitish.

Vertigo Anglica, Férussac (1822), *Tabl. Syst.* p. 64.
Turbo Anglicus, Wood (1828), *Ind. Test.* Supp. p. 19. pl. vi. f. 12.
Pupa Anglicus, Gray (1828), *Ind. Test. Supp.* p. 50.
Pupa ringens, Jeffreys (1830), *Trans. Linn. Soc.* vol. xvi. part 2. p. 356 (not of Michaud).
Vertigo (Isthmia) Anglica, Moquin-Tandon (1855), *Hist. Moll.* vol. ii. p. 404. pl. xxviii. f. 34 to 36.
Hab. In various parts of Britain, from Scotland to the Channel Islands. (Under stones, among dead leaves and moss.)

This species, the smallest of our *Pupæ*, but larger than any *Vertigo*, has been well observed by Mr. Jeffreys, although it is not entitled to the name by which he described it. The animal, he says, is furnished with short tentacles of a lighter shade than the upper part of the body, which is of a yellowish-grey or slate colour, with several dark lines or streaks along the sides, and underneath it is milk-white. "This is a shy little creature," adds Mr. Jeffreys, "although tolerably active when inclined to make its appearance. It has a singular habit of withdrawing slowly one of its eyes, which rolls backwards like a little ball until it reaches the neck, while the tentacle which supports it remains extended to its full length. This I have observed being done when there was no obstacle in the way. It also retracts occasionally and apparently without any reason, one of its horns, and not the other." The shell is of an ovate cylindrical form, with the aperture compressed into a characteristic triangular shape, furnished with as many as five parietal teeth, two of which are more internal than the rest, and have been observed by Mr. Alder to pass thread-like round the columella.

Pupa Anglica was discovered about forty years ago at Scarborough, by the now venerable Mr. Bean of that place, and named by De Férussac, who acknowledges having received specimens from him. No formal diagnose of characters is given in the 'Tableau Systématique' with the name, but it is accompanied by remarks, which leave no room for doubt on the species. After giving "Environs of Scarborough" as its habitat, M. De Férussac says, "This curious species may possibly be a *Pupa*; it is as stout as a *Pupa (Cochlodonta) muscorum*. It is consequently the largest *Vertigo* in Europe, if it belongs to that genus." Specimens were also deposited in the British Museum, and figured and named by Dr. Gray, *Pupa Anglica*, in the Supplement to Wood's 'Index Testaceologicus.'

The geographical range of the species has not been satisfactorily established. It has been collected in the west of Scotland and throughout Ireland, and in many scattered parts of England and the Channel Isles. Morelet met with *Pupa Anglica* in the neighbourhood of Cintra and Oporto, and the Abbé Dupuy gives Algeria as a habitat. M. Moquin-Tandon includes the species in his molluscan fauna of France, on the strength of having once found a single specimen of the shell in the mud of the river near Toulouse.

Vertigo striata. (*Considerably enlarged.*)

Genus XII. **VERTIGO**, *Müller*.

Animal; body rather short, acuminated towards the tail, carrying a blunt tumidly whorled shell, mostly dusky grey, pale towards the sides; upper tentacles rather short, no lower tentacles.

Shell; minute, sometimes sinistral, cylindrical, variously umbilicated, composed of from four to six semitransparent whorls, mostly smooth, rather obtuse at the apex, rounded at the base; aperture ovate, generally more or less toothed within, the teeth having a parietal form, winding in thread-like ridges into the interior.

The absence of lower tentacles is not the only character in which *Vertigo* differs from *Pupa*. The species are all minute, the largest *Vertigo* being smaller than the smallest *Pupa;* and they are characterized by the distinct habit of living in more watery places, generally about the roots of grass at the sides of lakes and rivers, though occasionally at a considerable elevation on the hills. The proportion in the number of species of *Vertigo* to *Pupa* in all parts of the world, so far as they have been eliminated from that genus, is about seventy to one hundred and fifty; but in Britain, we have nine species of *Vertigo* to only four of *Pupa*.

Our commonest and most widely distributed *Vertigo* is *V. pygmæa*. It is abundant throughout our islands, and throughout Europe, reaching to the Azores; it is, moreover, the only British species which has an extra-European range, passing into Siberia. Two species, *V. striata* and *edentula*, are found throughout Britain, but on the Continent they are confined chiefly to the central parts. *V. antivertigo* appears throughout the Continent, and in England and Ireland, but not in Scotland; while *V. minuta*, which appears nearly throughout Europe, is scattered and rare in Britain. *V. pusilla*, which is peculiar to Northern and Central Europe, is found in England and Ireland, but not in Scotland. *V. alpestris*, with a similar range on the Continent, has only been collected in the north of England, while *V. vertigo* and *Moulinsiana*, both natives of Central Europe, have been found in Britain, the first only in the west of England and Ireland, the second only in the west of Ireland.

The distribution of the species, generally, agrees pretty nearly with that of *Pupa*. Of about seventy hitherto described, twenty-five are European, two Syrian, eight Madeiran, three Indian, three from Teneriffe, one Mauritius, three South Africa. The proportion in the number of species of *Vertigo* to *Pupa*, is smaller in the West Indies than in Britain, but larger in the United States, where they include several species very closely allied to our own, but none in common. Three species of *Vertigo* have been recorded from Chili and Peru, and one from Brazil.

The British species are:—

1. **antivertigo.** Shell tumidly cylindrical, of four to five semitransparent fuscous whorls; aperture six to nine-toothed.
2. **Moulinsiana.** Shell tumidly ovate, of four to five very glossy semitransparent yellowish whorls; aperture two to four-toothed.
3. **pygmæa.** Shell ovately cylindrical, of four to five semitransparent reddish or fulvous whorls; aperture four to five-toothed.
4. **alpestris.** Shell oblong-cylindrical, of four to five transparent narrow fulvous-olive whorls; aperture four-toothed.
5. **striata.** Shell minute, shortly cylindrical, of four to four and a half transparent tumidly rounded strongly striated whorls; aperture four to six-toothed.
6. **pusilla.** Shell minute, stoutly cylindrical, of four to five semitransparent olive-horny whorls convoluted sinistrally; aperture six to seven-toothed.
7. **vertigo.** Shell minute, ovately cylindrical, of four to five semitransparent olive-horny whorls convoluted sinistrally; aperture four to five-toothed.
8. **edentula.** Shell minute, narrowly straightly cylindrical, of six to six and a half thin yellowish-horny whorls; aperture toothless.
9. **minuta.** Shell very minute, oblong-cylindrical, of five to five and a half strongly striated tumidly rounded whorls; aperture toothless.

1. Vertigo antivertigo. *Unreversed Vertigo.*

Shell; minute, tumidly cylindrical, compressly umbilicated, chestnut or fuscous horny, semitransparent, glossy, whorls four to five, convex, smooth; aperture triangularly heart-shaped, with from six to nine small teeth, lip thin, rather expanded.

Vertigo sexdentata, Studer (1789), *Faun. Helv. in Coxe, Trav. Switz.* vol. iii. p. 432 (without characters).
Pupa antivertigo, Draparnaud (1801), *Tabl. Moll.* p. 57.
Turbo sexdentatus, Montagu (1803), *Test. Brit.* p. 337. pl. xii. f. 8.
Vertigo sexdentatus, Férussac (1807), *Ess. Méth.* p. 124.
Odostomia sexdentata, Fleming (1814), *Edin. Encyc.* vol. vii. part 1. p. 76.
Pupa octodentata, Hartmann (1821), *Neue Alp.* vol. i. p. 219.
Vertigo septemdentata, Férussac (1822), *Tabl. Syst.* p. 68.
Alæa palustris, Jeffreys (1830), *Trans. Linn. Soc.* vol. xvi. p. 360 and 516.
Vertigo palustris, Leach (1831), *Turt. Man.* p. 104.
Vertigo antivertigo, Michaud (1831), *Comp. Drap.* p. 72.
Alæa antivertigo, Beck (1837), *Ind. Moll.* p. 85.
Vertigo (Isthmia) palustris, Gray (1840), *Turt. Man.* p. 204.
Pupa sexdentata, Fleming (1842), *Brit. Anim.* p. 262.
Stomodonta antivertigo, Mermet (1843), *Moll. Pyr.-Occid.* p. 54.
Vertigo (Isthmia) antivertigo, Moquin-Tandon (1855), *Hist. Moll.* vol. ii. p. 407. pl. xxix. f. 4 to 7.

Hab. Throughout Europe. Throughout England and Ireland. Channel Islands. (In marshy places, on water flags, etc.)

Although not more than about the twelfth of an inch in length, the shell of this species is, with the exception of the Toothless Vertigo (*V. edentula*) and *V. Moulinsiana,* which is scarcely yet established as British, the largest of the genus. It is of a shortly cylindrical tumid form of from four to five whorls, transparent and glossy, with from six to nine teeth. The aperture assumes a triangular form, and it has from two to three teeth at each angle. The animal is of a dusky slate-grey colour, and, according to Dr. Gray, has a couple of dots in place of a lower pair of tentacles. M. Moquin-Tandon describes this little mollusk as carrying its shell nearly straight when crawling, slightly balancing from right to left. The very sombre colour of the animal gives to the transparent shell a dark brown appearance.

2. Vertigo Moulinsiana. *Moulins' Vertigo.*

Shell; minute, tumidly ovate, rather openly umbilicated, yellowish horny, very glossy, semitransparent, whorls four to five, smooth, very tumid; aperture semioval, with from two to four small teeth, lip rather thin, reflected.

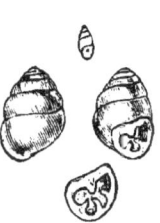

Pupa Moulinsiana, Dupuy (1849), *Cat. Extr. Gall. Test.* no. 284.
Pupa Charpentieri, Shuttleworth (1852), *Küst. Conch. Cab.* p. 129. pl. xvi. f. 41 to 43.
Vertigo (Isthmia) Moulinsiana, Moquin-Tandon (1855), *Hist. Moll.* vol. ii. p. 403. pl. xxviii. f. 31 to 38.
Vertigo Moulinsiana, Jeffreys (1862), *Brit. Conch.* vol. i. p. 255.
Hab. France. Switzerland. Heidelberg. West of Ireland. (Under stones in marshy places.)

This species is unknown to me, but it has been well observed in Switzerland by Mr. Shuttleworth and Mr. Jeffreys, and in France by the Abbé Dupuy and M. Moquin-Tandon. It is included in the British fauna only on the grounds of Mr. Jeffreys having lately discovered specimens among some shells which he collected nearly twenty years ago by the side of a small lake at Ballinahinch, near Roundstone, county Galway. "The situations," he says, "where I found it in Switzerland, were like that of the Irish habitat; and I have no doubt it will be discovered in this country by attention being thus drawn to it. The few districts of our eastern counties, as well as the wilds of Connemara, require to be more thoroughly searched."

The animal *V. Moulinsiana* is described as being of a dark-grey colour, lighter however, and more slender than that of *V. antivertigo*, with the tentacles more decidedly clavate. The shell is rather larger and more ventricose, and of lighter colour, with fewer teeth (from two to four) in the aperture.

Such is Mr. Jeffreys' description of the animal of Swiss specimens of *V. Moulinsiana*. M. Moquin-Tandon includes the species in his History of the Mollusks of France, but he is not able to give any description of the animal. He merely gives a drawing, on a considerably magnified scale, of the shell, of which the above figure is a reduced copy.

3. Vertigo pygmæa. *Pygmy Vertigo.*

Shell; minute, ovately cylindrical, slightly compressly umbilicated, reddish brown or fulvous horny semitransparent, glossy, whorls four to five, rounded, smooth; aperture semiovate, with from four to five teeth, only one of which is on the body whorl, lip thinly expanded.

Vertigo quinquedentata, Studer (1789), *Faun. Helv. in Coxe, Trav. Switz.* vol. iii. p. 432 (without characters).
Pupa pygmæa, Draparnaud (1801), *Tabl. Moll.* p. 57.
Vertigo pygmæa, Férussac (1807), *Ess. Méth.* p. 124.
Helix (Isthmia) cylindrica, Gray (1821), *Lond. Med. Repos.* vol. xv. p. 239 (not of Férussac nor Studer).
Vertigo similis, Férussac (1822), *Tabl. Syst.* p. 68.
Alæa vulgaris, Jeffreys (1830), *Trans. Linn. Soc.* vol. xvi. p. 359.
Vertigo vulgaris, Leach (1831), *Turt. Man.* p. 129.
Alæa pygmæa, Beck (1837), *Ind. Moll.* p. 95.
Vertigo (Isthmia) pygmæa, Gray (1840), *Turt. Man.* p. 199.
Stomodonta pygmæa, Mermet (1843), *Moll. Pyr.-Occid.* p. 55.
Hab. Throughout Europe. Siberia. Azores. (Under logs of wood or stones in wet places.)

Of all our nine species of *Vertigo,* this is the commonest and most widely diffused. Specimens may be collected in almost any part of Britain, sometimes by placing a log of wood upon the wet grass at night, and examining it in the morning. On the Continent it ranges from Sweden to Sicily and the Azores, and it appears also in Siberia. The animal is of the prevailing colour of the genus, dusky grey, speckled with black. The shell is extremely small, ovately cylindrical, and with from four to five teeth in the aperture, only one of which is on the body whorl.

"*V. pygmæa,*" says Mr. Jeffreys, "is a tolerably active and lively little creature, crawling by jerks, and carrying its shell nearly upright. It makes, like its congeners, a filmy epiphragm, but which is not iridescent. It may be in some degree considered a subalpine form, as it occurs at considerable heights. Dr. Johnston found it at the top of a mountain in East Lothian, at an elevation of 1200 feet, and M. Puton on the Vosges at 1640 feet."

FAMILY COLIMACEA.

4. Vertigo alpestris. *Alpine Vertigo.*

Shell; minute, oblong-cylindrical, narrowly umbilicated, fulvous olive, thin and transparent, very glossy, whorls four to five, increasing slowly, convex, smooth or finely striated; aperture semiovate, with four teeth, one of which is on the body whorl.

Vertigo alpestris, Alder (1830), *Trans. Nat. Hist. Soc. Northumb.* vol. ii. p. 340.
Vertigo (Isthmia) alpestris, Gray (1840), *Turt. Man.* p. 202.
Pupa Shuttleworthiana, Charpentier (1847), *Zeitsch. für Malak.* p. 148.
Pupa pygmæa, var., Forbes and Hanley (1853), *Hist. Brit. Moll.* vol. iv. p. 106. pl. cxxx. f. 6.

Hab. North of England. Northern and Central Europe. (Under stones among dead leaves on the hills.)

There is a difference of opinion among conchologists as to whether this species is anything more than a northern and alpine variety of *V. pygmæa*. The animal is, however, of a light straw-colour, with the tentacles and foot longer than in that species, and the shell, it will be seen, on a comparison of our figures, is more narrowly cylindrical and more distinctly striated; it is moreover paler in colour, and of a more glassy transparency.

---

5. Vertigo striata. *Striated Vertigo.*

Shell; minute, shortly cylindrical, compressly umbilicated, yellowish horny, thin, semitransparent, whorls four to four and a half, tumidly rounded, strongly obliquely striated, constricted at the sutures; aperture semiovate, a little contracted at the side, with four to six small teeth; lip thin, slightly reflected.

Alæa substriata, Jeffreys (1830), *Trans. Linn. Soc.* vol. xvi. p. 515.
Pupa substriata, Alder (1830), *Trans. Nat. Hist. Soc. Northumb.* vol. ii. p. 339.
Vertigo substriata, Alder (1837), *Mag. Zool. and Bot.* vol. ii. p. 112.
Vertigo curta, Held (1837), *Isis*, p. 304.
Vertigo (Isthmia) substriata, Gray, (1840), *Turt. Man.* p. 202, pl. vii. f. 84.

Hab. Central Europe (Bavaria). Throughout Britain. (Under stones or among decayed leaves, and at the roots of grass in wet places).

Six of the British species of *Vertigo* have smooth shells, merely finely striated with the lines of growth; in the remaining three the shells are sculptured with raised striæ. Of these *V. vertigo* is very tumidly cylindrical and sinistral, and *V. minuta* is oblong-cylindrical and toothless. To Mr. Jeffreys we are indebted for the discovery of a third strongly striated species, widely diffused throughout Britain, which is of a shortly cylindrical form, dextral in growth, and has from four to six small teeth in the aperture. The animal is of different shades of grey, tolerably active though timid, and carries its shell nearly upright.

Turton, Gray, and Forbes and Hanley refer this species to Montagu's *Turbo sexdentatus*, still, however, adopting Mr. Jeffreys' name *substriata;* but Montagu particularly describes his *T. sexdentatus* as being a smooth shell, and I incline to agree with Moquin-Tandon that it is Draparnaud's *P. antivertigo*. Mr. Jeffreys' name should be preserved for the species under consideration, dropping the auxiliary preposition. It is not consistent to continue the name "faintly striate" for a shell which it is necessary to characterize as being "very strongly striate."

V. striata has been collected in Bavaria, and is probably more widely diffused on the Continent than is at present supposed.

---

6. **Vertigo pusilla.** *Little Vertigo.*

Shell; minute, sinistral, stoutly cylindrical, compressly umbilicated, olive-horny, semitransparent, moderately glossy, whorls four to five, smooth, rounded; aperture triangularly ovate, with from six to seven, rather solid teeth, lip thinly expanded.

Vertigo pusilla, Müller (1774), *Verm. Hist.* part 2. p. 124.
Helix vertigo, Gmelin (1788), *Syst. Nat.* p. 3664.
Pupa vertigo, Draparnaud (1801), *Tabl. Moll.* p. 57.
Turbo vertigo, var., Montagu (1803), *Test. Brit.* p. 365.
Odostomia vertigo, Fleming (1814), *Edin. Encyc.* vol. vii. part 1. p. 77.
Vertigo heterostropha, Leach (1831), *Turt. Man.* p. 130.
Vertigo vertigo, Aleron (1837), *Bull. Soc. Phil. Perpig.* vol. iii. p. 92.
Vertigo (Vertilla) pusilla, Moquin-Tandon (1855), *Hist. Moll.* vol. ii. p. 409. pl. xxix. f. 12 to 14.

Hab. Northern and Central Europe. Throughout England, but chiefly in the southern and eastern counties, local and rare. More rare in Ireland. (In woods, among wet moss or under stones.)

It was the peculiarity of the sinistral or left-handed convolution of this species that induced the eminent Danish naturalist of the last century, Müller, to establish his genus *Vertigo*. It has a stoutly cylindrical moderately glossy shell, smooth, or only very faintly striated, with a triangularly ovate aperture, armed with at least six to seven solid teeth. The animal is of a dusky grey-slate colour, pale towards the sides, with a pair of widely diverging tentacles; and according to Forbes and Hanley it has a pair of almost obsolete lower tentacles. This observation has not, however, been confirmed by Moquin-Tandon.

V. pusilla is found in several parts of England, local and rare, and still more rare in the north and west of Ireland. On the Continent it ranges from Sweden to Switzerland and Lombardy.

7. **Vertigo vertigo.** *Reversed Vertigo.*

Shell; minute, sinistral, ovately cylindrical, attenuately contracted at the base, scarcely umbilicated, fuscous horny, semitransparent, glossy, whorls four to five, tumidly convex, obliquely rather strongly striated, constricted at the sutures; aperture small, compressly ovate, with four to five teeth, two of which are on the body-whorl, and only one on the opposite margin, lip rather thickened, but little reflected.

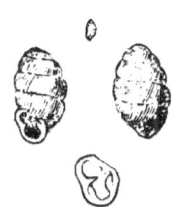

Turbo vertigo, Montagu (1803), *Test. Brit.* p. 363. pl. xii. f. 6.
Vertigo Venetzii, Charpentier (1822), *Féruss. Tabl. Syst.* p. 65 (without characters).
Vertigo angustior, Jeffreys (1830), *Trans. Linn. Soc.* 1830, vol. xvi. p. 361.
Vertigo nana, Michaud (1831), *Comp. de Drap.* p. 71. pl. xv. f. 24, 25.
Vertigo hamata, Held (1837), *Isis*, p. 304.
Vertigo plicata, August Müller (1838), *Weigm. Arch.* p. 210. pl. iv. f. 6.
Pupa nana, Deshayes (1838), *Anim. sans vert.* vol. viii. p. 190.
Pupa (Vertigo) Venetzii, Pfeiffer (1848), *Monog. Helic. Vic.* vol. ii. p. 364.
Vertigo (Vertilla) plicata, Moquin-Tandon (1855), *Hist. Moll.* vol. ii. p. 408, pl. xxix. f. 8 to 11.

Hab. Central Europe. West of England and Ireland. Rare. (At the roots of grass in wet places.)

It will be seen by our figure of this very rare species, that the shell of *V. vertigo* is distinguished from that of *V. pusilla*, by a more ovate form peculiarly constricted at the sutures, arising from a characteristic attenuate contraction of the last whorl on reaching maturity. Another peculiarity of the species consists in the unusual arrangement of the teeth, there being two on the body whorl to only one on the opposite margin. The animal as described by Mr. Jeffreys, who has observed it both in the west of England and in the south of France, is short and rather slow in its movements, black in front, and grey at the sides and underneath, with the tentacles thick and considerably diverging from each other.

The species was well observed by Montagu. It is unfortunate that he gave it a name which had been used by Müller to designate the genus; but *Vertigo vertigo* is no more objectionable than *Vertigo antivertigo*, and as conchologists are pretty well agreed on retaining the latter name in the nomenclature, we may as well do our eminent countryman the justice to retain also the former.

8. **Vertigo edentula.** *Toothless Vertigo.*

Shell; small, rather narrowly straightly cylindrical, minutely umbilicated, thin, yellowish horny, bright, glossy, whorls six to six and a half, rather flatly convex, smooth or faintly arcuately striated; aperture small, pyriformly rounded, toothless, lip thinly reflected.

Helix exigua, Studer (1789), *Faun. Helv. in Coxe's Trav. Switz.* vol. iii. p. 430 (without characters).
Pupa edentula, Draparnaud (1801), *Hist. Moll.* p. 52. pl. iii. f. 28, 29.
Vertigo edentula, Studer (1820), *Kurz. Verz.* p. 89.
Vertigo nitida, Férussac (1822), *Tabl. Syst.* p. 68.
Helix Offtonensis, Sheppard (1823), *Trans. Linn. Soc.* vol. xiv. p. 155.
Jaminia edentula, Risso (1826), *Hist. Nat. Europ. Mérid.* vol. iv. p. 89.
Turbo edentulus, Wood (1828), *Ind. Test.* Suppl. pl. vi. f. 14.
Alæa nitida and *revoluta*, Jeffreys (1830), *Trans. Linn. Soc.* vol. xvi. p. 358.
Alæa edentula, Beck (1837), *Ind. Moll.* p. 85.
Vertigo lepidula, Held (1837), *Isis*, p. 307.

FAMILY COLIMACEA.

Stomodonta edentula, Mermet (1843), *Moll. Pyr.-Occid.* p. 54.
Hab. Northern and Central Europe. Siberia. Throughout the British Isles. (Chiefly about the wet roots of grass and under dead leaves.)

The two remaining species of *Vertigo* have more whorls in the shell, more narrowly convoluted, and are toothless. *V. edentula*, the larger, is found in all parts of the British Islands, but it is local and not by any means common. On the Continent it ranges as far north as Lapland and Siberia; while its southern limit appears to be in Switzerland and Lombardy.

The animal of *V. edentula*, according to Moquin-Tandon, is timid and retires within its shell at the slightest touch. It is of the usual slate-grey colour, paler towards the sides, and is quite destitute of any rudimentary trace of lower tentacles. The shell is thin and rather elongately cylindrical, very delicately arcuately striated, with the simplest possible form of aperture.

---

9. **Vertigo minuta.** *Minute Vertigo.*

Shell; very minute, oblong cylindrical, rather conspicuously umbilicated, rather solid, fulvous horny, moderately glossy, whorls five to five and a half, tumidly rounded, constricted at the sutures, strongly striated; aperture rotundately ovate, toothless, lip expandedly reflected.

Pupa muscorum, Draparnaud (1801), *Tabl. Moll.* p. 56 (not *Turbo muscorum*, Linn.).
Pupa minuta, Studer (1820), *Kurz. Verz.* p. 89.
Pupa minutissima, Hartmann (1821), *Neue Alp.* p. 220. pl. ii. f. 5.
Vertigo cylindrica, Férussac (1822), *Tabl. Syst.* p. 28.
Pupa obtusa, Fleming (1828), *Brit. Anim.* p. 269 (not of Draparnaud).
Alæa cylindrica, Jeffreys (1830), *Trans. Linn. Soc.* vol. xvi. p. 359.
Vertigo muscorum, Michaud (1831), *Comp. de Drap.* p. 70.
Alæa minutissima, Beck (1837), *Ind. Moll.* p. 83.
Vertigo pupula, Held (1837), *Isis*, p. 308.
Vertigo (Isthmia) cylindrica, Gray (1840), *Turt. Man.* p. 201, pl. xii. f. 140.
Eruca muscorum, Swainson (1840), *Treat. Malac.* p. 334.
Stomodonta muscorum, Mermet (1843), *Moll. Pyr.-Occid.* p. 55.
Vertigo minutissima, Grüells (1846), *Cat. Moll. Esp.* p. 7.
Vertigo (Isthmia) muscorum, Moquin-Tandon (1855), *Hist. Moll.* vol. ii. p. 399. pl. xxviii. f. 20 to 24.

Hab. Throughout Europe. Rare in Britain. (Under stones in damp shady places on hills.)

The smallest of our land mollusks, collected only at rare intervals in widely scattered parts of our islands. On the Continent it is more plentiful in the northern and central parts, rare in the south. The animal is described by Moquin-Tandon, as being slightly narrow and rounded in front, very gradually attenuated and somewhat blunt behind, finely shagreened greyish-slate colour dotted with black. The shell, like that of *V. edentula*, is toothless, but very much smaller, and distinctly elevately striated.

Draparnaud mistook this for the Linnean *Turbo muscorum*.

Family III. AURICULACEA.

Respiratory and visceral organs distinct from the main contractile mass of the body, coiled within a spiral shell. Eyes at the base of the tentacles.

Before passing to the inoperculated Pond Snails (*Lymnæacea*) which are amphibious, breathing both water and air, we have to speak of a family breathing air, and regarded as land mollusks, but confined to wet places, and in most instances to localities within reach of the sea. Amid the hot swampy and saline marshes of South and Central America, and the islands of Australia and the Malayan and Polynesian archipelagos, are a number of mollusks of this kind, producing shells of comparatively solid growth and dark colour, enveloped, in most instances, by a brown horny epidermis, and having the columella mostly toothed or sculptured with winding plaits. Under the genera *Auricula*, *Scarabus*, *Melampus*, *Conovulus*, *Marinula*, *Pedipes*, and *Carychium*, about a hundred and fifty species have been described. In Britain, the family of *Auriculacea* is represented by only four species, one belonging to the genus *Carychium*, and three to *Conovulus;* and the family is scarcely more numerous in species in any part of Europe, Western Asia, or North America.

Carychium and *Conovulus* have very different shells, and they differ materially in habit, but the animals are very similar. They agree in the special characteristics, when compared with other air-breathing genera, of having the head produced into a ringed muzzle, and the eyes separate from the tentacles, simply as specks in the skin-surface, so to speak, of the head, presenting a transition in these respects to *Acme* and *Cyclostoma*, which are operculated air-breathers. The upper tentacles are short, thick, and rather approximated at the base; the lower tentacles are almost obsolete.

The particulars in which *Carychium* and *Conovulus* differ, are important. *Carychium*, in whatever country it has been observed, is a minute mollusk, carrying a transparent glassy shell toothed in the aperture somewhat after the manner of *Vertigo*, and living in inland places among moss or dead leaves wet with fresh water, or in crevices of the stems of freshwater plants. It appears throughout Europe, in Siberia, in North Africa, in the western sub-Himalayan mountains of India, and in the United States; yet it is doubtful whether more than ten species are known, including four of a particular globose form, inhabiting the subterranean caverns of Carniola, Austria, of which as many as thirteen species have been made by M. Bourguignat under the title of *Zospeum*. *Conovulus* is of much larger dimensions, and has a solid epidermis-covered shell of the characteristic exotic type already described. *Conovulus* is not found inland in the vicinity of fresh water. It inhabits mud-banks and sand-hills in places washed with brackish water, or even within range of the sea, especially in saline marshes. Mr. Cuming collected *Conovuli*, almost identical with the British species, at Lima, in saltwater ponds at least three miles from the sea, the intermediate land being chiefly salt, with the scantiest possible vegetation.

The British genera of *Auriculacea* are :—

1. **Carychium.** Animal pale yellowish, strongly bilobed in front, head produced into a ringed muzzle, eyes at the hinder base of the tentacles. Shell conically turbinated, of five to five and a half glassy whorls, one-toothed at the three angles of the aperture.

2. **Conovulus.** Animal violaceous yellow or grey, faintly bilobed in front, head produced into a ringed muzzle, eyes at the inner hinder base of the tentacles. Shell of six to seven fusiformly convoluted whorls, plaited and toothed on the columella, and sometimes toothed within the lip.

Carychium minimum. (*Considerably enlarged.*)

Genus I. **CARYCHIUM**, *Müller*.

Animal; pale yellowish, minutely black-and-white-speckled, strongly bilobed in front, carrying a minute glassy shell, head produced into a ringed muzzle, eyes conspicuous black specks at the hinder base of the upper tentacles, which are short and rather thick, lower tentacles almost obsolete.

Shell; subdiaphanous greenish white, of five to five and a half conically turbinated whorls, which are smooth or finely striated; aperture three-toothed.

Carychium, first distinguished by Müller, is a minute bright-eyed mollusk, carrying a conically turbinated glassy shell, with a rather superficially toothed aperture. It is widely distributed throughout the north temperate regions of both hemispheres, and it appears at an elevation of similar temperature in the Himalayas. M. Bourguignat remarks, in his monograph of the genus, that the *Carychia* are little sensible of cold or heat, since they inhabit the icy regions of Siberia and Lapland, as well as the countries of India, and the warmest parts of Italy and Spain. The range of temperature is undoubtedly very considerable between Siberia and Spain; but these countries are the confines of a province of distribution common to many land mollusks. The habitat of *Carychium* in India is on the shady side of mountains, at an elevation of from 5000 to 9000 feet. Only a few species are known, the most abnormal being a globose form (genus *Zospeum*, Bourguignat) inhabiting the great subterranean caverns of Carniola.

British specimens of *Carychium* are all referred to one species:—

1. **minimum.** Shell conically turbinated, of from five to five and a half smooth or faintly striated whorls; aperture three-toothed.

FAMILY AURICULACEA.

1. **Carychium minimum.** *Very small Carychium.*

Shell; conically ovate, scarcely umbilicated, subdiaphanous white or greenish, spire produced, whorls five to five and a half, rounded, rather constricted at the sutures, smooth or obliquely very faintly striated; aperture small, rather contracted, somewhat squarely ovate, three-toothed, a little expanded.

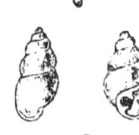

Carychium minimum, Müller (1774), *Verm. Hist.* part 2. p. 125.
Helix Carychium, Gmelin (1788), *Syst. Nat.* p. 3665.
Bulimus minimus, Bruguière (1789), *Enc. Méth. Vers*, vol. i. p. 310.
Auricula minima, Draparnaud (1801), *Tabl. Moll.* p. 54.
Turbo Carychium, Montagu (1803), *Test. Brit.* p. 339. pl. xxii. f. 2.
Odostomia Carychium, Fleming (1814), *Edin. Encyc.* vol. vii. p. 76.
Auricula Carychium, Klees (1818), *Diss. Test. Tubing.* p. 30.
Auricella Carychium, Hartmann (1821), *Syst. Gast.* p. 49.
Seraphia tridentata, Risso (1826), *Nat. Hist. Europ. Mérid.* vol. iv. p. 84.
Carychium nanum, Anton (1839), *Verz. Conch.* p. 48.
Carychium elongatum, A. and J. B. Villa (1841), *Disp. Syst.* p. 59.
Carychium (Auricella) minimum, Moquin-Tandon (1855), *Hist. Moll.* vol. ii. p. 413. pl. xxix. f. 15 to 26.

Hab. Throughout Europe. Siberia. Sicily. North Africa. (Among wet moss, at the roots of grass, or under stones in wet places.)

Authors are pretty well agreed that all the varieties of the little glassy shell known throughout Britain as *Carychium minimum* belong to one and the same species. Some specimens are smooth, others are obviously finely striated, and the teeth are more conspicuously developed in some specimens than in others, while the whorls vary a little in their shorter or more elongated mode of convolution. M. Bourguignat describes three additional species, *C. tridentatum, striolatum,* and *Rayianum,* as inhabiting France; and one is described by M. Morelet under the name *Auricula gracilis,* as a native of Portugal. A single species was discovered by Captain Hutton, in India, in Simla, and Landour, on the shady side of mountains, at an elevation of 5000 to 9000 feet; and the *Pupa exigua* of Say (*Bulimus exiguus*, Binney), common in all the Northern and Middle States of North America, is a *Carychium*.

Conovulus myosotis. (*Much enlarged.*)

Genus II. **CONOVULUS.**

Animal; violaceous yellow or grey, carrying a fusiformly oblong shell, head slightly lobed, produced into a broad ringed muzzle; upper tentacles rather stout, with the eyes a little behind their inner base, lower tentacles represented by a pair of closely approximating tubercles.

Shell; fusiformly oblong or ovate, opake-white or yellowish horny covered by a brown epidermis, whorls six to seven, smooth, sometimes faintly puckered at the sutures; aperture oblong, columella plaited and more or less toothed, lip sometimes toothed within on a thickened ridge.

The *Conovuli* live in brackish marshes, sometimes at a distance of two or three miles from the sea, but mostly inhabit the mouth of rivers, in the crevices of the stems or about the roots of water plants, or in the mud or sand of the seashore, within range of the spray of the sea.

They can scarcely, however, be said to be marine. They do not live in situations reached by the tide; they are not amphibious, but respire air, and are characterized by the same organization as the *Carychia*, which inhabit inland localities. In the saline swamps and marshes of intertropical countries, the *Conovuli* are more numerous; and in the nearest allied genus, *Auricula*, which is the type of the group, they attain a large size.

The animal of *Conovulus* has a broad, obtuse muzzle, an incipient development of that which is so conspicuous a feature in *Cyclostoma;* and the eyes are situated about the middle of the inner base of the tentacles, directed a little behind. The shell is of six or seven whorls, convoluted in a fusiformly oblong shape, with the columella plaited and more or less toothed. The inner lip is sometimes reflected, sometimes simple, and when simple it has a toothed inner ridge. On these two forms of *Conovulus* are founded M. Moquin-Tandon's subgenera *Ovatella* and *Phytia*, and Dr. Gray's genera *Alexia* and *Leuconia*. In the interior of the shell the septa of the whorls are generally absorbed towards the apex, and the visceral extremity of the animal losing its spiral contour becomes obtusely conical.

The British species of *Conovulus* are:—

1. **denticulatus.** Shell fusiformly ovate, of six to seven whorls, columella with four winding plaits and teeth, lip with an inner ridge of teeth.
2. **myosotis.** Shell fusiformly oblong, rather inflated, of six to seven whorls, columella strongly four-toothed, lip simple, slightly thickened within.
3. **bidentatus.** Shell fusiformly oblong, of six to seven whorls, columella with two teeth, the lower tooth being a nearly obsolete winding plait.

---

1. Conovulus denticulatus. *Toothed Conovulus.*

Shell; fusiformly ovate, rather solid, yellowish white, opake, covered with a chestnut horny epidermis, spire more or less produced, whorls six to seven, flatly convex, plicately striated next the sutures; aperture narrowly ovate, rounded at the base, lip simple, rather thickened and serrately toothed within, columella callous with four winding plaits and teeth.

Voluta denticulata, Montagu (1803), *Test. Brit.* p. 234. pl. xx. f. 5.
Voluta ringens and *reflexa,* Turton (1819), *Conch. Dict.* p. 250, 251.
Acteon denticulatus, Fleming (1828), *Brit. Anim.* p. 337.
Auricula denticulata, Jeffreys (1830), *Trans. Linn. Soc.* vol. xvi. p. 367.
Auricula tenella, Menke (1830), *Syn. Moll.* 2nd edit. p. 131.
Carychium personatum, Michaud (1831), *Comp. de Drap.* p. 73. vol. xv. f. 42, 43.
Pythia denticulata, Beck (1837), *Ind. Moll.* p. 103.
Auricula personata, Potiez and Michaud (1838), *Moll. Gal. de Douai,* vol. i. p. 105.
Conovulus (Ovatella) denticulatus, Gray (1840), *Man.* p. 225 pl. xii. f. 144.

Jaminia denticulata aud *quinquedens*, Brown (1845), *Illus. Conch.* p. 22. pl. viii. f. 6, 11.
Alexia denticulata, Gray (1847), *Pro. Zool. Soc.* p. 179.
Carychium (Ovatella) denticulatum, Moquin-Tandon (1855), *Hist. Moll.* vol. ii. p. 415. pl. xxix. f. 27 to 29.

Hab. Southern and Western Europe. South and south-west of England. (About the mouths of rivers in the mud left bare by the tide, in harbours under stones above high-water mark, or among the roots of water-plants, in brackish marshes.)

Conovulus denticulatus is distinguished by having the shell more or less toothed on both sides of the aperture. In addition to two very conspicuous winding plaits at the base of the columella, it has two small teeth on a callosity at the upper part, and a row of small teeth, sometimes however reduced to one only, on a callous ridge within the opposite lip. In the interior of the shell the septa of the whorls are generally absorbed.

M. Bouchard-Chantereaux, who has well observed this species in the Port of Wimereux, Department of Pas-de-Calais, mentions, in speaking of it under Michaud's name of *Carychium personatum*, that the animal, which shows little fear when lifted by the shell, and elongates itself and tries to find a place of attachment, lives under stones in that locality, but only under those above high-water mark, where it feeds on the detritus of marine plants and rotten wood. He never found the *Conovuli* in localities washed by the tide. They would live only a moderately long time on being immersed in sea-water, but would exist longer than when immersed in fresh water.

---

2. **Conovulus myosotis.** *Mouse-ear Conovulus.*

Shell; fusiformly oblong, yellowish-white, covered with a fulvous chestnut horny epidermis, spire moderately produced; whorls six to seven, flatly convex, plicately striated next the sutures, last whorl rather inflated; aperture somewhat squarely oblong, rounded at the base, lip simple, rather thickened within, columella callous, strongly four-toothed.

Auricula myosotis, Draparnaud (1801), *Tabl. Moll.* p. 53.

FAMILY AURICULACEA.

Carychium myosotis, Férussac (1807), *Ess. Méth. Conch.* p. 54.
Auricella myosotis, Jurine (1817), *Helv. Almanach*, p. 34.
Pythia myosotis, Beck (1837), *Ind. Moll.* p. 101.
Conovulus denticulatus, var. *myosotis*, Forbes and Hanley (1853), vol. iv. p. 195. pl. cxxv. f. 4, 5.
Carychium (Phytia) myosotis, Moquin-Tandon (1855), *Hist. Moll.* vol. ii. p. 417. pl. xxix. f. 33 to 39, and pl. xxx. f. 1 to 4.

Hab. Central and Southern Europe. South and south-west of England. (Under stones or in mud, about the mouths of tidal rivers.)

In this species the shell is of a more oblong square form, a little inflated, and there is no serrated ridge of small teeth within the lip, although it is a little swollen. M. Moquin-Tandon describes the animal as being rather rapid in its movements and irritable, emerging from its shell a little obliquely. The tentacles appear, he continues, like conical protuberances fixed on the muzzle; and in walking, which is rather rapid, the muzzle is used to assist locomotion as in *Cyclostoma*. The shell is carried nearly horizontally.

The principal British habitat of *Conovulus myosotis* is in brackish marshes about the estuaries of rivers on the south and south-western coasts. In the Thames it has been collected with *Assiminea Grayana* as high up as Greenwich, which fully entitles the genus to a place in this work.

---

3. **Conovulus bidentatus.** *Two-toothed Conovulus.*

Shell; fusiformly ovate, rather solid, white, covered with a light horny epidermis, spire rather acuminated, whorls six to seven, flatly convex, impressed round the upper part; aperture oblong-ovate, lip simple, slightly thickened within, columella callous, two-toothed, one tooth central and conspicuous, the other obliquely basal and rather obsolete.

Voluta bidentata, Montagu (1803), *Test. Brit.* p. 100. pl. xxx. f. 2.
Voluta alba, Turton (1819), *Conch. Dict.* p. 250.
Auricula bidentata, Férussac (1821), *Tabl. Syst.* p. 103.
Volvaria bidentata, Fleming (1828), *Brit. Anim.* p. 333.
Acteon bidentatus, Fleming (1828), *Brit. Anim.* p. 337.
Auricula erosa and *alba*, Jeffreys (1830), *Trans. Linn. Soc.* vol. xvi. p. 369.

Conovulus (Leuconia) bidentatus and *albus*, Gray (1840), *Man.* p. 227. pl. xii. f. 145.
Auricula dubia, Cantraine (1840), *Bull. Acad. Brux.* vol. ii. p. 383.
Auricula Micheli, Mittre (1841), *Rev. Zool.* p. 66.
Hab. Western and Southern Europe. North, south and south-west of England. Ireland. (In crevices of rocks and timber, among wet moss or under stones, on the banks of rivers near the sea.)

In this species the teeth are reduced in number to two, and only one of these is conspicuous. The other is an obscure winding basal plate. The shell is mostly paler than that of either of the preceding species, and there is a well-marked variety of shining diaphanous whiteness. It is found on widely remote parts of our coast in the crevices of rocks just above high-water mark, and generally throughout Ireland.

The testimony in favour of arranging the *Conovuli* with the Land and Freshwater rather than with the Marine genera, was ably confirmed about twelve years since by the observations on this species of Mr. Clarke. "I think it will be acceptable to malacologists," he wrote in a letter to the Annals of Natural History of that date, "to review my notes on the much-disputed point whether the animal respires free air or eliminates it from water by a pectinibranchous organ. Having submitted fourteen live animals to the powers of an excellent microscope, I am enabled to say that I found no traces of a regular pectinated membrane, but when the dissection turned out well, there appeared, as in the usual place of the *Helices*, what I considered the respiratory cavity, having its walls lined with an anastomosing network of vessels. The animal when put into water instantly escapes therefrom, apparently with the view of breathing air."

Order II. **PULMOBRANCHIATA**—BREATHING BOTH AIR AND WATER.

Respiratory organ a vascular sac for the respiration of air, with the addition of branchial lamellæ (gills), for the respiration also of water.

The Pulmobranchiate, or lung-gilled, inoperculated Cephals are possessed of a compound system of respiration, part lung, part gill, by which they are enabled to breathe both air and water.

Water is their natural element, but they can also live out of the water. For animals dwelling in shallow ponds and ditches, which are full or empty according to the amount of rainfall, and in summer are liable to be dried up for weeks together, such an arrangement of the breathing functions is necessary to existence. The respiratory organ is a vascular sac, over which the blood flows and is aerated in a network of minute vessels; and it is fitted with branchial plates, or lamellæ, for the respiration of water.

The British species are all contained in one family:—

1. **Lymnæacea.** Head produced into a short broad muzzle, tentacles sometimes slender and bristle-like, sometimes flatly triangular, with the eyes at the inner base. Shell extremely variable, sometimes flatly discoid, sometimes ovately or fusiformly spiral, sometimes limpet-shaped.

Family I. LYMNÆACEA.

Head produced into a short broad muzzle, tentacles sometimes slender and bristle-like, sometimes flatly triangular, with the eyes at their inner base. Shell extremely variable, sometimes flatly discoid, sometimes ovately or fusiformly spiral, sometimes limpet-shaped.

The *Lymnæacea*, or Pond Snails, are much less numerous over the globe in species, with far less variation of typical form, than the Land Snails; and they are fewer and less varied in the Eastern than in the Western Hemisphere. Yet they are comparatively numerous in Britain, plentiful in individuals, and very generally diffused. There are not many more species in all Europe than there are in the British Isles. They have no particular centre of creation in any part of their large area of distribution, which reaches from Greenland and Siberia to the Himalayas; and through all this wide range of latitude there is little difference between the species of any genus in form and colour. Below the Himalaya range to Tasmania, the *Lymnæa* and *Physa* appear in new forms, illustrative of a number of very interesting species; but the principal area of development of all our genera of freshwater mollusks is in Central and North America, extending southwards to Chili, and northwards to Vancouver's Island. Here the species are more abundant, and it will be seen that they are distributed in more distinct assemblages of types.

There are few groups of mollusks in which so variable a shell, generically speaking, is produced with so little variation in the animal as in the family of *Lymnæacea*, and the genera are unusually distinct from each other. In *Planorbis*, the shell is a slowly increasing tube, coiled rotariwise into a flattened disk; in *Physa* and *Lymnæa* it is an inflated spiral, sharply acuminated; in *Ancylus* it has only a tendency to coil in an early stage of growth, and enlarges without convolution like a limpet. The animal is characterized throughout by nearly a similar organization of parts. The respiratory sac, as we have already shown, is both pulmonary and branchial, the head is produced into an obtuse muzzle, and the eyes are situated at the inner base of the tentacles. The parts in which there are differences are the tentacles and the mantle. In *Planorbis* and *Physa*, the tentacles are slender and bristle-like; in *Lymnæa* they are flatly triangular; in *Ancylus* they are slender but short, widely separated, and triangular at the base. The mantle, like the shell which it secretes, is most variable. In *Planorbis*, it is retained within the shell; in *Physa* it is reflected over the shell on either side in a curiously digitate lobe; in a section of *Lymnæa* (*Amphipeplea*) it is reflected without digitation over nearly the whole of the shell; in *Ancylus* it lines the shell, which entirely covers the animal.

The genera of British *Lymnæacea* are:—

1. **Planorbis.** Animal carrying a horizontal rotary coiled tubular shell, tentacles slender and bristle-like. Shell horny or glassy, of from three to seven slowly increasing whorls.

2. **Physa.** Animal carrying an ovately inflated sinistrally coiled shell, over the edge of which the mantle is reflected in digitate lobes, tentacles slender. Shell transparent, shining, of five whorls.

3. **Lymnæa.** Animal carrying a variously shaped shell, over which the mantle in one instance is amply reflected, tentacles broadly flatly triangular. Shell varying from ventricosely ovate to elongately turreted, of from four to eight whorls.

4. **Ancylus.** Animal carrying a small limpet-shaped shell, which entirely covers it, tentacles short, slenderly triangular. Shell cap-like, with the vertex turned subspirally sometimes to the left, sometimes to the right.

Planorbis corneus.

Genus I. **PLANORBIS**, *Guettard*.

Animal; body slender, carrying a discoidly convoluted tubular shell, head obtuse, tentacles long and bristle-like, with the eyes at their base, mantle not reflected over the shell, foot small, narrow.
Shell; discoid, sometimes very thinly compressed, lens-shaped, apex sunk in the nucleus of the coil, whorls varying from three in number to seven, smooth or striated, sometimes keeled at the periphery.

Of the *Ammonite*-shaped *Lymnæacea* associated in this genus we have eleven species in Britain, nearly all that are known in Europe; and among them is one, the subject of our vignette, pre-eminent in size, the largest form of *Planorbis* in the world. The body of this mollusk is remarkably slim and tapering, enveloped by a mantle which secretes a rotary coiling tubular shell of horny or glassy substance increasing wonderfully slowly in diameter. The head of the animal is stout and obtusely proboscis-shaped, the tentacles slender and bristle-like, with the eyes small and black at their inner base, and the foot, as might be expected where the aperture of the shell is so contracted, is small, narrow, and very flexible. The most slenderly convoluted species are *P. carinatus, complanatus, vortex,* and *spirorbis,* the two first encircled at the periphery by a fine marginal keel; a sixth species, *P. contortus,* is chiefly remarkable for the closeness with which the whorls are pressed one upon another; only one species has any definite sculpture, *P. albus;*

the remainder have small glassy shells. *P. glaber*, allied to the last-mentioned, is smooth and very much smaller; and *P. fontanus*, of a peculiar quoit-like form, convex above, concave below, is smaller still. The last species, *P. nitidus*, better known to conchologists as *P. lacustris*, is distinguished from all others by having the whorls radiately divided internally by septa, and is fairly entitled to rank as a subgenus (*Segmentina*). Sometimes the shell of *Planorbis* is coiled to the right, sometimes to the left, but in a shell of discoid growth it is not easy to detect the difference. The best mode of observing it, is to notice the obliquity of the aperture. The upper disk of the shell is the side on which the margin of the aperture is the most advanced in growth. The animal of *Planorbis*, more particularly that of *P. corneus*, emits a purple fluid, probably blood, when irritated, retiring briskly into its shell.

Freshwater mollusks, as already observed, are much more evenly distributed over the globe, with much less variation of type, than land mollusks. *Planorbis* is the least varied of all. The shell is of the same typical form throughout the whole of the Eastern Hemisphere, and the species are fewer in number and less varied than in the Western. From Greenland to Tasmania, scarcely more than about thirty have been collected, one-half of which are European, and eleven of these are in Britain. The remainder, of moderate and small size, are scattered in Ceylon, India, Burmah, South Africa, and Australia. We have none from the Malayan Archipelagos. In America, between Chili and the Northern United States, there are not fewer than seventy species of *Planorbis* illustrative of two or three quite distinct types. Some of the Bahian and West Indian species are of large size, though not so large as the European *P. corneus*, and they have a peculiarly compressed refulgent shell. The most distinct form of *Planorbis* is that represented by *P. lentus*, *trivolvis*, and *bicarinatus*, of the United States, in which the whorls are broadly tumid and keeled on both sides, and deeply sunk on both sides in the centre. This, in a country where the *Lymnææ* are of the same type as the British and almost identical in species, is a point to be noted.

The British species of *Planorbis* are less local in their distribution than the *Helices*. Eight out of the eleven are present in all parts of Britain, and on the Continent range southwards into the islands of the Mediterranean, passing, with perhaps two exceptions, into North Africa. *P. corneus* is found only in our eastern, southeastern, and midland counties; *P. carinatus* belongs chiefly to the

FAMILY LYMNÆACEA.

eastern and south-eastern counties; and the little *P. nitidus* (*Segmentina*) common in the neighbourhood of London, is a South-England species, and ranges little south of the latitude of England on the Continent. *Planorbis* inhabits all kinds of stagnant pools and ditches and gently running brooks, chiefly adhering to flags and other water-plants. When left dry in the bed of a stream by the retiring of the water, the animal encloses itself within the shell by an epiphragm.

The British species of *Planorbis* are:—

1. **corneus.** Shell large and ventricose, of five to five and a half rudely striated faintly malleated whorls.
2. **albus.** Shell small, rather depressed, of four and a half to five longitudinally thread-striated whorls.
3. **glaber.** Shell very small, smooth, glossy, convex above, rather concave below, of four to four and a half rounded whorls.
4. **crista.** Shell very small, Nautiloid, of only three whorls, covered by a membranaceous epidermis, which is puckered and prolonged at the periphery into lashes.
5. **carinatus.** Shell moderate in size, very depressed, of five slowly enlarging whorls, keeled externally a little below the centre.
6. **complanatus.** Shell larger, very depressed, of five slowly enlarging whorls, keeled externally at the basal edge.
7. **vortex.** Shell rather small, very thinly depressed, of six to seven extremely slowly increasing whorls.
8. **spirorbis.** Shell small, very depressed, of five to six slowly increasing faintly two-angled whorls.
9. **contortus.** Shell rather small, depressed, of six to seven broader extremely closely coiled whorls.
10. **fontanus.** Shell minute, amber horny, of three and a half to four whorls convexly sloping on both sides towards the periphery.
11. **nitidus.** Shell minute, transparent amber horny, of three and a half to four whorls sloping on the upper side towards the periphery, divided internally at intervals by septa.

1. **Planorbis corneus.** *Horny Planorbis.*

Shell; stout, rather ventricose, livid ash, greyish olive, or white, more concave above than below, the spire being conspicuously immersed, whorls five to five and a half, glossy, densely transversely striated, longitudinally striated while young, then faintly malleated; aperture obliquely lunar-rounded, outer lip a little expanded, not reflected.

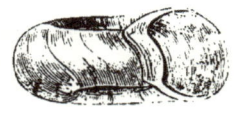

Helix cornea, Linnæus (1758), *Syst. Nat.* 10th edit. p. 770 (not of Draparnaud).
Planorbis purpura and *similis*, Müller (1774), *Verm. Hist.* part ii. p. 154 and 166.
Helix nana, Pennant (1776), *Brit. Zool.* ed. iv. vol. iv. p. 133. pl. lxxxiii. f. 125.
Helix cornu-arietis, Da Costa (1778), *Brit. Conch.* p. 60. pl. iv. f. 13.
Planorbis corneus, Poiret (1801), *Coq. de l'Aisne, Prod.* p. 87.
Planorbis Metidjensis, Forbes (1838), *Moll. Alg. Ann. Phil.* p. 254. pl. xii. f. 5.
Planorbis Dufouri, Graëlls (1846), *Cat. Moll. Espan.* p. 11. pl. i. f. 11 to 15.
Planorbis (Coretus) corneus, Moquin-Tandon (1855), *Hist. Moll.* vol. ii. p. 415. pl. 31. f. 32 to 38 and pl. xxxii. f. 1 to 6.
Planorbis Etruscus, elophilus, Nordenskioldi, anthracius, Banaticus, adelosius, and *aclopus*, Bourguignat (1859), *Rev. et Mag. de Zool.* no. 12 ; *Amén. Malac.* vol. ii. p. 127 to 135. pl. xvi., xvii.

Hab. Pretty general in Europe. Siberia. North Africa. South-eastern, eastern, and midland counties of England. (In muddy ponds and ditches.)

This well-known species stands conspicuously alone in the series on account of its colossal dimensions when compared with any other European form of *Planorbis*. It is indeed the largest in any part of the world, though nearly equalled in size by one or two Brazilian and West Indian species. The animal, as represented of the natural size in our vignette, is of dark reddish black, sluggish in its movements, mostly swimming at the under surface of the water in a half-contracted state, but very irritable; and when irritated or wounded it will emit a purple fluid. The shell is rather inflated in growth, and has the surface faintly malleated as well as densely transversely striated, and towards the apex, where the early growth of the shell is seen, it is strongly striated longitudinally. The malleated appearance arises from the enlargement and diffusion, after the first two or three whorls, of the longitudinal striæ. The spire is so much immersed that the inner edges of the whorls, usually forming the wall of the umbilicus in spiral shells, are pushed into a nearly flattened disk.

Planorbis corneus has a wide range on the Continent, passing into Siberia, and over the islands of the Mediterranean into North Africa; but it is curiously partial in Britain. In the ponds and ditches of the eastern, south-eastern, and midland counties of England, and of the south-eastern parts of Ireland, *Planorbis corneus* is not uncommon; but it does not appear in Cornwall or Devon, nor in Scotland. In colour, the shell is mostly of a livid ash-grey, passing into olive. Sometimes it is of a dark reddish olive, and sometimes entirely white.

2. **Planorbis albus.** *White Planorbis.*

Shell; rather depressed, thin, whitish horny, covered with a scarcely perceptible hairy epidermis, less concave above than below, lower concavity a broadly excavated umbilicus, spire moderately immersed, whorls four and a half to five, longitudinally very finely thread-ridged, interstices faintly decussated with transverse striæ; aperture obliquely lunar ovate.

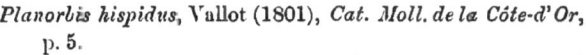

Planorbis albus, Müller (1774), *Verm. Hist.* part 2. p. 164.
Planorbis villosus, Poiret (1801), *Coq. de l'Aisne, Prod.* p. 95.
Planorbis hispidus, Vallot (1801), *Cat. Moll. de la Côte-d'Or*, p. 5.
Helix Draparnaudi, Sheppard (1825), *Trans. Linn. Soc.* vol. xiv. p. 158.
Planorbis reticulatus, Risso (1826), *Hist. Nat. Europ. Mérid.* vol. iv. p. 98.
Planorbis Draparnaldi, Jeffreys (1830), *Trans. Linn. Soc.* vol. xvi. p. 386.
Gyraulus hispidus, Hartmann (1841), *Erd. und Süssw. Gast.* pl. xxv.
Planorbis (Gyraulus) albus, Moquin-Tandon (1855), vol. ii. p. 440. pl. xxxi. f. 12 to 19.

Hab. Throughout Europe. Siberia. North Africa. (Everywhere common on water-plants.)

This is a thinly compressed form of *Planorbis*, in which the whorls of the shell are longitudinally thread-striated, somewhat after the manner of the early whorls of the preceding species, and they are covered with a slight hairy epidermis. The striæ do not, however, become obsolete or irregular with age; they present a neatly defined sculpture throughout, and on the periphery of the outer whorl the central stria assumes a prominence inclining to the magnitude of a keel. The spire, as seen on the upper disk of the shell, is only moderately immersed. On the lower disk the whorls form an even superficially concave umbilicus.

The animal of *P. albus* varies in colour from grey to reddish or yellowish-brown. It is common on water-plants in all parts of the British Isles. Mr. Macgillivray describes having collected it abundantly in the Aberdeen Canal, on different species of *Potamogeton*, and ascertained, by keeping specimens for many days alive, that it fed on that plant with great voracity. "The animal," he adds, "in walking in the water, bears the shell inclined at an angle of from seventy to eighty degrees. Out of the water, it drags the shell, laid flat, by sudden jerks." Abroad it has the same extended range as the preceding species.

3. Planorbis glaber. *Smooth Planorbis.*

Shell; rather convex, depressed above, concave below, thin, glossy, greyish horny, sometimes iridescent, whorls four to four and a half, rounded, smooth; aperture obliquely lunar-rounded.

Planorbis glaber, Jeffreys (1830), *Trans. Linn. Soc.* vol. xvi. p. 387.

Planorbis lævis, Alder (1830), *Trans. Nat. Hist. Soc. Northumb.* vol. ii. p. 337.

Hab. Throughout Europe. North Africa. Madeira. (On water-plants in marshes, lakes, and ponds.)

This little glossy species, unnoticed by Continental writers, has been well observed by Mr. Jeffreys. It is obviously distinct from *P. albus*, with which species it is the nearest allied. The whorls are tumidly rounded and smooth, and the shell is uniformly small. Mr. Jeffreys thinks that *P. cornu* of Ehrenberg, *P. Rossmasleri* of Auerswald, and *P. gyrorbis* of Seckendorf, are referable to *P. glaber*. The animal is described by the same author as being yellowish-grey, with rather short cylindrical tentacles, ending in a blunt point; foot rather broad, especially in front.

4. **Planorbis crista.** *Crest Planorbis.*

Shell; elegantly discoid, rather depressed, greenish horny, transparent, covered by a fine membranaceous epidermis, disposed in transverse ridges which become puckered, and serrate the periphery of the last whorl, upper disk flattened, with the sutures rather deeply impressed, lower disk excavately umbilicated in the centre, whorls three, slightly angled, a little keeled at the angle; aperture obliquely rounded, openly expanded.

Nautilus crista, Linnæus (1758), *Syst. Nat.* 10th edit. p. 709.
Turbo nautileus, Linnæus (1767), *Syst. Nat.* 12th edit. p. 1241.
Planorbis imbricatus, Müller (1774), *Verm. Hist.* part ii. p. 165.
Helix nautilea, Walker (1784), *Test. Minut.* f. 20, 21.
Planorbis imbricatus and *cristatus*, Draparnaud (1805), *Hist. Moll.* p. 44. pl. i. f. 49 to 51, and pl. ii. f. 1 to 3.
Planorbis nautileus, Fleming (1814), *Edin. Encyc.* vol. vii. part 1. p. 69.
Planorbis (Gyraulus) nautileus, Moquin-Tandon (1855), *Hist. Moll.* vol. ii. p. 438. pl. xxxi. f. 6 to 11.

Hab. Throughout Europe. North Africa. (On water-flags in ponds and ditches.)

The minute semitransparent horny shell of this species, more generally known to collectors by the second name which Linnæus gave to it, is composed of only three whorls, necessarily more rapidly enlarging than if it consisted of a coil of a greater number; and the aperture expands like that of a French horn. But the character which chiefly arrests attention is its crested coating of epidermis. At frequent intervals during the growth of the shell it is puckered on the periphery of the outer whorl, and a kind of crest is formed of serrated lashes. The animal as described by M. Moquin-Tandon is greyish-brown, minutely speckled with black, slow in its movements and irritable, carrying its shell a little inclined when crawling.

It may be observed, on reference to the synonymy, that Linnæus and Draparnaud both made two species of this. In adult specimens, the puckered epidermis is sometimes conspicuous, while the crest of lashes on the periphery is nearly obliterated; and sometimes both are obliterated. The names *crista* and *cristatus* have been given to young specimens, and *nautileus* and *imbricatus* to adult specimens.

Planorbis crista is found on water-flags in ponds and ditches in all parts of our islands, and it extends on the Continent from the extreme north to south, passing into Algeria.

5. **Planorbis carinatus.** *Keeled Planorbis.*

Shell; very depressed, concave above, the spire being a little immersed, concavely flattened below, pale horny, smooth, glossy, whorls five, rounded above, convex below, strongly keeled rather below the middle; aperture obliquely ovate, somewhat rhomboidal.

Helix planorbis, Linnæus (1758), *Syst. Nat.* 10th edit. p. 769 (not of Da Costa).
Planorbis carinatus, Müller (1774), *Verm. Hist.* part ii. p. 175 (not of Studer).
Helix limbata, Da Costa (1778), *Test. Brit.* p. 63. pl. iv. f. 10 (not of Draparnaud).
Planorbis acutus, Poiret (1801), *Coq. de l'Aisne, Prod.* p. 91.
Helix carinata, Montagu (1803), *Test. Brit.* p. 450.
Helix planata, Maton and Rackett (1807), *Trans. Linn. Soc.* vol. viii. p. 109. pl. v. f. 14.
Planorbis umbilicatus, Studer (1820), *Kurz. Verz.* p. 92 (not of Müller).
Planorbis lutescens and *disciformis*, Jeffreys (1830), *Trans. Linn. Soc.* vol. xvi. p. 385 and 521.
Planorbis planatus, Turton (1831), *Man.* p. 110. f. 92.
Planorbis (Gyrorbis) carinatus, Moquin-Tandon (1855), *Hist. Moll.* vol. ii. p. 431. pl. xxx. f. 29 to 33.
Hab. Europe, widely diffused but partial. Siberia. England and Ireland. Chiefly in eastern and south-eastern counties. (In stagnant marshes.)

This and the following species require to be examined together. They resemble each other so closely as scarcely to show any perceptible difference; but there are differences and they are constant. On comparing our figures of the shells, which are roughly drawn but characteristic, it will be seen that *P. carinatus* is smaller and more finely striated than *P. complanatus*, and that the encircling keel is not so close to the basal edge. The lower disk in the upper figure of *P. carinatus* is seen just below the keel, which is not the case in the corresponding figure of *P. complanatus*, and the keel by a correlation of growth winds more conspicuously into the aperture. The shell of *P. carinatus* is moreover paler in colour and more glossy. There appears to be scarcely any appreciable difference in the animals of the two species: both are violet-red, more or less speckled with black, with long and slender tentacles.

P. carinatus is much less generally diffused than *P. complanatus*. It occurs, however, in many wide-spread localities in England, becoming gradually more plentiful towards the eastern and south-eastern counties.

6. **Planorbis complanatus.** *Smooth Planorbis.*

Shell; depressed, moderately concave above, the spire but little immersed, concavely flattened below, dingy horny, rather rough, whorls five, rounded above, convex below, strongly keeled at the lower edge; aperture obliquely ovate, somewhat rhomboidal.

Helix complanata, Linnæus (1758), *Syst. Nat.* 10tL edit. p. 769 (not of Montagu).
Planorbis umbilicatus, Müller (1774), *Verm. Hist.* part ii. p. 160.
Planorbis complanatus, Studer (1789), *Faun. Helv.* in *Coxe, Trav. Switz.* vol. iii. p. 435 (not of Poiret nor Draparnaud).
Helix lacustris, Razoumowsky (1789), *Hist. Nat. Jorat*, vol. i. p. 273.
Planorbus marginatus, Draparnaud (1805), *Hist. Moll.* p. 45. pl. ii. f. 11 to 13.
Helix cochlea, Brown (1818), *Mem. Wern. Soc.* vol. ii. p. 501. pl. xxiv. f. 10.
Helix rhombea, and *terebra* Turton (1822), *Conch. Dict.* p. 47.
Planorbis rhombeus, Turton (1831), *Man.* p. 108.
Planorbis Sheppardi, Leach (1831), *Turt. Man.* p. 140.
Planorbis (Gyrorbis) complanatus, Moquin-Tandon (1855), *Hist. Moll.* vol. ii. p. 428. pl. xxx. f. 18 to 28.

Hab. Throughout Europe. Siberia. North Africa. England and Ireland. (Everywhere present in marshes, ponds, canals, ditches.)

The shells of this and the preceding species, though resembling each other closely in general appearance, were well distinguished by Linnæus. In *P. carinatus* (*Helix planorbis*, Linn.), the keel is nearly central; in *P. complanatus* (*Helix complanatus*, Linn., but more generally known to collectors as *P. marginatus*, Drap.), the keel is quite on the basal edge of the whorl. The name *complanatus* is not the most appropriate, for the species is the less smooth of the two, and it is darker in colour. It is much the commoner, being present in all parts of Britain, excepting Scotland, in stagnant marshes, ponds, canals, and ditches. Dr. Gray remarks that *P. complanatus* breeds very rapidly in ponds of warm water that is emitted from steam-engines in Yorkshire; and the specimens found in such situations have a curious tendency to assume a more spiral form. The range of *P. complanatus* abroad is similar to that of *P. carinatus*, and the two species are very closely connected by an intermediate form, *P. submarginatus* of Cristofori and Jan, *P. intermedius* of Charpentier.

7. Planorbis vortex.   *Whirl Planorbis.*

Shell; extremely depressed, fulvous horny, thin, upper disk moderately concave with the sutures rather indented, lower disk flatly concave, with the whorls sloping to the lower margin, where they are obtusely angled; whorls six to seven, very narrow, increasing very slowly, faintly obliquely striated; aperture obliquely lunar, lip simple.

Helix vortex, Linnæus (1758), *Syst. Nat.* 10th edit. p. 772.
Planorbis vortex, Müller (1744), *Hist. Verm.* part 2. p. 158.
Helix planorbis, Da Costa (1778), *Test. Brit.* p. 65. pl. iv. f. 12 (not of Linnæus).
Planorbis tenellus, Studer (1820), *Kurz. Verz.* p. 92.
Planorbis compressus, Michaud (1831), *Compl.* p. 81. pl. xvi. f. 6 to 8.
Planorbis (Gyrorbis) vortex, Moquin-Tandon (1855), *Hist. Moll.* vol. ii. p. 433. pl. xxx. f. 34 to 37.

Hab. Throughout Europe. Siberia. North Africa. (On plants in shallow stagnant water.)

P. vortex and *spirorbis* are very similar, and should also be studied together. Their shells are of all land shells the most thinly, slowly, and flatly coiled. That of *P. vortex* is composed of from six to seven whorls, sloping towards the lower margin, where it is obtusely keeled, and the lower disk of the shell is consequently less concave than the upper. In *P. spirorbis* the shell is smaller, and has from one to two fewer whorls, the whorls are less angled at the periphery, and more truly discoid, so that the two disks are nearly equally concave. In both species the animals are purplish reddish-brown, minutely speckled with black, and are in all respects similar. When the animal of *P. vortex* retires into its shell in a torpid state from lack of moisture while the shallow pools which it inhabits are dried up, it forms a callous ridge within the margin of the aperture.

Mr. Macgillivray, who carefully observed the habits of this species from specimens collected in the neighbourhood of Aberdeen, remarks that the animal slowly glides along the surface of a stem or leaf under water by a series of undulations. Every now and then the shell is jerked forward by a sudden movement, which is not performed by the foot, but by the muscles which pass from it, or along the neck into the body. It is abundantly diffused throughout the British Isles, and in all parts of Europe, passing into Siberia and North Africa.

8. **Planorbis spirorbis.** *Roll Planorbis.*

Shell; extremely depressed, dingy horny, thin, upper and lower disk nearly equally concave, each showing impressed sutures; whorls five to six, finely striated, convex, slightly angled at the margin of both disks; aperture lunar ovate, mostly ridged within.

Helix spirorbis, Linnæus (1758), *Syst. Nat.* 10th edit. p. 770.
Planorbis spirorbis, Müller (1774), *Verm. Hist.* part ii. p. 161.
Planorbis rotundatus, Poiret (1801), *Prod.* p. 93.
Planorbis leucostoma, Millet (1813), *Moll. Maine et Loire*, p. 16.
Planorbis septemgyratus, Ziegler (1835), *Rossm. Icon.* vol. i. p. 106. f. 64.
Planorbis Perezii, Graëlls (1850), *Dupuy, Hist. Moll.* vol. iv. p. 441. pl. xxv. f. 6.
Planorbis fragilis, Millet (1854), *Moll. Maine et Loire*, 3rd edit. p. 43.
Planorbis (Gyrorbis) spirorbis and *rotundatus*, Moquin-Tandon (1855), *Hist. Moll.* vol. ii. p. 435 and 437. pl. xxx. f. 38 to 46 and pl. xxxi. f. 1 to 5.

Hab. Throughout Europe. Siberia. North Africa. (On plants in shallow stagnant water.)

This species is very closely allied to the preceding, and it has a similar distribution. Both were, however, separately distinguished by Linnæus and Müller, and they have been regarded as separate species by all subsequent writers on the subject. Typical specimens of *P. spirorbis* are composed of fewer whorls than *P. vortex*, and they are more truly discoid, the upper and lower disk of the shell being almost equally concave, and each showing impressed sutures. The aperture varies in outline according to the greater or less development of keel on the edge of either disk.

---

9. **Planorbis contortus.** *Twisted Planorbis.*

Shell; depressed, brownish horny, upper disk flat with the sutures impressly channelled, lower disk broadly excavately umbilicated; whorls six to seven, extremely closely coiled one upon the other, finely striated, rounded at the periphery, gently sloping to the edge which is obtusely angled; aperture small, compressly lunar.

Helix contorta, Linnæus (1758), *Syst. Nat.* 10th edit. p. 770.
Planorbis contortus, Müller (1774), *Verm. Hist.* part ii. p. 162.
Helix crassa, Da Costa (1778), *Brit. Conch.* p. 66. pl. iv. f. 11.
Helix umbilicata, Pulteney (1799), *Cat. Dorset.*
Planorbis (Bathyomphalus) contortus, Moquin-Tandon (1855), *Hist. Moll.* vol. ii. p. 443. pl. xxxi. f. 24 to 31.
Hab. Throughout Europe. Siberia. (In ponds and ditches.)

The shell of this small *Planorbis* has the whorls more closely coiled one upon the other than that of any other mollusk. Although of truly discoid growth, the under surface is distinctly and largely umbilicated; the upper surface is very little concave, and the sutures of the very narrow whorls are impressly channelled throughout. The animal is dusky brown about the head and neck, black at the foot; and the tentacles are rather elongately setaceous, tapering to a point. It lives among the roots and stems of water-plants, in ponds and ditches throughout Britain, but it is local; on the Continent it is also generally diffused, passing in an easterly direction into Siberia and southwards to the islands of the Mediterranean.

Planorbis contortus is sluggish and irritable, says M. Moquin-Tandon, allowing itself to sink in the water on being touched. It then rises with a slightly jerking movement, carrying its shell horizontally, and on reaching the surface manages to let it float.

---

10. **Planorbis fontanus.** *Fountain Planorbis.*

Shell; depressed, bright amber horny, upper disk with the spire a little immersed towards the apex, lower disk with a moderate-sized sunk umbilicus; whorls three and a half to four, convexly sloping on both sides towards the periphery, which is angularly rather produced; aperture horizontally triangularly lunar.

Helix fontana, Lightfoot (1786), *Phil. Trans. Roy. Soc.* vol. lxxvi. p. 165. pl. ii. f. 1.
Planorbis complanatus, Draparnaud (1805), *Hist. Moll.* p. 47. pl. ii. f. 20 to 22 (not of Studer nor Poiret).
Helix lenticularis, Alten (1812), *Syst. Abhandl. Augsb.* p. 35. pl. ii. f. 4.
Planorbis fontanus, Fleming (1814), *Edin. Encyc.* vol. vii. p. 69.
Planorbis lenticularis, Sturm (1829), *Deutsch. Faun.* vol. viii. f. 16.
Planorbis nitidus, Gray (1840), *Turt. Man.* p. 268. pl. viii. f. 93 (altered in edit. of 1857 to *P. fontanus*).
Hippeutis lenticularis, Hartmann (1842), *Erd. und Süss. Gasterop.* p. 51.
Planorbis (Hippeutis) fontanus, Moquin-Tandon (1855), *Hist. Moll.* vol. ii. p. 426. pl. xxx. f. 10 to 17.

Hab. Throughout Europe. Siberia. (In ponds and clear streams, on water-plants.)

Forbes and Hanley and Mr. Jeffreys have named this species *Planorbis nitidus* on the ground that it is in chief part the *P. nitidus* of Müller, but Moquin-Tandon has clearly shown that our next species is Müller's *P. nitidus;* and he has been followed in that opinion by Dr. Gray. *P. fontanus*, the smallest of the genus, has a bright transparent amber horny shell of only three and a half to four whorls, sloping in a peculiar quoit-like form towards the periphery, and the aperture assumes a horizontal, triangularly lunar shape. The umbilicus is small compared with its ordinary dimensions in *Planorbis*, and instead of being superficial it has a sunken aspect. The animal is of a dusky grey, sometimes tinged with yellow and finely dark-speckled, timid and slow in its movements. The tentacles are long and filiform, and but little flexuous. The eyes are rather large, round, and black. The species is very generally and widely diffused over the same range of latitude as the preceding.

11. **Planorbis nitidus.** *Shining Planorbis.*

Shell; depressed, semitransparent amber horny, upper disk with the spire a little immersed, lower disk convexly flattened with a rather narrow sunken umbilicus, whorls three and a half to four, convexly sloping on the upper part towards the periphery which is obtusely angled, last whorl marked with a few distant radiating lines indicating internal septa.

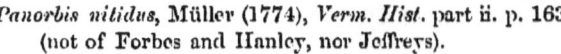

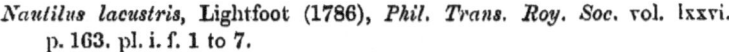

Panorbis nitidus, Müller (1774), *Verm. Hist.* part ii. p. 163 (not of Forbes and Hanley, nor Jeffreys).
Helix lineata, Walker (1784), *Test. Minut. Rar.* p. 8. pl. i. f. 28.
Nautilus lacustris, Lightfoot (1786), *Phil. Trans. Roy. Soc.* vol. lxxvi. p. 163. pl. i. f. 1 to 7.
Helix nitida, Gmelin (1788), *Syst. Nat.* p. 3624.
Planorbis complanatus, Poiret (1801), *Prod.* p. 93.
Planorbis clausulatus, Férussac (1820), *Journ. Phys.* p. 240.
Planorbis nautilus, Sturm (1823), *Deutsch. Faun.* vol. vi. pl. xv.
Segmentina lineata, Fleming (1828), *Brit. Anim.* p. 279.
Segmentina nitida, Fleming (1830), *Edin. Encyc.* vol. xii.
Hemithalamus lacustris, Leach (1831), *Turt. Man.* p. 137.
Segmentaria lacustris, Swainson (1840), *Treat. Malac.* p. 338.
Planorbis (Segmentina) nitidus, Moquin-Tandon (1855), vol. ii. p. 424. pl. xxx. f. 5 to 9.

Hab. Central and (perhaps) Northern Europe. Chiefly in south of England; rare in Scotland and Ireland. (In ponds and ditches, on duckweed and other water-plants.)

The shell of this species is distinguished in a very characteristic manner by the presence of three or four septa within the last whorl, and when seen externally they give the whorl an appearance of being radiately divided into segments. They were described by Müller, the founder of the species, as "streaks like ligaments," which he supposed to be the repairs of fractures. Mr. Jeffreys considers them as being somewhat analogous to the parietal teeth of *Pupa* and *Vertigo*, which is not improbable. The shell of *P. nitidus* is rather larger and duller in aspect than that of the preceding species, and the whorls do not slope into so prominent an angle at the periphery, whilst they are more flattened around the umbilicus, which is smaller.

It has been observed that *Planorbis nitidus* is rather more active in its movements than others of the genus, and carries its shell quite horizontally. Its geographical range is not very clearly established. It is not uncommon among duckweed and dead leaves in the neighbourhood of London; and it is recorded as having been collected in Ireland and in the south of Scotland. On the Continent it does not appear to have been noticed south of the Pyrenees. Central Europe seems to be its chief area of habitation, and it is said to have been collected in Sweden.

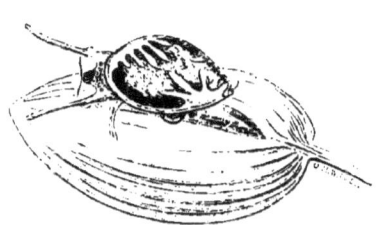

Physa fontinalis.

Genus II. **PHYSA**, *Draparnaud*.

Animal; body ocellated, carrying a transparent sinistrally coiled shell, over the edge of which the mantle is sometimes reflected, fringed with digitate lobes, foot lanceolate and pointed behind, obtuse in front, yellow, clouded more or less with very minute dusky specks, head rather truncate, furnished with two contractile filiform tentacles which are broad at their base, at the inner corner of each base are situated the eyes, black and conspicuous.

Shell; sinistral, ovate and ventricose or fusiformly oblong, transparent, glassy, of four whorls with the spire short, or of five whorls with the spire attenuated; aperture ovate, lip simple, columella thinly callously twisted.

This pretty mollusk, with its ocellated colouring seen through a transparent sinistrally convoluted shell, and with its curiously digitate fringe-lobed mantle, is the European representative of a genus belonging more especially to Australia and Central America. In Australia and New Zealand, some thirty species of *Physa* have been collected, many of them covered with an epidermis and of large size, and nine of them are of a form (*Ameria*) found in no other part of the world, in which the upper, or hinder, portion of the shell is sharply angled. As many as fifty species of *Physa* inhabit the continent and islands lying between Bahia and the Northern United States, three of which, *P. Maugeræ, influviata*, and *aurantia*, natives respectively of Jamaica, Guatemala, and Mazatlan, are conspicuous for their size and dark shining surface. In Europe there are but four species of *Physa*, and only two of these pass into Britain. *P. fontinalis* is plentifully distributed in our ponds and ditches and running brooks. *P. hypnorum* is more partially diffused and scarcer, and on the Continent it is one of the few mollusks whose range of habitation is limited to the north side of the Pyrenees.

Our two *Physæ* are very distinct from each other. *P. fontinalis* is of a rather inflated oval form, having the mantle reflected over the shell on either side in a digitate fringe-like lobe. Sometimes it is lobed pretty evenly along each side, sometimes, as in the variety represented in our vignette, more in the direction of the columella and spire, the lobe on the columellar side being the larger. The portion of the animal seen through the transparent shell, is delicately ocellated. *P. hypnorum* is of a fusiformly oblong shape, and has a more attenuately convoluted shell with no lobed overlapping of the mantle. For this reason it has been set apart as a separate genus (*Aplexa*). In both species, the foot of the mollusk is lanceolately pointed behind, the respiratory and other important organs are on the left side of the animal, the shell is therefore convoluted to the left, and the tentacles are filiform, or bristle-like, as in *Planorbis*, which genus it resembles also in having the eyes at their inner base. The difference in the structure of the shell in *Planorbis* and *Physa*, one producing a discoidly convoluted, the other an inflated oval one, contrasts remarkably with their similarity in other respects.

The *Physæ*, especially *P. hypnorum*, are active in habit, whether swimming, foot uppermost, on the surface of the water, holding themselves stationary at different depths in the water, or gliding through it in sudden jerks by an hydraulic action of the foot. By bringing the lateral margins of this organ into contact, the animal constructs a tube for inhaling and suddenly expelling the water either upwards or downwards. Montagu stated, and the statement has been repeated by Jeffreys, that the animal spins a mucous thread for letting itself down in the water and rising again for respiration, but I have not succeeded in confirming this observation, and have great doubts of its accuracy.

One or two species of *Physa* of the ordinary *P. fontinalis* type have been collected in Syria, China, Singapore, and the Philippine and Sandwich Islands, and a single curiously turriculate species with a delicate transparent glassy shell has been lately discovered in Egypt, and at Benguela, south of the Congo River. At Natal, in the same range of habitation, another singularly opposite form of *Physa* occurs, short and globose, with the spire not at all exserted, having the columella truncated at the end (*Physopsis*); and a second species of this form has been lately collected at Port Essington, North Australia.

The British *Physæ* are:—

1. **fontinalis.** Shell sinistral, ovate, transparent, of four smooth whorls, the last of which is ventricosely inflated.
2. **hypnorum.** Shell sinistral, fusiformly oblong, transparent, of five smooth whorls, convoluted attenuately.

---

1. **Physa fontinalis.** *Stream Physa.*

Shell; sinistral, ovate, mostly subglobose, yellowish horny, very thin and transparent, almost glassy, spire short, rather obtuse, sometimes produced, whorls four, convex, smooth, moderately shining, faintly irregularly striated in the direction of the lines of growth, opakely margined at the sutures, last whorl inflated; aperture pyriformly ovate, columella slightly callously twisted.

Bulla fontinalis, Linnæus (1758), *Syst. Nat.* 10th edit. p. 727.
Planorbis bulla, Müller (1774), *Verm. Hist.* part ii. p. 167.
Turbo adversus, Da Costa (1778), *Test. Brit.* p. 96. pl. v. f. 6.
Bulimus perla, Müller (1781), *Naturf.* vol. xv. p. 6. pl. i.
Bulimus fontinalis, Bruguière (1789), *Enc. Méth. Vers*, vol. i. p. 306.
Physa fontinalis, Draparnaud (1801), *Tabl. Moll.* p. 52.
Helix bullaoides, Donovan (1803), *Brit. Shells*, vol. v. pl. clxviii. f. 2.
Bulla fluviatilis, Turton (1819), *Conch. Dict.* p. 27.
Limnea fontinalis, Sowerby (1823), *Gen. Limn.* f. 8.
Bulimus fontinalis, Beck (1837), *Ind. Moll.* p. 117.

Physa (Bulimus) fontinalis, Moquin-Tandon (1855), *Hist. Moll.* vol. ii. p. 451. f. 9 to 13.

Hab. Throughout Europe. (On aquatic plants, both in stagnant and running water.)

Common as this sinistral water-snail is throughout Britain and the whole of Europe, not only in tranquil ponds, ditches, and canals, but also in running brooks, it is not often seen with the mantle so neatly developed as in our vignette of a living specimen crawling on a leaf of *Potamogeton*, from a drawing obligingly contributed by Mr. Berkeley. The animal has not been observed in this country with the care it merits. Dr. Gray has given a woodcut of it, and it is figured by Forbes and Hanley, but Mr. Jeffreys' figures of this plentifully distributed mollusk are a copy of those of Moquin-Tandon. More original observation of the animal of *Physa fontinalis* is needed, for there are certainly two very distinct forms of the species. The mantle is distinguished in a very characteristic manner, by being reflected on either side of the shell in a digitate lobe. In the short subglobose form figured by the above-named authors, the lobes are quite lateral, but in the oblong form, with a more produced spire represented in our vignette, it will be seen that the right lobe is towards the front, and the left lobe is towards the apex of the shell, the right lobe being always more digitate than the left. Dr. Gray relates, in his edition of 1840 of Turton's 'Manual,' that Mr. James Sowerby sent him a specimen of the long-spired variety of *Physa fontinalis* under the name of *Physa acuta?*, which he received from Anglesey, and continued to breed in his waterbutt. Mr. Sowerby named it *acuta?*, as differing from the common species in the following particulars:—"One of the lobes," he says, "covers the columella and is five-parted, the other is turned upon the spire, and is three-parted." This interesting variety is the mollusk of our vignette, but it is not the *Physa acuta*, which is quite another species, a native of Central and Southern France, in which the mantle is not reflected.

The shell of *Physa fontinalis* is extremely thin and transparent, and of a bright yellowish horny substance, moderately glossy. It is composed of four whorls, always convoluted sinistrally, and has a pyriformly ovate aperture, with the columella slightly callously twisted. Its range of habitation appears to be very general throughout Europe. Near London, it may be found abundantly in such places as the marshy parts of Battersea and the Isle of Dogs.

2. **Physa hypnorum.** *Moss Physa.*

Shell; sinistral, fusiformly oblong, very thin, transparent, brittle, shining, spire attenuately produced, whorls five, convex, faintly irregularly striated in the direction of the lines of growth, opakely margined at the sutures; aperture obliquely ovate, rather narrow, columella callously edged.

Bulla hypnorum, Linnæus (1758), *Syst. Nat.* 10th edit. p. 727.
Planorbis turritus, Müller (1774), *Verm. Hist.* part ii. p. 169.
Bulla turrita, Gmelin (1788), *Syst. Nat.* p. 3428.
Helix marmorata, Gmelin (1788), *Syst. Nat.* p. 3665.
Bulimus hypnorum, Bruguière (1789), *Enc. Méth. Vers,* vol. i. p. 301.
Physa hypnorum, Draparnaud (1801), *Tabl. Moll.* p. 52.
Physa turrita, Studer (1823), *Kurz. Verz.* p. 92.
Limnea turrita, Sowerby (1853), *Gen. Limn.* f. 10.
Aplexa hypnorum, Fleming (1828), *Brit. Anim.* p. 276.
Nauta hypnorum, Leach (1831), *Turt. Man.* p. 152.
Physa cornea, Massot (1845), *Soc. Agr. Pyr.-Orient.* vol. vi. part 2. p. 236. f. 4.
Physa (Nauta) hypnorum, Moquin-Tandon (1855), *Hist. Moll.* vol. ii. p. 453. pl. xxiii. f. 11 to 15.

Hab. Northern and Central Europe to the Pyrenees. Arctic Siberia. Throughout Britain. (In ponds and ditches, upon blades of grass and other plants.)

This more elongated species is very distinct from the preceding, and it is not without good grounds that it has been regarded as a separate genus, *Aplexa.* In *Physa fontinalis* the mantle is reflected over the shell on either side in a fringed or digitate lobe; in *Physa hypnorum* it is not reflected beyond the merest lodgment on the edge of the shell; and the shell, notwithstanding, is the more brittle and glossy of the two. Another peculiarity in the species under consideration, consists in the tentacles being more slenderly subulate, whilst the animal is altogether darker in colour. There is, too, an additional whorl in the shell. The mantle of *P. hypnorum,* according to the observations of M. Desmoulins, is ocellated with yellow or golden spots, like that of *P. fontinalis.*

Its range of habitation is rather partial in Britain, and in Europe its place is taken below the Pyrenees by *Physa acuta.* It prevails more towards the north; and was collected by Middendorf in Arctic Siberia.

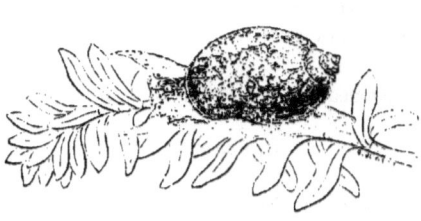

Lymnæa limosa.

Genus III. **LYMNÆA**, *Bruguière*.

Animal; body ovate, carrying a horny shell varying greatly in structure, but mostly ventricosely inflated, mantle reticulated, sometimes capaciously reflected over the shell, foot obtusely attenuated behind, truncate in front, with the margin slightly undulate, head obtuse, furnished with two flat obliquely triangular tentacles, with the eyes on obscure tubercles at their inner base.

Shell; varying from ventricosely ovate of four whorls to elongately turreted of eight whorls, with the spire proportionally sharply acuminated, yellowish horny, sometimes extremely thin and transparent, smooth or malleated, columella arcuately twisted, lip simple, columellar lip more or less appressly dilated on the body whorl.

On comparing the *Lymnæa limosa* of our vignette with the *Physa fontinalis* figured at the head of the preceding genus, it will be seen that not only do the shells differ in their manner of convolution, the first-named being coiled to the right, the second to the left, but that there is an important difference in the animal's tentacles. As in *Physa*, the foot of *Lymnæa* is pointed behind, the head is truncate and the eyes are similarly placed; the tentacles are different. Those organs are no longer filiform, or bristle-like, but flatly broadly triangular. Along with this difference in the tentacles, there is a difference in the structure and composition of the shell. It is extremely variable in form, sometimes sharply fusiform, sometimes rotundately inflated, and it is of a dull horny striated or malleated texture, never vitrified to anything approaching a porcelain gloss. This remark applies, however, only to the *Lymnææ* proper, of which we have six species in Britain. The two remaining species, *L. glutinosa* and *involuta*, have an extremely thin bright transparent shell, covered almost entirely by a reflected extension of the mantle; and many authors consider them entitled to rank as a separate genus, *Amphipeplea*.

Lymnæa is represented in its distribution over the globe in three very distinct forms. In India, neighbourhood of Calcutta, the shell is of a characteristic cylindrically oblong form. In the Malayan islands and Punjaub districts of India, it is of a peculiar silvery-horny substance marked with opake-white linear streaks. In Western Asia north of the Himalayas, over the whole of Europe extending to Greenland, and over all the United States, the *Lymnææ*. excepting *L. glabra* and the section *Amphipeplea*, produce a dull horny malleated shell, such as we have already described. The inland waters of Central America and Australia have few *Lymnææ*; they are chiefly inhabited by *Physæ*. In the Patagonian and Chilian waters of South America the two genera are mainly dispensed with in favour of another type peculiar to that locality, *Chilina*.

The *Lymnææ* of Western Asia and Europe are plentiful in individuals, but comparatively limited in species; and, unlike other Caucasian genera, which diminish in species as they are further removed from the central area of the general molluscan fauna, they all pass into Britain. Other species have been described by Continental authors, but they do not appear to me to be tenable. Another peculiarity in the distribution of the Caucasian *Lymnææ* is, that nearly all the species appear to pass in a modified form into the United States. The species of the two countries, if not identical, present a marked degree of parallelism,—*L. limosa* with *L. catascopium*, *L auricularia* with *L. macrostoma*, *L. stagnalis* with *L. jugularis*, *L. palustris* with *L. elodes*, and *L. truncatula* with *L. desidiosa*. Our *L. glabra* and the two *Amphipepleæ*, *L. glutinosa* and *involuta*, have no representative in America; and, concomitant with this, there is a peculiarity in their geographical distribution in Europe. Like *Physa hypnorum*, they are not found south of the Pyrenees. *L. involuta*, it should be mentioned, is a doubtful species, perhaps a variety of *L. glutinosa* in which the spire is immersed. It has only been collected in the south of Ireland. The only other species of *Amphipeplea* at present known are three of fine dimensions, collected respectively at the island of Luzon, Philippines, and in the vicinities of Moreton Bay and of Melbourne, Australia.

The animal of *Lymnæa* varies in colour and marking. Sometimes the body is speckled with dusky sometimes with opake-white dots; the mantle is generally cloudedly reticulated. *L. stagnalis* and *limosa*, as observed in the vivarium, are decidedly carnivorous. Mr. Jones, the active Secretary of the Cotteswold Club, informs me that he has frequently given them dead bleak and other small fish, which they have reduced to a skeleton in a very short time, although abundance of vegetable food was at hand.

The British species of *Lymnæa* are:—

1. **limosa.** Shell obliquely ovate, of four whorls, of which the last is ventricosely inflated.
2. **auricularia.** Shell squarely semiglobose, of three to four whorls, of which the last is very largely widely expanded.
3. **stagnalis.** Shell ovately turreted, with the spire acuminately produced, of five to six whorls, of which the last is globosely inflated.
4. **palustris.** Shell oblong-ovate, rather solid, of five to six whorls, rough grey, fuscous red in the interior.
5. **truncatula.** Shell smaller, acuminately ovate, rather solid, of five to six roughly striated whorls.
6. **glabra.** Shell small, elongately turreted, with a subcylindrical spire, of eight rather closely convoluted smooth silky whorls.
7. **glutinosa.** Shell globosely ovate, extremely thin and transparent, of only three whorls, having the columellar lip largely membranaceously reflected over the body whorl.
8. **involuta.** Shell subtruncately oval, extremely thin and transparent, with the spire immersed, of three to four whorls, having the columellar lip largely membranaceously reflected over the body whorl.

FAMILY LYMNÆACEA.

1. **Lymnæa limosa.** *Mud Lymnæa.*

Shell; obliquely ovate, compressly minutely umbilicated, rather thin, yellowish horny, spire short, acuminated, whorls four, convex, irregularly striated in the direction of the lines of growth, the last much the largest, ventricosely inflated; aperture ovate, columella arcuately twisted, lip broadly appressed over the umbilicus.

Helix limosa, Linnæus (1758), *Syst. Nat.* 10th edit. p. 774.
Buccinum peregrum, Müller (1774), *Verm. Hist.* part ii. p. 130.
Helix putris, Pennant (1777), *Brit. Zool.* p. 139. f. 137 (not of Linnæus).
Helix peregra, inflata, and *teres*, Gmelin (1788), *Syst. Nat.* p. 3659, 66, 67.
Buccinum medium and *rivale*, Studer (1789), *Faun. Helv. Coxe, Trav. in Switz.* vol. iii. p. 433, 434.
Bulimus pereger, Bruguière (1789), *Enc. Méth. Vers*, p. 301.
Bulimus limosus, Poiret (1801), *Coq. de l'Aisne*, p. 39.
Limneus pereger, Draparnaud (1801), *Tabl. Moll* p. 48.
Helix lutea, Montagu (1803), *Test. Brit.* p. 380. pl. xvi. f. 6.
Limneus ovatus, Draparnaud (1805), *Hist. Moll.* p. 50. pl. ii. f. 30, 31.
Limneus fontinalis, Studer (1820), *Kurz. Verz.* p. 93.
Limnæus vulgaris, C. Pfeiffer (1821), *Deut. Moll.* vol. i. p. 89. pl. iv. f. 22.
Limnea intermedia, Férussac (1822), *Lam. Anim.* vol. vi. part 2. p. 162.
Lymnæa limosa, Fleming (1828), *Brit. Anim.* p. 274.
Lymnæa putris, Fleming (1830), *Edin. Encyc.* vol. vii. p. 77.
Gulnaria peregra, Leach (1831), *Turt. Man.* p. 146.
Limnea marginata, Michaud (1831), *Comp. de Drap.* p. 88. pl. xvi. f. 15, 16.
Limnea thermalis, Boubée (1833), *Bull. Hist. Nat.* p. 28.
Limnea lineata, Bean (1834), *Mag. Nat. Hist.* vol. vii. p. 493. f. 62.
Gulnaria ovata, Beck (1837), *Ind. Moll.* p. 114.
Lymnea lacustris, Potiez and Michaud (1838), *Moll. de Douai*, vol. i. p. 219. pl. xxii. f. 11, 12.
Lymnæa peregrina, Mauduyt (1839), *Moll. de la Vienne*, p. 95.
Amphipeplea lacustris, Brown (1845), *Illus. Conch.* p. 30. pl. xv. f. 24.
Limnæa Burnetti, Alder (1848), *Ann. and Mag. Nat. Hist.* Ser. 2. vol. ii. p. 396. pl. xi. f. 1 to 3.
Limnæa glacialis, Dupuy (1849), *Cat. Extr. Test.* no. 199.
Limnæa Trencaleonis and *Nouletiana*, Gassies (1849), *Moll. de l'Agénais*, p. 163 and 166.
Lymnæa Hookeri, Reeve (1850), *Pro. Zool. Soc Lond.* part 18. p. 49.

Lymnæa Boissii, Dupuy (1851), *Hist. Moll.* vol. v. p. 479. pl. xxv. f. 9.
Lymnæa teres, Bourguignat (1853), *Voy. Mer Morte*, p. 58.
Lymnæa (Gulnaria) ovata and *limosa*, Moquin-Tandon (1855), *Hist. Moll.* vol. ii. p. 465 to 470. pl. xxxiv. f. 11 to 16.

Hab. Throughout Europe. Siberia. Thibet. Afghanistan. (In ponds, lakes and ditches, and in springs at various elevations.)

Water snails, as we have already remarked, are distributed over the globe, with less variation of typical character than land snails. *Lymnea limosa*, more generally known to collectors by the name *peregra* than by its Linnean name, is perhaps the most abundantly and widely diffused of all our mollusks. It is the common pond snail both of the Germanic and Lusitanian regions of Europe, it ranges over Western Asia from Siberia to Afghanistan and Thibet, and it is so nearly approached in the United States by Mr. Say's *L. catascopium* that doubts are entertained whether the two alleged species are really distinct. It has also a near representative in Greenland in *L. Vahlii*. The shell of *L. limosa* is yellowish horn-colour, generally of a uniform tint, but it is extremely variable in form. The variability is, however, clearly a modification of the same specific type. There is no fear of mistaking the most widely inflated form of *L. limosa* for *L. auricularia*. All the numerous variations of the species arise simply out of a more or less elongately spired plan of convolution; the remaining differences follow in the order of correlation of growth. Where the spire is shortest, as in the variety which has been named *L. Burnettii*, the whorls are most inflated; where it is longest, as in a Thibetan variety named *L. Hookeri*, they are the most restricted in diameter.

The animal of *L. limosa* varies chiefly in colour. It is described by different observers as being dark-grey or brown, or greenish-brown or olive-yellow, mottled with grey or black, and sometimes speckled with opake yellow or white. The tentacles are flatly triangular, broad at the base, with the eyes, small clear and bright, between them. Sometimes the body and shell are convoluted sinistrally; it is however the exception, not as in *Physa* the rule, of growth. Specimens in this state Mr. Metcalfe informs me he once collected in the neighbourhood of Scarborough, but never in any other locality. The largest specimens he ever found of *L. limosa* of the regular dextral growth were in some ponds on the north side of Hampstead Heath.

FAMILY LYMNÆACEA.

 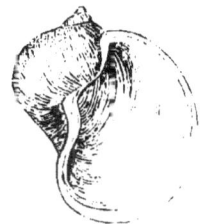

2. **Lymnæa auricularia.** *Ear-shaped Lymnæa.*

Shell; somewhat squarely semiglobose, compressly umbilicated, rather thin, pallid horny, spire very small, sharp, whorls three to four, convex, irregularly striated in the direction of the lines of growth, sometimes evanescently obscurely irregularly ridged and malleated in the opposite direction, extremely rapidly enlarging, last whorl abruptly widely auricularly inflated; aperture very large, outer lip thinly expanded, columella callously twisted, lip appressly dilated over the umbilicus.

Helix auricularia, Linnæus (1758), *Syst. Nat.* 10th edit. p. 774.
Buccinum auricula, Müller (1774), *Verm. Hist.* part 2. p. 126.
Turbo patulus, Da Costa (1774), *Test. Brit.* p. 95. pl. v. f. 17.
Bulimus auricularius, Bruguière (1789), *Enc. Méth. Vers,* vol. i. p. 304.
Limnæus auricularius, Draparnaud (1801), *Tabl. Moll.* p. 48.
Radix auriculatus, De Montfort (1810), *Conch. Syst.* vol. ii. p. 207.
Limneus acronicus and *Hartmanni,* Studer (1820), *Kurz. Verz.* p. 93.
Limneus acutus, Jeffreys (1830), *Trans. Linn. Soc.* vol. xvi. p. 373.
Gulnaria auricularia, Leach (1831), *Turt. Man.* p. 148.
Limneus ampullaceus, Rossmässler (1835), *Icon. Moll.* vol. ii. p. 19. f. 121.
Gulnaria ampla, Monnardii, and *Hartmani,* Hartmann (1842), *Erd. und Süss. Gast.* p. 69 to 72. pl. v. to vii.
Limnæa canalis, Villa (1851), *Dup. Hist. Moll.* vol. v. p. 482. pl. xxii. f. 12.
Limnæa (Gulnaria) auricularia, Moquin-Tandon (1855), *Hist. Moll.* vol. ii. p. 462. pl. xxxii. f. 21 to 31. and pl. xxxv. f. 1 to 10.

Hab. Nearly throughout Europe, but local. Siberia. Cashmere. Central and southern England and Ireland. (In ponds, lakes, marshes, or on aquatic plants.)

The shell of this species is composed of a whorl less than that of the preceding species, and it is especially distinguished by its rapidly enlarging growth; the last whorl is so abruptly and widely outspanned as almost entirely to cover the animal. It was well distinguished from *L. limosa* by Lister, Linnæus, and Müller, and it appears to differ from all its varieties. The surface of the shell is frequently malleated by an irregular decussation of spiral ridges, which become faint and finally disappear.

The geographical range of *L. auricularia* abroad, is very extended. I give Siberia on the authority both of Middendorf and Gerstfeldt, and Cashmere on that of M. Morelet, who states that there are specimens in the Museum of Paris, collected in that locality by the unfortunate Jacquemont. In Britain, the species has not been found north of the north-midland counties of England.

3. **Lymnæa stagnalis.** *Pond Lymnæa.*

Shell; ovately turreted, compressly umbilicated, rather thin, yellowish horny, spire produced and sharply acuminated, whorls five to six, slopingly convex round the upper part, then ventricose, striated in the direction of the lines of growth, sometimes evanescently obscurely irregularly ridged and malleated in the opposite direction; aperture moderate, somewhat squarely ovate, columella callously twisted, lip broadly appressly dilated over the umbilicus.

Helix stagnalis and *fragilis*, Linnæus (1758), *Syst. Nat.* 10th edit. p. 774.
Buccinum stagnale, Müller (1774), *Verm. Hist.* part ii. p. 132.
Turbo stagnalis, Da Costa (1778), *Test. Brit.* p. 93. pl. v. f. 11.
Buccinum fragile, Studer (1788), *Faun. Helv. Coxe, Trav. in Switz.* vol. ii. p. 434.
Bulimus stagnalis, Bruguière (1789), *Enc. Méth. Vers*, vol. i. p. 303.
Lymnæa stagnalis, Lamarck (1801), *Syst. Anim. sans vert.* p. 91.
Lymnus stagnalis, De Montford (1810), *Conch. Syst.* vol. ii. p. 263.
Lymnæa fragilis, Fleming (1814), *Edin. Encyc.* vol. vii. p. 77.
Bulimus fragilis, Lamarck (1822), *Anim. sans vert.* vol. vi. part 2. p. 23.
Buccinum roseo-labiatum, Wolf (1823), *Sturm, Deutsch. Faun.* sect. 6. part 1.
Limneus major, Jeffreys (1830), *Trans. Linn. Soc. Lond.* vol. xvi. p. 375.
Stagnicola elegans, Leach (1831), *Turt. Man.* p. 145.
Stagnicola vulgaris, Hartmann (1842), *Erd. und Süss. Gast.* pl. viii. f. 12.
Limnæa (Lymnus) stagnalis, Moquin-Tandon (1855), *Hist. Moll.* vol. ii. p. 471. pl. xxxiv. f. 17 to 20.
Limnæa raphidia, Bourguignat (1860), *Rev. et Mag. de Zool.* no. 1, 2. *Amén. Malac.* vol. ii. p. 184. pl. xviii. f. 6, 7.

Hab. Nearly throughout Europe. Siberia. Cashmere. England, chiefly midland and southern counties, and Ireland. (In slow streams, canals, ponds, marshes, etc.)

This fine species stands alone among the *Lymnæacea* of the Eastern Hemisphere for the conspicuous prominence of its size. In the Western Hemisphere, it is represented in a remarkable degree of parallelism by the *Lymnæa jugularis* of Lake Superior, distinguished by the same prominent assemblage of characters. It ranges in this country with *L. auricularia*, not being found in Scotland, and appearing extremely rare and local in England, north of the midland counties. In the Danube and other parts of Southern Europe, *L. stagnalis* attains a larger size than in any part of Britain. It is recorded by Gerstfeldt and Middendorf in their lists of Siberian species, and, according to Morelet, it has been collected by Jacquemont in Cashmere.

The animal of *L. stagnalis* is of a dusky fawn or yellowish-grey colour, variously speckled with brown and opake-white dots. The shell is remarkably sharply acuminated, composed of only six whorls convoluted necessarily in a constricted manner towards the sutures. Its variations are limited, the most distinct being a delicate and rather slender form, which has been named *L. fragilis*, resulting probably from the circumstance of its inhabiting more tranquil waters. Mr. Jeffreys remarks while speaking of this mollusk's habit of floating on the water, that "before descending to the bottom, it withdraws its body into the shell, and in so doing, disengages the air from its pouch, which escapes with a perceptible noise."

4. **Lymnæa palustris.** *Marsh Lymnæa.*

Shell; oblong-ovate, compressly umbilicated, rather solid, fuscous horny, purplish-brown, sometimes opake violet-grey, spire moderately produced, whorls five to six, convex, roughly striated in the direction of the lines of growth, evanescently obscurely irregularly ridged and malleated in the opposite direction; aperture rather small, interior fuscous red, columella moderately twisted, lip appressed over the umbilicus.

 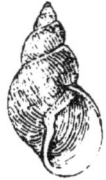

Buccinum palustre, Müller (1774), *Verm. Hist.* part ii. f. 131.
Helix palustris and *corvus,* Gmelin (1788), *Syst. Nat.* p. 3658, and 3665.
Bulimus palustris, Bruguière (1789), *Enc. Méth. Vers,* vol. i. p. 302.
Helix crassa, Razoumowsky (1789), *Hist. Nat. Jorat,* vol. i. p. 276 (not of Da Costa).
Helix striatula, Olivi (1792), *Zool. Adriat.* p. 178 (not of Linnæus nor Gray).
Limneus palustris, Draparnaud (1801), *Tabl. Moll.* p. 50.
Helix fontinalis, Donovan (1803), *Brit. Shells,* vol. v. p. 175. f. 2.
Lymnæus fuscus, C. Pfeiffer (1821), *Deutsch. Moll.* vol. i. p. 92. pl. iv. f. 25.
Limneus communis and *tinctus,* Jeffreys (1830), *Trans. Linn. Soc.* vol. xvi. p. 376 and 378.
Stagnicola communis, Leach (1831), *Turt. Man.* p. 142.
Limnophysa palustris, Fitzinger (1833), *Syst. Verz.* p. 113.
Lymnea Vogesiaca and *disjuncta,* Puton (1847), *Moll. des Vosges,* p. 58 and 60.
Lymnæa corrus, Dupuy (1849), *Cat. Ext. Gall. Test.* no. 195.
Limnæa (Lymnus) palustris, Moquin-Tandon (1855), *Hist. Moll.* vol. ii. p. 475. pl. xxxiv. f. 25 to 35.

Hab. Throughout Europe. Siberia. North Africa. (In shallow, muddy waters).

From its habit of living in more shallow and muddier waters than the three preceding species, *Lymnæa palustris* has a rather more solid, roughly coated shell. The whorls, which are more symmetrically convoluted, increase with the regularity of a Peruvian *Bulimus*, and the shell is not unlike one in general aspect. The outer surface is of a dull fuscous-horny hue, sometimes glazed with a clean opake violet-grey coating, promiscuously freckled with dots of a darker grey, while the interior is lined with a thin enamel of warm fuscous red. The animal is mostly dark violet tinged with green, variegated with black and opake white or yellowish specks. It is described by Moquin-Tandon as being sluggish and irritable, withdrawing its tentacles on the slightest touch, and very voracious. It crawls out of the water, carrying its shell almost horizontally.

L. palustris abounds in every part of Europe, reaching southwards to North Africa, and it appears to be equally common with *L. limosa* in Siberia, but we have no record as yet of its presence in Cashmere or Thibet. In America it is closely represented by *L. elodes* of the Northern States, and by *L. umbrosa* and *reflexa* of the Western and Middle States. Dr. Gould, while speaking of *L. elodes* in his 'Report of the Invertebrata of Massachusetts,' has some curious remarks on its economy. "The animal," he says, "attains maturity and dies about the end of June. At this time the young may be seen with the old, about an eighth of an inch in length, and these continue to grow rapidly during the season. But after the early part of July, it is rare to find an adult shell containing a living animal. At this time the exterior of the shell is much eroded. In fact, the animals, as they cluster together, actually devour each others' shells; the aperture becomes white and sometimes chalky, and the brown submarginal callus of the outer lip is thus covered over." Its European analogue, says Dr. Gould, is *L. palustris*; and while alluding to these and to *L. umbrosa* and *reflexa*, he adds, "After all, these species are so nearly allied that no description, and, perhaps, no figure will enable any person to determine any one of them by itself. They must be learned by comparison and by interchanging specimens. But the difficulty does not end here. It is no easy matter to assign the limits of a species. No one presents a greater variety. The length of mature shells varies from half an inch to an inch, and it is remarkable that the largest specimens are usually the most fragile. The surface usually has an uneven aspect, coated with mud."

5. Lymnæa truncatula. *Truncate Lymnæa.*

Shell; acuminately ovate, distinctly umbilicated, rather solid, yellowish horny, spire rather produced, whorls five to six, convex, truncate next the sutures, roughly and often densely striated in the direction of the lines of growth; aperture rather small, ovate, columella callous, but little twisted, lip elongately expandedly reflected round the circumference of the umbilicus.

Buccinum truncatulum, Müller (1774), *Verm. Hist.* part ii. p. 130.
Helix truncatula, Gmelin (1788), *Syst. Nat.* p. 3659.
Buccinum fossarum, Studer (1789), *Faun. Helv. Coxe, Trav. in Switz.* vol. iii. p. 433.
Bulimus truncatus, Bruguière (1789), *Enc. Méth. Vers*, vol. i. p. 310.
Bulimus obscurus, Poiret (1801), *Coq. de l'Aisne*, p. 35 (not of Draparnaud).
Limneus minutus, Draparnaud (1801), *Tabl. Moll.* p. 51.
Helix fossaria, Montagu (1803), *Test. Brit.* p. 372. pl. xvi. f. 9.
Lymnæa fossaria, Fleming (1814), *Edin. Encyc.* vol. vii. p. 77.
Limnæus truncatulus, Jeffreys (1830), *Trans. Linn. Soc.* vol. xvi. p. 377.
Stagnicola minuta, Leach (1831), *Turt. Man.* p. 143.
Limnophysa minuta, Fitzinger (1833), *Syst. Verz.* p. 113.
Limnophysa truncatula, Beck (1837), *Ind. Moll.* p. 112.
Limnæa Doublieri, Requien (1845), *Moq.-Tand. Hist. Moll.* vol. ii. p. 474.
Lymnea oblonga, Puton (1847), *Moll. des Vosges*, p. 60.
Limnæa microstoma, Drouët (1852), *Baudon, Moll. de l'Oise*, p. 14.
Limnæa (Lymnus) truncatula, Moquin-Tandon (1855), *Hist. Moll.* vol. ii. p. 473. pl. xxxiv. f. 21 to 24.

Hab. Throughout Europe. North Africa. Siberia. Afghanistan. (On the muddy margins of stagnant and slow running waters.)

Lymnæa truncatula, so named by Müller from the whorls of the shell being rather truncate round the upper part next the sutures, is a mud-dwelling species, plentifully diffused about the margin of stagnant and gently running waters in all parts of Europe. Its foreign distribution is similar to that of the preceding species, extending in addition into Afghanistan. It prefers to live out of the water rather than in it, and is found at a considerable elevation on the mountain. The animal of *L. truncatula* is of a rusty black, varying to light grey, profusely speckled with black dots. The shell partakes of the form of *L. palustris*, differing, however, in the sutures being more constricted, which gives the upper part of the whorls a truncately swollen character, while it is of very much smaller size.

6. Lymnæa glabra. *Smooth Lymnæa.*

Shell; elongately turreted, almost subulate, minutely umbilicated, yellowish horny, thin, smooth and silky, spire rather cylindrical, whorls eight. slopingly flatly convex, densely extremely finely striated, margined at the suture; aperture very small, pyriformly ovate, sometimes ribbed within, columella arcuately twisted, lip reflected over the umbilicus.

Buccinum glabrum, Müller (1774), *Verm. Hist.* part ii. p. 135.
Helix octona, Pennant (1777), *Brit. Zool.* ed. 4. vol. iv. p. 138. pl. lxxx. f. 135 (not of Linnæus).
Helix glabra, Gmelin (1788), *Syst. Nat.* p. 3658 (not of Studer).
Bulimus glaber, Bruguière (1789), *Enc. Méth. Vers.* vol. i. p. 312.
Bulimus leucostoma, Poiret (1801), *Coq. de l'Aisne*, p. 37.
Helix octanfracta, Montagu (1803), *Test. Brit.* p. 396. pl. ii. f. 8.
Limneus elongatus, Draparnaud (1805), *Hist. Moll.* p. 52. pl. iii. f. 3, 4.
Lymnæa oetanfracta, Fleming (1814), *Edin. Encyc.* vol. vii. p. 78.
Lymnæa leucostoma, Lamarck (1822), *Anim. sans vert.* vol. vi. part 2. p. 62.
Lymnæa elongata, Sowerby (1823), *Gen. Shells, Lim.* f. 6.
Limneus subulatus, Kickx (1830), *Syn. Moll. Brabant*, p. 60. f. 13, 14.
Stagnicola octanfracta, Leach (1831), *Turt. Man.* p. 141.
Omphiscola glabra, Beck (1837), *Ind. Moll.* p. 110.
Limnæus glaber, Gray (1840), *Turt. Man.* ed. 2. p. 242. pl. ix. f. 106.
Leptolimnea elongata, Swainson (1840), *Treat. Ma'ac.* p. 338.
Limnæa variabilis, Millet (1854), *Moll. Maine-et-Loire*, p. 51.
Limnæa (Lymnus) glabra, Moquin-Tandon (1855), *Hist. Moll.* vol. ii. p. 478. pl. xxxiv. f. 36 to 37.

Hab. Northern and Central Europe. England. (Partially diffused in drains, ditches, and shallow pools.)

This is one of the few European inland mollusks which, like *Physa hypnorum*, is peculiar to the Germanic portion of the Continent. There is no record of its having been found on the Lusitanian side of the Pyrenees. It commences to appear north of that range, and is diffused partially and more sparingly up to Scandinavia. In France, it is partial in its distribution, and becomes scarcer towards the north. Draparnaud made no mention of it in his 'Tableau,' but it was described in the 'Histoire' published after his decease. In Scotland, it is unknown. The latest record of its habitats in England (and local habitats are important in the case of a species like the present) is that given by Mr. Jeffreys :—"Northumberland, Durham, York, Salop, Norfolk, Suffolk, Essex, Oxon, Wilts., Dorset, Cornwall, Guernsey, Jersey, Cork, Belfast," to which I am able to add Surrey and Sussex, on the authority of an able observer, Mr. C. H. Gatty, of Felbridge Park, East Grinstead.

Lymnæa glabra, with its slender, cylindrically elongated shell of eight whorls, presents a curious antithesis in form to *L. auricularia*, with its wide, expandedly inflated shell of scarcely four whorls. The structure of the shells of water snails differs in a most remarkable degree among genera associated in the same locality; as, for example, of *Planorbis*, *Lymnæa*, and *Ancylus*. And wherever the same genera occur in other parts of the globe, they are with little exception typically the same. Descending to species, we have among the *Lymnææ* great contrasts in the plan of convolution, accompanied by very little difference in details of parts. The growth of *L. auricularia* is limited to four whorls, and the tubular increase of the shell is rapid; the mollusk opens out its shell to the utmost degree of expansion, as though striving, so to speak, to make the most of its brief sum of volutions. In *L. glabra* we have the reverse of these conditions and results. The allotted growth of the species is eight volutions, and the increase is accordingly slow. The shell is drawn out into an elongately subulate spire, not sharply turned as in the first growth of the larger species, which have a largely inflated whorl to construct on arriving at maturity, but a leisurely constructed equable cylindrical range of whorls, with little or no inflation of the aperture. The parts of the shell in detail—the columella, lip, surface of sculpture, etc.—are the same in both species, and the mollusk is the same dusky slate, black-speckled creature. "*L. glabra*," says M. Moquin-Tandon, "is a very sluggish mollusk, carrying its shell a little horizontally, sometimes floating on the surface of the water, sometimes crawling on the sides of a vase out of the water; at other times it is quite sunk in its shell."

7. Lymnæa glutinosa. *Glutinous Lymnæa.*

Shell; globosely ovate, bright amber horny, extremely thin and transparent, submembranaceous, spire small, whorls three, longitudinally finely plicated, impressed round the upper part, thin, convex, last whorl ventricosely inflated; aperture ovate, large, columella thinly callous, lip very largely membranaceously reflected over the body whorl.

Buccinum glutinosum, Müller (1774), *Verm. Hist.* part ii. p. 129.
Helix glutinosa, Gmelin (1788), *Syst. Nat.* p. 3659.
Bulimus glutinosus, Bruguière (1789), *Enc. Méth. Vers*, vol. i. p. 306.
Limneus glutinosus, Draparnaud (1805), *Hist. Moll.* p. 50.
Amphipeplea glutinosa, Nilsson (1822), *Moll. Suec.* p. 58.
Myxas Mülleri, Leach (1831), *Turt. Man.* p. 149.
Limnæa (Amphipeplea) glutinosa, Moquin-Tandon (1855), *Hist. Moll.* vol. ii. p. 461. pl. xxxiii. f. 16 to 20.

Hab. Europe, from Sweden to the Pyrenees. Syria. England. (Local in stagnant pools and ditches.)

The two remaining species of *Lymnæa* represent a quite distinct form (*Amphipeplea*), in which the animal is much larger in proportion to the shell, and has the mantle, a viscid sulphur-spotted mass, capaciously reflected. In young specimens the shell, which is of the thinnest possible consistency, membranaceous and flexible, with the columella lip largely appressed on the body whorl, is entirely concealed from view. In adult specimens a small oval space in the middle of the back remains uncovered. The shell is a bright amber colour, so transparent that the columellar axis of the whorls is seen throughout as in a glass bubble. Yet notwithstanding its flexibility, tenuity, and extreme transparency, it is convoluted with the utmost symmetry, and is a remarkably delicate and beautiful object in the cabinet.

L. glutinosa is described as being a very active and sensitive mollusk, withdrawing the reflected extension of its mantle in the water immediately on being touched. It is pretty generally distributed throughout the Continent north of the Pyrenees, and appears, according to Dr. Gray, in Syria. In Britain it has only

8. Lymnæa involuta. *Involute Lymnæa.*

Shell ; subtruncately ovate, bright yellowish horny, extremely thin and transparent, submembranaceous, spire small, immersed, whorls three to four, longitudinally finely plicated, contracted round the upper part, thin, convex, last whorl moderately inflated; aperture oblong-pyriform, reaching to the plane of the apex, columella thinly callous, lip very largely membranaceously reflected over the body whorl.

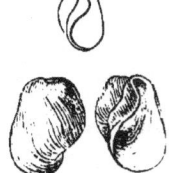

Limneus involutus, Harvey (1834), *Trans. Linn. Soc.* vol. xvii. p. 559.

Amphipeplea involuta, Thompson and Goodsir (1840), *Ann. and Mag. Nat. Hist.* vol. v. p. 22. pl. i.

Hab. Ireland. (In a small lake on Cromaylaun mountain, near the lakes of Killarney.)

This very interesting *Bulla*-like *Lymnæa* was discovered thirty years ago by Professor Harvey in a mountain lake near the lakes of Killarney. It is still, we believe, found in that locality, specimens from which we have before us, and it has not been found elsewhere. The animal has been described anatomically by Mr. Goodsir. It is the same as that of *L. glutinosa*, and it produces a shell in all respects similar except in the very peculiar contraction and immersion of the spire, which gives a more than usually sinuous elevation to the aperture. It is of the same transparent, membranaceous horny substance, similarly finely longitudinally plicated, and it is characterized by the same broad and thinly appressed reflection of the columellar lip amalgamated, as it were, with the body whorl.

There is no satisfactory record of *L. glutinosa* ever having been found in Ireland. Whether, therefore, *L. involuta* is a variety of that species or not, its presence in that island is of the highest interest.

Ancylus fluviatilis. (*Considerably Enlarged.*)

Genus IV. **ANCYLUS**, *Geoffroy*.

Animal; ovoid and conical, carrying an overlapping light cap-shaped shell, incurved at the posterior summit into a small subspiral hook, which inclines either to the right or to the left according to the position of the respiratory orifice. Mantle very thin, not reflected over the edge of the shell. Head broad, obtuse, with two rather distant slenderly triangular tentacles, having the eyes at their inner base.

Shell; orbicularly ovate and elevately convex, more or less radiately ridged and striated, incurved at the posterior summit into a hooked vertex, turning subspirally to the right; or, oblong-ovate, moderately convex and smooth, incurved a little posterior to centre into a hooked vertex turning subspirally to the left.

The existence in Europe of a small limpet adhering to stones and to the stems and leaves of water plants in ponds, lakes, ditches, and shallow streams, made it necessary to introduce the genus *Ancylus* long before the animal had been observed. Lamarck placed the four species known to him, two European, two West Indian, at the end of his marine water-breathing family of *Calyptraciens*, but " only provisionally," said the acute conchologist. "According to the observations of M. De Férussac, the animal rises to the surface of the water to respire air." The researches of subsequent naturalists have shown that the animal has a similar organization of parts to the animal of *Physa*, although taking the form of an ovoid conical mass, secreting a simple *Patella*-like shell, with, however, a tendency to coil in a very early stage of growth.

There are two very distinct forms of *Ancylus*, one (*Ancylastrum*, Moquin-Tandon), having the respiratory and other vital organs situated on the left side of the animal, in which case the vertex of the shell inclines to the right; another (*Velletia*, Gray), having the respiratory and other organs on the right side of the animal, and the vertex of the shell inclines to the left. The animal is wholly covered by the shell. In our vignette, from a drawing made by the Rev. M. J. Berkeley, the front of the shell is lifted to show the animal, which it may be seen has the broad head and slender subtriangular tentacles of *Physa*, with the eyes situated as in that genus at their inner base.

The geographical distribution of *Ancylus* is curious. In Europe we have probably only a single species of each of the groups, *A. fluviatilis* and *lacustris*. More than a dozen species have been described as European, but they appear to me quite untenable. Local modifications of form, substance, and colour cannot be sustained as specific characters. The only other species of *Ancylus* known in the Eastern Hemisphere are one in Siberia, one in Tasmania, one in New Zealand, one in Natal; and one in Teneriffe appears to be distinct from the Madeiran form of the European species. In the Western Hemisphere *Ancylus* has a well-established centre of geographical diffusion in the West Indies and Central America. Southwards it ranges to Venezuela and Bahia, and to Mexico and Chili. Northwards it reaches over California and the United States to Oregon and Newfoundland. About thirty species, reducible apparently to twenty-four, have been described from this wide-spread region.

The dextral form of *Ancylus* (*Velletia*) has the same range of geographical distribution as the sinistral (*Ancylastrum*), with the important additional habitat of Bengal (*A. Baconi*), but it is scarcer in species and more limited in individuals. It has been stated that the *Velletia* form of *Ancylus* has not been found out of Europe, but I consider the *A. Barilensis*, inhabiting Lake Baril, Bahia, *A. Haldemanni*, from Massachusetts, *A. Verreauxi*, from the Cape of Good Hope, and an undescribed species from Mexico in Mr. Cuming's collection, as belonging to this group. No *Ancylus* has been found among the *Lymnæaceà* of the Malayan streams.

The British *Ancyli* are :—

1. **fluviatilis.** Shell rotundately ovate, elevated, radiately striated, with the vertex quite posterior, incurved to the right.
2. **lacustris.** Shell oblong-ovate, rather depressed, smooth, with the vertex posterior to central, incurved to the left.

1. **Ancylus fluviatilis.** *River Ancylus.*

Shell; somewhat rotundately ovate, more or less elevated, radiately wrinkled and striated, the striæ being sometimes neatly defined, sometimes obscure, bluish white, covered with an olivaceous epidermis; vertex sharply incurved to the right, quite behind.

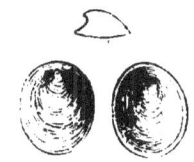

Ancylus fluviatilis, Müller (1774), *Verm. Hist.* part ii. p. 201.
Patella fluviatilis, Gmelin (1778), *Syst. Nat.* p. 3711.
Patella cornea, Poiret (1801), *Coq. de l'Aisne*, p. 101.
Ancylus riparius, Desmarest (1814), *Bull. Soc. Philom.* p. 19. pl. i. f. 2.
Ancylus capuloides, Jan (1838), *Porr. Mal. Terr. e Fluv.* p. 87. pl. i. f. 7.
Ancylus, vitraceus, strictus, and *obtusus*, Morelet (1845), *Moll. du Port*, p. 87, 88. pl. viii. f. 3, 4, 5.
Ancylus deperditus, Dupuy (1851), *Hist. Moll.* vol. v. p. 494. pl. xxvi. f. 4.
Ancylus Janii, simplex, gibbosus, Deshayesianus, and *cyclostoma*, Bourguignat (1853), *Journ. Conch.* p. 183 to 187.
Ancylus (Ancylastrum) fluviatilis, Moquin-Tandon (1855), *Hist. Moll.* vol. ii. p. 484. pl. xxxv. f. 5 to 38, and pl. xxxvi. f. 1 to 49.

Hab. Throughout Europe. Algeria. Madeira. (Adhering to stones, and, less frequently, to plants, in shallow streams and running brooks.)

In this species the respiratory and more important organs of the animal are situated on the left side, and the subspiral vertex of the shell,—showing that the mollusk commenced life with a tendency to follow the spiral plan of growth,—is turned to the right. The body is of a finely speckled slate-colour, and the head, as represented in our vignette, is broad with the tentacles slenderly triangular and widely separated, the eyes being small and distinct at their inner base. The shell of *A. fluviatilis*, though thin, is of a firm symmetrical growth, rayed with fine ridges and striæ, orbicularly ovate at the base, and rising, cap-like, into a broad cone, is sharply hooked at the posterior summit, with its embryonic nucleus turned subspirally to the right. It chiefly inhabits gently running streams.

The geographical range of this species appears to be very general throughout Europe, and it extends to North Africa and Madeira. More than a dozen species have been made of its different varieties. Mr. Cuming possesses a well-authenticated series of these alleged species, but they will scarcely bear the test of comparison when viewed with a due consideration of the conditions arising out of differences of latitude and the corresponding results of the law of correlation of growth. Gerstfeldt describes a Siberian species, *A. Sibiricus*, which is very like *A. fluviatilis*, but it is of a remarkably conoid form.

---

2. **Ancylus lacustris.** *Lake Ancylus.*

Shell; oblong, rather depressed, compressed at the sides; very thin, transparent, smooth, with the vertex a little posterior to central, hooked obliquely to the left.

Patella lacustris, Linnæus (1758), *Syst. Nat.* 10th edit. p. 783.
Ancylus lacustris, Müller (1774), *Verm. Hist.* part ii. p. 199.
Patella oblonga, Lightfoot (1786), *Phil. Trans. Roy. Soc.* vol. lxxvii. p. 168. pl. iii.
Acroloxus lacustris, Beck (1838), *Ind. Moll.* p. 124.
Velletia lacustris, Gray (1840), *Turt. Man.* p. 50. f. 126.
Ancylus Moquinianus, Bourguignat (1853), *Journ. Conch.* p. 197. pl. vi. f. 9.
Ancylus (Velletia) lacustris, Moquin-Tandon (1855), *Hist. Moll.* vol. ii. p. 488. pl. xxxvi. f. 50 to 55.

Hab. Throughout Europe. (Adhering to stems and leaves of plants in ponds, lakes and canals.)

A. lacustris is not only distinguished from *A. fluviatilis*, in having the principal vital organs on the right side of the animal, and the vertex of the shell turned to the left, but in being of a slighter flatter growth, devoid of radiating striæ, with the vertex not far posterior to the centre. There is also a difference in the habit of the two species. *A. fluviatilis* mostly adheres to stones in gently running brooks; *A. lacustris* mostly adheres to the stems and leaves of plants in still waters, in ponds, canals, lakes. It is not without reason, therefore, that the genus *Ancylus* has been subdivided into two.

As in the case of *Ancylastrum* so in *Velletia*, there is to my mind only one European species, and they have much the same geographical distribution; *Ancylastrum* is, however, the scarcer of the two. The animal is of rather a lighter colour, and of a more livid hue.

TRIBE II. **OPERCULATA.**—WITH OPERCULUM.

The Operculated Freshwater Cephals are much less numerous in kind than the Inoperculated. There are only eleven species in Britain. The operculum is an apparatus used chiefly by water-breathing mollusks; and in those inhabiting the sea it is developed in great variety. Freshwater mollusks are comparatively few in number, much fewer indeed in Europe than in the same isothermal latitudes of North America; and the proportion that are gilled and operculated, is smaller still. Our eleven species illustrate seven very characteristic generic forms, two of which, *Cyclostoma* and *Acme*, while possessing rudimentary gills, breathe air.

The operculum is sometimes spiral, with the nucleus inclined towards the side, sometimes it is enlarged by concentric additions with the nucleus in the centre; and it is either horny or superficially testaceous.

ORDER I. **PULMONIFERA**—AIR-BREATHING.

Respiratory organ a vascular sac for the respiration of air.

When describing the Pulmoniferous division of the Inoperculated Cephals, we noticed that it embraced one-half of our entire series of Land and Freshwater mollusks. Of the Pulmoniferous Operculated Cephals, we have only two species in Britain. The pulmonary organ, or lung, is a sac, over the surface of which the blood flows in minute vessels for the purpose of being aerated; and it has been observed that in the present Order they are closer together, in parts forming flexuous ridges, which are supposed to be rudimentary gills.

The Pulmoniferous Operculated Cephals are contained in one Family:—

1. **Cyclostomacea.** Animal having the head produced into a ringed or ridged proboscis, tentacles bristle-like, with the eyes at their outer base, foot bearing on the upper posterior surface a shelly or horny operculum.

Family I. **CYCLOSTOMACEA.**

Animal having the head produced into a ringed or ridged proboscis, tentacles bristle-like, with the eyes at their outer base, foot bearing on the upper posterior surface a shelly or horny operculum.

Except in intertropical countries, from whence more than a thousand species of this family have been described, the operculum is an apparatus belonging almost exclusively to the water-dwelling mollusks. We have in Britain only two operculated mollusks respiring and dwelling in air, each the type of a distinct genus. They live in damp places or places near the sea, and have a certain affinity with the *Auriculacea* of similar habit among the inoperculated tribe. The head, which in the *Auriculacea* is enlarged into a ringed muzzle, is in this family produced into an absolute proboscis, and the tentacles are also bristle-like, with the eyes at their outer base. The genera *Cyclostoma* and *Acme* present much the same contrast in their resemblance as *Conovulus* and *Carychium*. One is comparatively of large size, carrying a solid testaceous shell, the other is very small, carrying a transparent glassy shell.

Our British *Cyclostoma* is the northernmost member of a group which, like *Physa*, is abundantly represented in the West Indies, but throughout all Europe has scarcely more than two other species nearly allied, and a few small species (*Pomatias*) in which the relationship is more removed. Unlike *Physa* and *Lymnæa*, which so abound in the United States, *Cyclostoma* does not appear in the New World north of Florida, and then only in a single species, very closely allied, if not identical with, one belonging to Cuba. The most striking example of the presence of this genus in the Eastern Hemisphere, is at Madagascar and the neighbouring islands of Bourbon and Rodriguez. Here *Cyclostoma*, with an operculum corresponding in structure with that of our own species, produces a bold globosely turbinated shell, richly coloured and variously banded and keeled, numerous in species, but all stamped with a well-defined local peculiarity of character. The operculum of these is, indeed, more like that of our own species than the operculum of the more closely resembling, in other respects, West Indian species,—a flat rotary constructed calcareous plate, with a central nucleus, like the apical nucleus of the shell, enlarged spirally as the shell enlarges.

The other British genus of this family, *Acme*, is an extremely small mollusk, with a turriculate glassy shell, scarcely the twelfth of an inch in length, of which the operculum is horny and of fewer whorls than the shell. One or two other species have been described on the Continent, but their specific value is a little doubtful. The genus is not known out of Europe.

The British genera of *Cyclostomacea* are :—

1. **Cyclostoma.** Animal carrying an ovately turbinated solid shell of five whorls. Operculum shelly.
2. **Acme.** Animal carrying a minute turriculate glassy shell of six whorls. Operculum horny.

Cyclostoma elegans.

Genus I. **CYCLOSTOMA,** *Lamarck.*

Animal; oblong, carrying an ovate tubularly turbinated shell, mantle free, head produced into a prominent ringed proboscis, tentacles slender, wrinkled, with the eyes on superficial swellings at their outer base, foot large and bilobed in front, short behind, bearing an operculum.

Shell; ovately turbinated, of five rather solid tubular whorls, having a nearly circular aperture. Operculum calcareous, spiral, smooth.

This conspicuously proboscis-snouted mollusk is the solitary representative in Britain of a group inhabiting in great abundance the islands of the West Indies. According to the generally received subdivision of the numerous foreign species of this family, founded on the characters of the operculum, it is associated with a number of globosely turbinated forms of larger size, peculiar to Madagascar and the adjacent islands; but taking the generality of its characters into consideration, it has a more direct affinity with the West Indian and Central American *Cyclostomacea*, arranged under the head of *Chondropoma*. It can hardly be said that there are more than two other species in Europe, though more have been described,—one, *C. costulatum*, in the neighbourhood of the Caspian Sea, and another, *C. sulcatum*, which is diffused abundantly throughout the south of France, Italy, Spain, and the islands of the Mediterranean, passing into North Africa. There are, however, five or six small species, of a form allied to *Cyclostoma*, known as *Pomatias*, which are tolerably abundant in France, but do not appear on this side of the English Channel.

There is no *Cyclostoma* indigenous to the United States. Although our land and freshwater mollusks are mostly generically represented in North America, either by species nearly allied, or by species locally distinct, the only *Cyclostoma* that has found its way into that country is the *C. dentatum*, a Cuban species of a pupoid West Indian type, which has become naturalized in woods and open places about Key West, Florida.

The animal of *Cyclostoma* is distinguished in a remarkable manner by this proboscis-like extension of the head, of which it not unfrequently avails itself when crawling; and the foot is peculiar in being obtusely lobed in front. It was observed by Rossmässler that the foot is composed of two longitudinal portions, one of which advances, when the animal is in motion, while the other remains adherent; and Mr. Binney notices that it is contracted in an undulatory manner, the shell being suddenly jerked forward with each contraction. Attached to the hinder part of the foot is the operculum, which has a nucleus in the centre analogous to the apex of the shell, and is composed of five horizontal coils corresponding to the shell's five whorls. We have said that the operculum is an apparatus belonging more especially to the water mollusk. *Cyclostoma* lives chiefly in the vicinity of water; not in wet places, but at the roots of shrubs near the sea-coast, and it has been observed, that the pulmonary vessels are here and there closer together, forming flexuous ridges, which are supposed to be rudimentary gills. The position of the eyes is more outward than in the inoperculated water snails, and in the operculated water snails, between which groups *Cyclostoma* is intermediate, it is more outward still.

Our single British *Cyclostoma* is—

1. **elegans.** Shell ovately turbinated, solid, of five finely corded whorls, frequently articulated with livid red-brown dots.

1. Cyclostoma elegans. *Elegant Cyclostoma.*

Shell; ovately turbinated, solid, compressly narrowly umbilicated, violet-tinged or yellowish drab, sometimes articulated with somewhat square livid red-brown dots, whorls five, corded throughout with close-set spiral ridges, the interstices between which are minutely reticulated with longitudinal striæ, sutures of the whorls linearly canaliculately impressed; aperture pyriformly rounded, tinged with orange, sinuated at the upper part, lip simple, continuous, operculum flat, five-whorled.

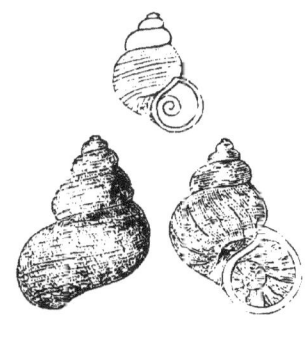

Nerita elegans, Müller (1774), *Hist. Verm.* vol. ii. p. 177.
Turbo tumidus, Pennant (1777), *Brit. Zool.* ed. iv. vol. iv. p. 128. pl. lxxxii. f. 110.
Turbo striatus, Da Costa (1778), *Test. Brit.* p. 86. pl. v. f. 9.
Turbo elegans, Gmelin (1788), *Syst. Nat.* p. 3606.
Pomatias elegans, Studer (1789), *Faun. Helv., Coxe Trav. in Switz.* vol. iii. p. 432.
Turbo reflexus, Olivi (1792), *Zool. Adriat.* p. 170.
Cyclostoma elegans, Draparnaud (1801), *Tabl. Moll.* p. 38.
Cyclostoma marmoreum, Brown (1829), *Edin. Journ. Sci.* October.
Cyclostoma (Ericia) elegans, Moquin-Tandon (1855), *Hist. Moll.* vol. ii. p. 496. pl. xxxvii. f. 3 to 23.

Hab. Central and Southern Europe. Canary Islands. Britain, south of Yorkshire. (Under stones and about the roots of shrubs.)

In some of our inland counties, but chiefly near the seacoast, in chalk districts, *Cyclostoma elegans* may be collected in great abundance in the spring of the year, among the roots of bushy shrubs. In dry weather it buries itself in the soil by the aid of its muscular proboscis and lobed foot. The shell is of an elegantly turbinated oval form of solid substance, finely corded throughout, generally of a livid drab colour, more or less articulated with red-brown dots. The operculum is composed of five whorls, corresponding with the five whorls of the shell, enlarging in a rotary manner as the aperture of the coiling shell enlarges.

C. elegans is not found in Scotland, nor in Northern Europe. It is most abundant, along with *C. sulcatum*, in the centre and south of Europe, and in the islands of the Mediterranean; and it has been collected in the Canary Islands.

Acme lineata. (*Considerably enlarged.*)

GENUS II. **ACME**, *Hartmann*.

Animal; rather elongated, carrying a slender transparent cylindrically turriculate shell, head proboscis-like, tumid and then attenuated, transversely wrinkled, tentacles bristle-like, not swollen at the tip, eyes at the base of the tentacles inclining outwardly, foot bearing a horny subspiral operculum.

Shell; cylindrically elongated, fuscous horny, transparent, glossy, of six rather slender closely lineated whorls, aperture small, more or less sinuated at the upper part.

Acme is a minute form, carrying a transparent amber shell, which, though elongately turreted, is barely the twelfth of an inch in length. Its affinity with the preceding genus is shown by the proboscidiform head, and slender tentacles with the eyes at the base inclining outwardly, and by the presence of an operculum, which in this instance is horny and subspiral.

Acme lineata is diffused pretty generally throughout Europe, and the genus is not found elsewhere. M. Moquin-Tandon not only makes three species of the *Acme lineata*, but he characterizes those in which a marginal cleft in the aperture of the shell is more apparent than in others, as subgenera, *Auricella* and *Platyla*. He has studied the living animal with care, and describes its habits with much interesting detail.

Our British species of *Acme* is:—

1. **lineata.** Shell of six transparent flatly convex whorls, more or less linearly striated.

1. Acme lineata. *Lineated Acme.*

Shell; minute, subcylindrically elongated, obscurely compressly umbilicated, transparent, horny, fuscous amber, glossy, whorls six, rather flatly convex, obtusely angled at the base, longitudinally closely linearly striated, striæ sometimes obsolete; aperture small, pyriformly rounded, more or less sinuated at the upper part, columellar lip thinly reflected.

Turbo, Walker and Boys (1784), *Test. Minut. Rar.* p. 112. pl. ii. f. 42 (without specific name).
Helix cochlea, Studer (1789), *Faun. Helv. Coxe, Trav. in Switz.* vol. iii. p. 430 (without characters).
Bulimus lineatus, Draparnaud (1801), *Tabl. Moll.* p. 67.
Turbo fuscus, Montagu (1803), *Test. Brit.* p. 330.
Auricula lineata, Draparnaud (1805), *Hist. Moll.* p. 57. pl. iii. f. 20, 21.
Auricella lineata, Jurine (1817), *Helv. Alman.* p. 34.
Carychium cochlea, Studer (1820), *Kurz. Verz.* p. 89.
Acicula lineata, Hartmann (1821), *Neue Alp.* vol. i. p. 215.
Acme lineata, Hartmann (1821), *Syst. Gast.* p. 49.
Carychium lineatum, Férussac (1822), *Tabl. Syst.* p. 104.
Cyclostoma lineatum, Férussac (1822), *Dict. Hist. Nat.* vol. ii. p. 90.
Carychium fuscum, Fleming (1828), *Brit. Anim.* p. 270.
Pupula lineata, Charpentier (1837), *Moll. Suiss.* p. 22.
Acme fusca, Gray (1840), *Turt. Man.* p. 223. pl. vi. f. 66.
Truncatella lineata and *polita*, Hartmann (1840), *Erd. und Süss. Gast.* p. 1. pl. i. and p. 5. pl. ii.
Cyclostoma fuscum, Moquin-Tandon (1843), *Moll. Tolouse*, p. 14.
Acme Moutonii, Dupuy (1849), *Cat. Extr. Test.* no. 4.
Acme (Auricella) lineata and *Moutonii*, Moquin-Tandon (1855), *Hist. Moll.* vol. ii. p 508, 509. pl. xxxviii. f. 3 to 7.
Acme (Platyla) fusca, Moquin-Tandon (1855), *Hist. Moll.* vol. ii. p. 509. pl. xxxviii. f. 8 to 16.
Acicula fusca, Gray (1857), *Turt. Man.* p. 39. pl. vi. f. 66.

Hab. Throughout Europe. Siberia. (In damp places, under stones, and among moss.)

When objects are minute, and have to be delineated on a magnified scale, naturalists are apt to exaggerate characters, and give substance to forms that are in great measure the result of fancy. This is especially the case with minute shells. *Acme lineata* has a cylindrically elongated shell of six flatly convex whorls, which, according to the observations of British conchologists, are sometimes closely lineated, sometimes nearly smooth. French writers consider that there are three, if not four, species of *Acme*, and that a marginal cleft in the upper corner of the aperture, which in some is deeply cut, in others almost obsolete, is of sufficient importance to be selected as a basis of subdivision. We cannot satisfy ourselves that there is more than a single species, unless exception be made in favour of a more elongated eight-whorled species, *A. spectabilis* of Rossmässler. *Acme lineata* is pretty generally diffused throughout Europe in wet places, under stones in drains, or under moss or decayed leaves.

M. Moquin-Tandon describes *Acme* as being rapid in its movements and very irritable. Mr. Jeffreys remarks that a living specimen which he observed in the North of Ireland did not seem to be shy or inactive while kept in the shade, but when exposed to the glare of the sun it immediately shut up and disappeared. Dr. Gray says, that the animal walks with its shell nearly perpendicular, twisting it round in a very odd manner, and then letting it suddenly fall again. The operculum, which is horny, is composed of fewer whorls than the shell, only about two and a half, and is sunk rather deep within the aperture.

Order II. **BRANCHIFERA.**

Respiratory organ gills of filaments or lamellæ for the respiration of water.

The water-breathing operculated Cephals are more numerous than those which respire air. Of the latter, as we have just seen, there are only two British species, of two genera, *Cyclostoma* and *Acme*; of the former we have nine, illustrative of five genera, *Assiminea, Bythinia, Paludina, Valvata,* and *Neritina*. The branchial chamber occupies its usual position on the right side of the animal, near the head. In *Assiminea* and *Bythinia*, the branchiæ form a row of wrinkled lamellæ. In *Paludina*, they consist of three rows of filaments, and the entrance to the chamber is by a tubular fold of a fleshy lobe immediately behind the right tentacle; on the opposite side there is a corresponding rudimentary lobe, which, in *Ampullaria*, an amphibious genus of this family belonging to a warmer climate, developes into a siphon for the conveyance of air to a pulmonary chamber. Mr. Berkeley has stated his conviction that there is a pulmonary as well as a branchial chamber in *Assiminea*, and it is certainly the habit of that mollusk to live as much out of water, always, however, in its vicinity, as in it. In *Valvata*, the branchiæ take the form of a plume of filaments, and are external. In *Nerita* they form a long, partially free, acute triangular membrane, and are internal.

The branchiate, or gilled, Operculated Cephals, are comprised in three Families :—

1. **Littoracea.** Head produced into a ringed muzzle. Branchiæ lamellar, internal.
2. **Peristomata.** Head produced into a proboscis, tentacles sometimes filiform, sometimes cylindrical, with the eyes sessile or pedunculated, on swellings at their outer base, branchiæ mostly lamellar or filamentary and internal, sometimes plumose and external.
3. **Neritacea.** Head short, tentacles slender with the eyes sessile at their outer base. Branchiæ internal, in the form of a long acute triangular membrane.

FAMILY I. **LITTORACEA.**

Head produced into a ringed muzzle. Branchiæ lamellar, internal.

The *Littoracea*, or Periwinkles, are a family dwelling on the margin of the sea, between tide-marks, or within reach of the spray. They are moderately numerous in species and plentiful in individuals. With comparatively little speciality of geographical distribution, they fringe the shore throughout both hemispheres, marking the rocks with riband-like zones ; and the zones indicate the resting-place of particular varieties of species at particular heights of the tide. Some of the *Littoracea* extend their habitat into brackish water, mud, and swamp, and the respiratory apparatus is modified to the necessities of such a change of existence. In Borneo, *Littoracea* dwell in the mangrove-trees, not only about the roots but among the branches. In Britain we have an abnormal member of this family inhabiting, in countless numbers, the banks and inlets of the Thames, sometimes in the water, but more frequently out of it, as high up as Greenwich. It is the only British species of the family that is not absolutely marine.

The genus established for its reception is :—

1. **Assiminea.** Animal carrying a pyramidally conical shell, head produced into a ringed muzzle, tentacle and eye-stalk united on each side in one, with the eye at their summit. Operculum horny, composed of a few rapidly enlarging whorls.

Assiminea Grayana. (*Much enlarged.*)

Genus I. ASSIMINEA, *Leach.*

Animal; body small, carrying a rather solid pyramidally conical shell, head produced into a ringed muzzle, notched in front, tentacles short, united with the eye pedicles, and bearing the eye at their summit, foot ample, broad in front, short and rather obtuse behind, carrying a slight horny few-whorled operculum.

Shell; pyramidally conical, rather solid, with a minute nearly closed umbilicus, of six smooth slopingly convex whorls somewhat obtusely angled at the base; aperture with the columellar lip thinly callously reflected. Operculum horny, composed of a few rapidly enlarging whorls.

About the banks and inlets of the Thames, in the neighbourhood of Greenwich and Woolwich, where the water commences to be brackish, a small periwinkle exists in profusion, sometimes in the water, but more frequently out of it, under stones, or on the mud, or about the roots of water-flags. The animal partakes of the characters of the marine *Littorina*, and of a form of *Bythinia* (*Hydrobia*, Hartmann), in having the horny operculum composed of a few rapidly enlarging whorls; but it differs from both in the very important feature of having the tentacle and eye-stalk on each side united in one, with the eye at its summit as in the retractile tentacle of the *Colimacea*. A little marine species first discovered by

Delle Chiaje in the Bay of Naples, and named by him *Helix littorea*, since collected by Mr. Metcalfe in the Fleet, at Weymouth, and by Professor Forbes in Whitcliff Bay, Isle of Wight, has been pronounced by the last-named author to be an *Assiminea*, with the same condition of the eyes and tentacles. It has been referred indifferently to *Rissoa Truncatella* and *Paludinella*.

The nearest approach to our brackish water *Assiminea*, so far as regards the shell, appears in India and China. Several allied forms, all of similar habit, have been described from the vicinity of Bombay, under the generic title of *Optodiceros*; and others not very far removed, inhabiting tanks, and clumps of damp tree-moss, have been described under the names *Tricula* and *Hydrocena*. Of none of these has the structure of the tentacle and eye-stalk been observed, and it is not, therefore, possible to determine their exact relationship. *Assiminea* of the British type is not found in the New World. There is a globose form at Valdivia, Chili, encircled by a linear groove below the suture; and in North America the genus is represented to some extent by certain species of *Amnicola*.

Our British *Assiminea* is:—

1. **Grayana.** Shell pyramidally conical, of six comparatively solid slopingly convex whorls obtusely angled at the base.

---

1. **Assiminea Grayana.** *Gray's Assiminea.*

Shell; pyramidly conical, rather solid, with a minute nearly closed umbilicus, fulvous red, translucent, whorls six, slopingly convex, obtusely angled at the base; aperture rather small, ovate, outer lip simple, columellar lip moderately callously expandedly reflected.

Assiminea Grayana, Leach (1816), *Moll. Brit. Syn.* ined. (*fide* Gray).
Nerita (Syncera) hepatica, Gray (1821), *Lond. Med. Repos.* p. 239.
Assiminea Grayana, Fleming (1828), *Brit. Anim.* p. 275.
Limneus Grayanus, Jeffreys (1830), *Trans. Linn. Soc.* vol. xvi. p. 378.
Paludina Grayana, Potiez and Michaud (1838) *Moll. Gall. de Douai*, vol. i. p. 251. pl. xxv. f. 23, 24.
Truncatella Grayana, Clarke (1855), *Ann. Nat. Hist.* ser. ii. vol. xvi. p. 115.

Hab. Banks of the Thames, between Greenwich and Woolwich. (In great profusion, under stones and on mud, about the roots of the water-flag.)

This peculiar form of periwinkle, with its tentacle and eye-stalk curiously united in one, and bearing the eye at its summit, was discovered on the banks of the Thames, in the neighbourhood of Greenwich, by Dr. Leach, in 1816. It abounds under stones, or in the mud, or about the roots of water-flags, and it does not appear to have been collected in any other locality. Forty years later, when a controversy sprang up between Dr. Gray and Mr. Clarke in consequence of the latter strangely asserting that the animal is a *Truncatella*, the same locality was had recourse to for living specimens. Mr. Jeffreys, who undertook to procure them, wrote word to Mr. Clarke:—"I went to Greenwich last Saturday, and have the pleasure of sending you some lively examples of this curious mollusk as well as a few *Littorina (?) anatina*. The shell of the latter is closely allied to *Bythinia*, but the operculum is that of *Littorina*. I found both of them more or less distributed along the banks of the Thames, from a little below Greenwich Hospital to the upper pier at Woolwich, a distance of about three miles. I met with them occasionally in the same localities, but their habitats are somewhat different. The *Littorina* (*Bythinia similis*, Dupuy, *Hydrobia similis*, Hartmann) inhabits muddy ditches and their banks, and it is gregarious. The other mollusk (*Assiminea Grayana*) inhabits muddy places, but seldom occurs under water. It is in countless profusion at and about the roots of the water-flag, and is more generally dispersed. It is associated with *Limneus palustris* and *truncatulus*."

The shell of *Assiminea Grayana* is of a well-defined pyramidally conical form, rather obtusely angled at the base. It is rather solid, but yet translucent, of a reddish fawn colour.

Family II. **PERISTOMATA.**

Animal with the head produced into a ringed proboscis, carrying a turbinated shell with the aperture entire.

The name *Peristomata*, signifying 'Entire-mouths,' was given to this family by Lamarck, for the sake of distinguishing the operculated freshwater Cephals, *Bythinia*, *Paludina*, and *Valvata*, in which the shell is of a tubularly turbinated structure, and has the aperture entire, from the *Lymnæacea*, which have no operculum, and whose shell is formed of whorls resting so far one upon the other, that the margins of the aperture are divided. Along with this similarity in the structure of the shell, there is a characteristic similarity in the animal. The head is in each case produced into a ringed proboscis, and the foot is furnished with an operculum. But there are important generic differences. *Bythinia* has the tentacles thread-like and flexible, with the eyes at their outer base, on sessile swellings, the operculum being sometimes concentric, sometimes paucispiral. *Paludina* has the tentacles rather stout and cylindrical, with the eyes raised on stalks to make room for a pair of lobes, of which the right-hand one is folded into a tube, in place of a simple orifice, for conveying water to the branchial chamber. *Valvata* has the eyes at the inner base of the tentacles, and the branchiæ are external, in the form of a plume of spiral filaments. In the intertropical parts of both hemispheres there is a genus belonging to this family, *Ampullaria*, larger in dimensions than either of the foregoing, and more abundant in species, of which we have no representative in Britain. The *Bythiniæ* and *Paludinæ* are scattered throughout the Eastern Hemisphere, excepting Australia; and they are comparatively plentiful in North America. *Valvata* is confined to the temperate and north-temperate latitudes of the Old World, and it occupies the same latitudes in the New World, reaching southwards to the West Indies. The *Peristomata* are essentially of mud-dwelling habits. Many of them, when exposed for hours daily on the river's brink at the fall of the tide, will burrow into the mud and remain embedded until its return. We have seven species in all.

The British genera of *Peristomata* are :—

1. **Bythinia.** Animal with flexible filiform tentacles, having the eyes placed externally at their base on sessile swellings, carrying a conically turbinated shell, of which the operculum is sometimes concentric, sometimes paucispiral.
2. **Paludina.** Animal with cylindrical tentacles, having the eyes raised externally on stalks, to make room for a pair of lobes, of which the right one forms a branchial siphonic tube.
3. **Valvata.** Animal with rather slender tentacles, having the eyes at their inner base, and the branchiæ external in the form of a plume of spiral filaments. Shell sometimes turbinated, sometimes discoid.

Bythinia tentaculata.

Genus I. **BYTHINIA**, *Gray*.

Animal; elongated, subcylindrical, carrying a light turbinated shell, head produced into a lengthened muzzle, cleft in front, tentacles slenderly elongated and flexible, with the eyes sessile at their outer base, foot oblong-triangular, broad in front, attenuated behind, bearing a subtestaceous operculum, which is sometimes paucispiral, with a lateral nucleus, sometimes concentric, with the nucleus in the centre.

Shell; conically turbinated, minutely umbilicated, fulvous green, semitransparent, smooth, of five to seven more or less rounded whorls, covered with a slight horny epidermis, aperture pyriformly ovate, with the margins continuous.

Before treating of the large and well-known *Paludinæ* of our canals and rivers, we have to speak of three much smaller mollusks formerly included in the same genus, until separated on very sufficient grounds by Dr. Gray, under the title of *Bythinia*. The *Paludinæ* bring forth their young alive, and the tentacles are cylindrical and firm, with the eyes raised upon stalks. The *Bythiniæ* are oviparous, and the tentacles are long and slender, almost filiform, with the eyes sessile; the operculum differs also in being superficially testaceous. In *B. tentaculata* and *Leachii*, the operculum is formed of concentric additions round a central nucleus, but in *B. similis*, a species associated on the banks of the Thames with *Assiminea Grayana*, the operculum is spiral, and two-whorled, with a lateral nucleus like that of its associate. Mr. Jeffreys includes it in a separate genus *Hydrobia*, along with the marine *Rissoa ventrosa* of authors. M. Moquin-Tandon refers it, along with nine other species, natives of the brooks and rivers of France, to the genus *Bythinia*, but as a separate section, *Bythinella*. The shell of

Bythinia is an elegantly turbinated cone of five to seven smooth more or less rounded whorls of a semitransparent fulvous green colour, covered by a thin horny epidermis. The aperture being contracted at the upper, or, more strictly speaking, hinder, part, has a pear-shaped tendency, with the margins continuous and generally dark-edged. Sometimes the dark peristome of an immature aperture may be noticed across the penultimate whorls, in adult specimens in the form of a varix.

Dr. Gray notices the presence of a small veil on the right side of the neck of *Bythinia*, and the Messrs. Adams speak of a small lobe, but this appears to be an unimportant fold of the mantle, quite distinct from the tubular lobe in *Paludina* used for conveying the water to the branchial chamber, and for the free development of which, at the base of the tentacles, the eye is raised on a conjoined stalk.

The *Bythiniæ*, which are now being examined by M. Frauenfeld, of Vienna, the intelligent naturalist of the Austrian Novara expedition, have much the same distribution in both hemispheres as *Paludina*. They are scattered, more or less sparingly, throughout Europe, Asia, North Africa, and North America. The presence of twelve species of *Bythinia* in France, is a tolerable indication of their being more plentifully diffused in Europe than has been hitherto supposed. In Asia, there are three described by Gerstfeldt from Siberia, and several have been collected in India, China, and the Philippine Islands. In North Africa, *Bythinia* has been collected in the tributaries of the Nile. In North America, twenty-three species are recorded in Mr. Binney's 'Descriptive Catalogue' of the *Peristomata* of that country, now passing through the press, nineteen of which, having a spiral-whorled operculum, are referred, in the proof-sheets he has been good enough to favour me with, to Gould and Haldeman's genus *Amnicola*.

The British *Bythiniæ* are:—

1. **similis.** Shell very small, of five to six convex whorls channelled at the sutures. Operculum horny, of two whorls, nucleus lateral.
2. **tentaculata.** Shell comparatively large, of five rather ventricose whorls. Operculum somewhat shelly, enlarged concentrically, nucleus central.
3. **Leachii.** Shell smaller, of five scalariform whorls. Operculum somewhat shelly, enlarged concentrically, nucleus central.

1. **Bythinia similis.** *Like Bythinia.*

Shell; very small, conically ovate, minutely umbilicated, fulvous horny, semitransparent, whorls five to six, convex, smooth, channelled at the sutures, aperture obliquely ovate, columellar lip thinly callously reflected. Operculum horny, of two whorls, nucleus lateral.

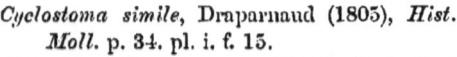

Cyclostoma simile, Draparnaud (1805), *Hist. Moll.* p. 34. pl. i. f. 15.
Valvata similis, Hartmann (1821), *Syst. Gast.* p. 57.
Paludina similis, Michaud (1831), *Comp. de Drap.* p. 93.
Bythinia similis and *Moutonii*, Dupuy (1849), *Cat. Extram.* p. 45 and 48.
Hydrobia similis, Dupuy (1851), *Hist. Moll.* vol. v. p. 552. pl. xxvii. f. 9.
Rissoa anatina, Forbes and Hanley (1853), *Hist. Brit. Moll.* vol. iii. p. 134. pl. lxxxvii. f. 3, 4 (not *Cycl. anatium*, Drap.).
Bythinia (Bythinella) similis, Moquin-Tandon (1855), *Hist. Moll.* vol. ii. p. 526. pl. xxxix. f. 18, 19.

Hab. Central and Southern Europe. Corsica. Siberia. England. (In ditches near the Thames at Greenwich.)

This is the little mollusk referred to by Mr. Jeffreys in the 'Annals of Natural History' for 1855, under the name *Littorina (?) anatina*, when speaking of the habitat, in muddy ditches, occasionally overflowed by the tide, on the banks of the Thames between Greenwich and Woolwich, of *Assiminea Grayana*. Along with the general characters of *Bythinia* it has the few-whorled operculum of its associate *Assiminea*. Moquin-Tandon refers it on this account to a subgenus, *Bythinella*, together with nine other similarly operculated species, natives of France; and Dupuy places it in Hartmann's genus *Hydrobia*, as lately adopted by Mr. Jeffreys. Gerstfeldt, while adopting the genus *Hydrobia* for a new Siberian species, *H. Angarensis*, retains the present in *Bythinia*.

The animal is described as being of a dark grey or fulvous brown hue, speckled with white. The shell is distinguished by its very small size and channelled suture. M. Moquin-Tandon describes it as being widely diffused in France, chiefly in the vicinity of the Pyrenees, and extending southwards to the island of Corsica. Its appearance in Siberia leads to the conclusion that it is pretty generally diffused throughout the Continent.

2. **Bythinia tentaculata.** *Tentacled Bythinia.*

Shell; conically ovate, with a minute nearly closed umbilicus, fulvous green, subtransparent, apex rather sharp, whorls five, smooth, convex, the last rather ventricose; aperture somewhat pyriformly ovate, lip dark-edged, scarcely reflected. Operculum subtestaceous, striated concentrically around a central nucleus.

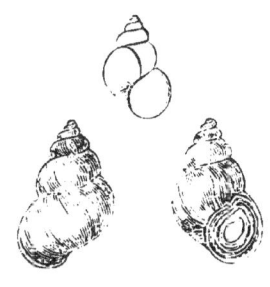

Helix tentaculata, Linnæus (1758), *Syst. Nat.* ed. 10. p. 774.
Nerita jaculator, Müller (1774), *Verm. Hist.* part ii. p. 185.
Turbo nucleus, Da Costa (1778), *Brit. Conch.* p. 91. pl. v. f. 12.
Bulimus tentaculatus, Poiret (1801), *Coq. de l'Aisne*, p. 61.
Cyclostoma impurum, Draparnaud (1801), *Tabl. Moll.* p. 41.
Turbo janitor, Vallot (1801), *Exerc. d'Hist. Nat.* p. 6.
Cyclostoma jaculator, Férussac (1807), *Ess. Méth. Conch.* p. 66.
Lymnæa tentaculata, Fleming (1814), *Edin. Encyc.* vol. vii. p. 78.
Paludina impura, Brard (1815), *Coq. Paris*, p. 183. pl. vii. f. 2.
Paludina jaculator, Studer (1820), *Kurz. Verz.* p. 91.
Turbo tentaculatus, Sheppard (1823), *Trans. Linn. Soc.* vol. xiv. p. 152.
Bithynia jaculator, Risso (1826), *Hist. Nat. Europ. Mérid.* vol. iv. p. 100.
Paludina tentaculata, Fleming (1828), *Brit. Anim.* p. 315.
Bithinia tentaculata, Gray (1840), *Turt. Man.* p. 93. pl. x. f. 120.
Bithinia (Elona) tentaculata, Moquin-Tandon (1855), *Hist. Moll.* vol. ii. p. 528. pl. xxxix. f. 23 to 44.

Hab. Throughout Europe. (In gentle streams and still waters.)

Bythinia tentaculata, the largest and most generally diffused of our British species, is well distinguished by its filiform, irregularly spreading tentacles. It was not improbably this peculiarity of structure that suggested to the mind of Linnæus the specific name *tentaculata*. It will be seen also on reference to our vignette, that they have a very flexible appearance, with sessile eyes at the base, all of which characters go to show their distinctness from *Paludina*, and the correctness of Dr. Gray's views in having founded the genus.

The shell of *B. tentaculata* is conically turbinated and ventricose, delicately coloured with a semitransparent fulvous green;

3. Bythinia Leachii. *Leach's Bythinia.*

Shell; conically turbinated, somewhat scalariform, minutely umbilicated, pale semitransparent fulvous green, whorls five, rather narrow, rounded, smooth; aperture rather small, rounded, but little pyriform. Operculum spirally striated around a central nucleus.

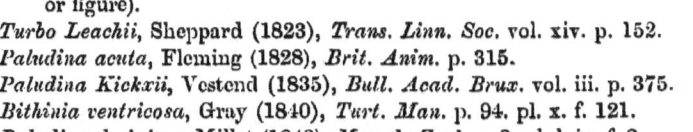

Paludina ventricosa, Gray (1821), *Lond. Med. Repos.* vol. xv. p. 239 (without characters or figure).
Turbo Leachii, Sheppard (1823), *Trans. Linn. Soc.* vol. xiv. p. 152.
Paludina acuta, Fleming (1828), *Brit. Anim.* p. 315.
Paludina Kickxii, Vestend (1835), *Bull. Acad. Brux.* vol. iii. p. 375.
Bithinia ventricosa, Gray (1840), *Turt. Man.* p. 94. pl. x. f. 121.
Paludina decipiens, Millet (1843), *Mag. de Zool.* p. 2. pl. lxiv. f. 2.
Paludina ventricosa, Brown (1845), *Illus. Conch. Brit.* p. 27. pl. xiv. f. 74, 75.
Paludina Michaudi, Dural (1845), *Rev. Zool. Soc. Cuv.* p. 211.
Bithinia Kickxii and *Michaudi*, Dupuy (1849), *Cat. Extram. Test.* no. 41 and 43.
Bythinia (Elona) Leachii, Moquin-Tandon (1855), *Hist. Moll.* vol. ii. p. 527. pl. xxxix. f. 20 to 22.
Hab. Throughout Europe. North Africa. (In gentle streams and still waters.)

This species is distinguished from the preceding, in being very much smaller and less ventricose. The whorls of the shell are rather narrow, rounder, and more loosely convoluted than in *B. tentaculata*, and they have a scalariform appearance. It is equally widely diffused, but is rather scarcer and more local. Southwards it passes into Algeria. The animal is speckled with black and yellow upon a light grey ground.

Dr. Gray mentions that the eggs of this species are disposed in a double row of six or seven pairs, on a tough strap-shaped green membrane attached to aquatic plants.

FAMILY PERISTOMATA. 191

Paludina contecta.

Genus II. **PALUDINA,** *Lamarck.*

Animal; oblong, broadly dilated in front, rounded behind, carrying a conically turbinated shell, head produced into a proboscis, lobed on each side at the base, the right lobe forming the orifice to the respiratory cavity, tentacles cylindrical, rising from between the lobes and proboscis with the eyes raised on a short conjoined stalk. Operculum horny, formed concentrically around a sublateral nucleus.

Shell; conically turbinated, of five to six whorls, sometimes umbilicated, sometimes with the umbilicus nearly closed, aperture moderate, pyriformly rounded, with the lip simple and slightly thinly expanded.

The canals, river-inlets, and ponds of Central Europe possess among their numerous small molluscan population already described, two of large size and characteristic structure, called *Paludinæ*, or Marsh Snails; and both appear, a little dwarfed in dimensions, in England, more especially in the southern half of our island. They are not found in Scotland or Ireland, although the genus, as we shall presently see, has a much more northerly range, in Asia, even in a more largely developed form. The *Paludinæ* constitute an important feature in our English freshwater fauna. They are distinguished by the habit of being ovo-viviparous. The eggs are hatched in the ovary, and at the end of about two months the young are ejected, three or four at a time, alive.

The animal of *Paludina*, it may be observed on reference to our vignette, has a largely dilated foot or disk, broad in front and attenuately rounded behind; the head is produced into a proboscis, and the tentacles, unlike those of *Bythinia*, are rather stout and cylindrical, with the eyes raised on short conjoined stalks. This modification in the position of the eyes, is apparently designed to make room for an organ at the outer base of the right tentacle, used for conveying water to the branchial chamber. It is in the form of a tubular lobe, and on the left side of the neck there is a corresponding lobe, to which no particular use is assigned. In the large *Ampullariæ* of the tropics, affecting situations where the water is more liable to be dried up, the animal is provided with a double system of respiration, having, in addition to the branchial chamber on the right side, a pulmonary chamber on the left side. The dormant lobe of *Paludina* is then developed into an elongated siphonic tube, for the passage to this chamber of the air. In most of the marine water-breathing Cephals, the siphonic lobes, which appear in such a rudimentary form in *Paludina*, are combined into a conspicuous central tube adapted to the same use, and the shell is either notched in front for its reception, as in *Buccinum*, or extended into a long canal for its special protection, as in *Murex*.

The *Paludinæ* adhere strongly by their long and wide-spread foot to the places selected for attachment, but they are sensitive to the touch, and readily fall. Our two British species were for a long time thought to be varieties of one and the same. Lister inclined to the fancy that the differences between them are merely differences of sex, that *P. vivipara* is the female of *P. contecta*. Their true specific characters were first noticed in France by Millet, and in England, a few years subsequently, by Dr. Gray. In *P. contecta*, the shell is composed of five and a half prominently tumid whorls, convoluted so loosely as to leave a deep umbilicus in the centre; in *P. vivipara*, the shell is composed of a whorl less, and the whorls are moderately ventricose, and more constrictedly convoluted, the umbilicus being reduced to a mere compressed chink. When very young, the shell is encircled with lines of extremely delicate ciliary bristles. Lamarck originally named this genus *Vivipara*, but on feeling the impropriety of using a specific adjective in the sense of a generic noun, he changed it to *Paludina*.

The general distribution of *Paludina* over the globe, does not range with that of the inoperculated freshwater genera *Planorbis*, *Lymnœa*, or *Physa*. We have no record of its existence in the waters of Australia, New Zealand, or Polynesia; nor does it appear in the West Indies or South America. About sixty species are known, of which rather more than two-thirds belong to the Eastern Hemisphere, and the remainder to North America. India and China, the chief area of habitation of this genus in the Old World, contribute about fifteen species, embracing several very characteristic varieties of form, sculpture, and painting. In some, the shells are conspicuously keeled, in others they are prettily banded, but green or olive-green is in all the predominant ground colour. The Philippine Island rivers contribute only half-a-dozen comparatively insignificant species of *Paludina*, their place being occupied to a great extent by *Melaniæ*, remarkable for the size and exquisitely acuminate proportions of their shells; and in Ceylon the rapids, chiefly those flowing from Adam's Peak, are inhabited by a very characteristic group having peculiarly globose shells, *Tanalia* and *Paludomus*.

The remainder of the Old World *Paludinæ*, so far as they are at present known, are three from Borneo and Celebes, three from Burmah and Siam, (in which latter country the land and freshwater mollusks all present strikingly new specific forms,) and six from more northern latitudes, of which two are European, two Siberian, and two Japanese. Temperature, which has so marked an influence in the development of molluscan life among the land snails, as may be seen by comparing the West African and Philippine *Helices* and *Bulimi* with the European, or those of Bolivia and Venezuela with the North American, has comparatively little influence among the water snails. We have no such large *Physæ* in Britain as inhabit Australia and the West Indies, but in *Lymnœa stagnalis*, which ranges eastward and northward to Siberia, we have the largest species known of that genus. It is the same with *Paludina*. The largest variety of the *P. gigantea* of Bengal is exceeded in size by *P. Ussuriensis*, a species collected by Gerstfeldt at the mouth of the Ussuri, a tributary of the Amoor, in Siberia, which is in the isothermal latitude of Lapland, little south of the line of permanent ground frost. The European *Paludina vivipara*, at least a light inflated form of similar structure and colour referred to that species, appears in North America, but the *Paludinæ* of that continent consist of sixteen species, chiefly of a peculiarly solid type, forming the genus *Melantho* of Bowditch.

o

The British *Paludinæ* are:—
1. **contecta.** Shell scalariformly turbinated, of five to six tumidly produced whorls, convoluted around a deep umbilicus.
2. **vivipara.** Shell obtusely ovately turbinated, of four to five moderately ventricose whorls, closely convoluted, leaving merely a compressed umbilical chink.

1. **Paludina contecta.** *Covered Paludina.*

Shell; scalariformly turbinated, openly deeply umbilicated, rather thin, dark olive-green, encircled with three purple-red bands, whorls five to six, longitudinally densely plicately striated, tumidly produced, obtusely angled round the upper part, with the sutures impressed; aperture somewhat pyriformly rounded, lip simple, a little expanded.

Nerita vivipara, Müller (1774), *Verm. Hist.* part ii. p. 182.
Cochlea vivipara, Da Costa (1778), *Test. Brit.* p. 81. pl. vi. f. 2.
Helix vivipara, Schröter (1779), *Gesch. Fluss-Conch.* p. 330. pl. viii. f. 2.
Cyclostoma viviparum, Draparnaud (1801), *Tabl. Moll.* p. 40.
Natica vivipara, Férussac (1801), *Mém. Soc. Méd.* vol. iv. f. 395.
Cyclostoma contectum, Millet (1813), *Moll. Maine et Loire*, p. 5.
Lymnæa vivipara, Fleming (1814), *Edin. Encyc.* vol. vii. p. 77.
Paludina vivipara, Studer (1820), *Kurz. Terz.* p. 91 (not of Say).
Paludina crystallina, Gray (1821), *Lond. Med. Repos.* vol. xv. p. 239.
Vivipara communis, Dupuy (1851), vol. v. p. 537. pl. xxvii. f. 5.
Paludina Listeri, Forbes and Hanley (1853), *Hist. Brit. Moll.* vol. iii. p. 8. pl. lxxi. f. 16.

Hab. Europe, chiefly in central parts. England, chiefly midland and southern counties, not in Scotland or Ireland. (In canals, ponds, and gently running waters.)

FAMILY PERISTOMATA. 195

An attentive comparison of the shells of this and the following species, will show that they are distinguished from each other by well-marked differences. *Paludina contecta* has a larger shell than *P. vivipara*, composed of a whorl more. The whorls are tumidly produced, obtusely angled round the upper part, which gives them a somewhat narrow appearance, and being coiled around a broader axis, a deep umbilicus is left in the centre. The colour is a dark olive-green, with the bands darker still, and more defined. The animal of *P. contecta* is a dingy brown, crowded with minutely speckled yellow dots.

The range of both species of *Paludina* is curiously confined in Britain to England. One or two stray dead shells of the genus have been found in Scotland, but there is no record of the animal having been taken alive; nor has it been taken alive in Ireland. The northern limit of *Paludina* in Britain, is in Yorkshire and Lancashire. On the Continent, *Paludina contecta* and *vivipara* range as far north as Finland. Their chief area of habitation is in Central Europe, extending, more scattered, to the Mediterranean. It has been stated that *P. contecta* is not found south of the Pyrenees, but we have little doubt that M. Charpentier's *P. ampullacea*, a native of Italy, is a variety of it.

2. **Paludina vivipara.**   *Viviparous Paludina.*

Shell; obtusely ovately turbinated, minutely compressly umbilicated, rather stout, fulvous olive, encircled with three purple-red bands, whorls four to five, longitudinally densely plicately striated, moderately ventricose, aperture somewhat pyriformly rounded, lip simple, a little expanded.

Helix vivipara, Linnæus (1758), *Syst. Nat.* ed. 10. p. 772.
Nerita fasciata, Müller (1744), *Verm. Hist.* part ii. p. 182.

Helix fasciata, Gmelin (1788), *Syst. Nat.* p. 3646.
Helix ventricosa, Olivi (1792), *Zool. Adriat.* p. 178.
Helix compactilis, Pulteney (1799), *Cat. Shells, Dorset*, p. 48.
Bulimus viviparus, Poiret (1801), *Coq. de l'Aisne*, p. 61.
Cyclostoma achatinum, Draparnaud (1801), *Tabl. Moll.* p. 40.
Viviparus fluviorum, De Montford (1810), *Conch. Syst.* vol. ii. p. 247.
Paludina achatina, Studer (1820), *Kurz. Verz.* p. 91.
Paludina vulgaris, Gray (1821), *Lond. Med. Repos.* vol. xv. p. 239.
Turbo achatinus, Sheppard (1823), *Trans. Linn. Soc.* vol. xiv. p. 152.
Paludina vivipara, Say (1830), *Amer. Conch.* pl. x.
Paludina fasciata, Deshayes (1838), *Lam. Anim. sans vert.* vol. viii. p. 512.
Vivipara fasciata, Dupuy (1851), *Hist. Moll.* vol. v. p. 540. pl. xxvii. f. 6.

Hab. Europe, chiefly in central parts. England, chiefly midland and southern counties. Not in Scotland, Ireland, or United States. (In canals, ponds, and gently running waters.)

The shell of this species has a more constricted closely coiled growth than that of the preceding, the whorls do not bulge out into an angular shelf as in *P. contecta*, and the apex is more obtuse. Its range of habitation is the same both in England and the Continent, and it appears, lighter and more inflated in growth, in North America, chiefly in the States of Georgia and Indiana. Mr. Jeffreys has remarked that the animal of *P. vivipara* is of a darker colour than that of *P. contecta*. The tentacles are bluish-black with bright yellow spots, the difference of their size in the male being very perceptible.

Valvata cristata. (*Much enlarged.*)

Genus III. **VALVATA**, *Müller*.

Animal; oblong, carrying a light heliciform or discoid shell, head produced into a ringed proboscis, tentacles cylindrical, moderate, approximating at the base, with the eyes on inner tubercular swellings, foot divided in front into two crescent-shaped segments, scooped out at the sides, branchiæ plumose, sometimes exserted, accompanied by a filiform tentacular appendage. Operculum horny, concave, rather sunk in the aperture, concentrically spiral, deeply ridged.

Shell; globosely heliciform or discoid, of from three to five rounded whorls of a semitransparent straw-colour; aperture small, with the margin continuous.

Included in the family of *Peristomata* by Lamarck, Gould, Moquin-Tandon, and others, is another freshwater mollusk, which partakes only to a very moderate extent of the characters of *Bythinia* and *Paludina*. *Valvata* has the same proboscis-like head, and it is provided with an operculum, but in other respects both the animal and its shell are distinguished by characters peculiarly their own. The eyes of *Valvata* at the base of the tentacles are on inner tubercular swellings, the foot is cleft in front into a pair of crescent-shaped segments, the branchiæ, most important feature of all, appear in the form of an external plume, composed of spirally twisted filaments, and accompanying this is a tentacular thread, designed apparently both for its protection when exserted, and for drawing the currents of water to it. The proboscis of *Valvata* is almost as conspicuously developed as that of *Cyclostoma*. In a beautifully executed drawing of *V. piscinalis* by Mr. Berkeley, now before me, but in which the branchial plume is not sufficiently exserted for illustration, the proboscis is rigidly protruded to half the length of the tentacles. The eyes are placed more inwardly than in *Cyclostoma*.

Valvata, as may be seen on reference to our outline figures of the shells, is small in Britain, and it is almost equally small in all its places of habitation. *V. piscinalis* and *cristata* are very generally diffused throughout Europe and Western Asia. Both species appear in Siberia, and as far southwards as Lycia and the islands of the Mediterranean. Several other forms of *Valvata*, collected over the same extended range, have been described as species,—*Baicalensis*, Gerstfeldt, *Amoorensis*, Bourguignat, *prasina*, Parreyss, *trochlea*, Dunker, *contorta*, Menke, *alpestris*, Shuttleworth; of none of which have I been able to satisfy myself, yet it would be premature to include them in the list of synonyms. The North American species are limited by Mr. Binney, in his 'Descriptive Catalogue,' now passing through the press, to four: *V. tricarinata*, *sincera*, and *humeralis*, Say, and *pupoidea*, Gould. The first of these has the shell encircled in a very prominent manner by three strongly developed keels, the others are European forms. We have no record of the existence of *Valvata* in eastern or southern Asia, or in South Africa, Australia, or South America. The genus was supposed, until lately, to be confined to the temperate and north temperate regions of the globe. *Valvata* appears, however, in its European form, as in the case of *Physa* and *Neritina*, in the West Indies. *V. depressa*, Say, and *V. pygmæa*, C. B. Adams, from that locality, are almost identical with our *V. piscinalis* and *cristata*. It is not improbable that *Valvata*, or a genus nearly allied to it, exists in other parts of Central America. Mr. Cuming collected several specimens of a shell very like that of *Valvata*, of a solid keeled type, near Real Llejos, Guatemala, in the bed of a mountain rapid, about four miles from the sea.

Our two British *Valvatæ* have very different shells. *V. piscinalis* has a globosely heliciform shell, faintly spirally ridged, while *V. cristata* has a smooth, discoid, *Planorbis*-like shell. The operculum is in both species concentrically spiral, rather deeply sutured, and its position is somewhat sunk in the aperture.

The British species of *Valvata* are:—
1. **piscinalis.** Shell small, of four and a half to five pale, faintly ridged, moderately enlarging whorls, convoluted globosely.
2. **cristata.** Shell minute, of three and a half to four smooth, slowly enlarging whorls, convoluted discoidly.

---

1. **Valvata piscinalis.** *Fish-pond Valvata.*

Shell; somewhat globosely heliciform, deeply narrowly umbilicated, pale straw-colour, semitransparent but solid, whorls four and a half to five, depressed at the apex, longitudinally densely very finely striated, spirally faintly ridged, ridges sometimes obsolete; aperture somewhat pyriformly rounded.

Nerita piscinalis and *pusilla*, Müller (1774), *Verm. Hist.* part ii. p. 171 and 172.

Trochus cristatus, Schröter (1779), *Gesch. Fluss-Conch.* p. 280. pl. vi. f. 11.

Helix piscinalis and *fascicularis*, Gmelin (1788), *Syst. Nat.* p. 3627 and 3641.

Nerita obtusa, Studer (1789), *Coxe, Trav. in Switz.* vol. iii. p. 436.
Turbo fontinalis, Pulteney (1799), *Cat. Shells, Dorset.* p. 45.
Turbo cristatus, Poiret (1801), *Coq. de l'Aisne*, p. 29 (not of Maton and Rackett).
Cyclostoma obtusum, Draparnaud (1801), *Tabl. Moll.* p. 39.
Valvata minuta, Draparnaud (1805), *Hist. Moll.* p. 42. pl. i. f. 36 to 38.
Valvata piscinalis, Férussac (1807), *Ess. Syst. Conch.* p. 75.
Lymnæa fontinalis, Fleming (1814), *Edin. Encyc.* vol. vii. p. 78.
Valvata obtusa, Brard (1815), *Coq. Paris*, p. 190. pl. vi. f. 17.
Turbo thermalis, Dillwyn (1817), *Desc. Cat. Shells*, p. 852.
Valvata depressa, C. Pfeiffer (1821), *Deuts. Moll.* vol. i. p. 100. pl. iv. f. 33.
Valvata Moquiniana, Reyniés (1851), *Dupuy, Hist. Moll.* vol. v. p. 586. pl. xxviii. f. 15.

Hab. Throughout Europe. Siberia. Asia Minor. (Adhering to stones, submerged sticks, etc., in tranquil and gently running waters.)

Valvata piscinalis is almost milk-white, showing its bright blue-black eyes very conspicuously on the inner base of each tentacle. The shell is of a gradually enlarging turbinated growth, of a pale semitransparent straw-colour, faintly spirally ridged. Sometimes the ridges are very apparent, sometimes they are almost obsolete, and the whorls of the shell being more or less closely convoluted, it follows that the umbilicus is more or less open. There is much reason to believe that more of the species of Continental authors have been founded on these variations of growth, than appear in our list of the synonymy. *V. piscinalis* is very generally diffused throughout Europe and Western Asia. The most northern habitat is that recorded by Gerstfeldt in Siberia; the most southern that of specimens in my possession, collected by Captain Spratt, in Lycia. The species is found in all parts of the British Isles, and it is very closely represented in the West Indies by *V. depressa*.

M. Moquin-Tandon describes *V. piscinalis* as being slow in its movements, and very irritable, sometimes crawling on solid bodies, sometimes swimming on the surface of the water, moving itself in all directions by the two crescent segments in the front part of the foot, and continually agitating its protruded trunk, while it falls on the least touch to the bottom of the water. Mr. Jeffreys notices that the eggs of this mollusk are deposited on various substances, sometimes on the shell of a *Planorbis*, enclosed in a globular capsule, having a short stalk by which it is attached. About the twelfth day, the capsule distends and bursts, and the young fry are ejected alive.

2. Valvata cristata. *Crested Valvata.*

Shell; minute, discoid, very largely superficially umbilicated, dull horny, whorls three and a half to four, rounded, narrow, increasing slowly, smooth or minutely striated, spire rather sunk, with the sutures impressed; aperture small, round, with the lip thinly reflected.

Valvata cristata, Müller (1774), *Verm. Hist.* part ii. p. 198.
Nerita valvata, Gmelin (1788), *Syst. Nat.* p. 3675.
Valvata pulchella, Studer (1789), *Coxe, Trav. in Switz.* vol. iii. p. 436.
Valvata planorbis, Draparnaud (1801), *Tabl. Moll.* p. 42.
Helix cristata, Montagu (1803), *Test. Brit.* vol. ii. p. 460. vign. i. f. 7 and 8.
Valvata spirobis, Draparnaud (1805), *Hist. Moll.* p. 45. pl. i. f. 32 and 33.
Trochus cristatus, Maton and Rackett (1807), *Trans. Linn. Soc.* vol. viii. p. 169.
Turbo cristatus, Turton (1819), *Conch. Dict.* p. 227.
Valvata branchialis, Gruithuisen (1821), *Nov. Act. Nat. Cur.* vol. x. p. 437. pl. xxxviii. f. 13.

Hab. Throughout Europe. Siberia. (Adhering to stones and submerged sticks, in tranquil and gently running waters.)

The name *cristata* given to this *Valvata*, refers to the branchial plume, not to the shell, and, as may be seen by our list of synonyms, it has been given by different authors to both species. *V. cristata*, the subject of our vignette of the living animal, is much smaller than *V. piscinalis*, and very much darker in colour; while the shell is constructed on a different plan, being formed of only three and a half to four slowly enlarging whorls lodged one upon the other, in a nearly discoidal manner, as in *Planorbis glaber*.

The distribution of *V. cristata* is similar to that of *V. piscinalis*, and it is represented in the same manner in the West Indies, by *V. pygmæa* of Professor C. B. Adams. In habit it is also the same, making active use while swimming, of its trunk and foot segments.

Family III. NERITACEA.

Head produced into a short proboscis, eyes on short slender stalks detached from the tentacles, branchiæ parallel, rounded plates set on a long triangular membrane. Shell few-whorled, rather solid.

The *Neritacea* are a numerous group of mollusks, part marine, part fluviatile, of essentially tropical origin. One small species prevails throughout the temperate fresh waters of the Eastern Hemisphere; there is no strictly marine species north of Senegal. In the Western Hemisphere there is no species of either habit north of California. The animal is small, nearly covered by a comparatively massive shell, of rarely more than three whorls, brilliantly painted, and often enveloped by a rather fibrous epidermis. The head is produced into a short proboscis, and the eyes are peculiar in being raised on short slender stalks detached from the tentacles, while the branchiæ appear in the form of parallel rounded plates, set on a long triangular membrane, which is partially free. The animal encloses itself firmly in the shell, by means of an operculum which hinges at one corner on the columella by a kind of apophysis. The *Neritacea* are about two hundred and fifty in number, comprised in three genera, *Nerita*, *Neritina*, and *Navicella*, inhabiting chiefly the seashore and rivers of the Polynesian and Malayan Archipelagos. Of the river *Neritacea*, which are the more numerous, we have one species in Britain.

It belongs to the genus:—

1. **Neritina.** Animal carrying an obliquely ovate shell, of three whorls, having a tranverse semilunar aperture.

Neritina fluviatilis. (*Moderately Enlarged.*)

Genus I. **NERITINA.** *Lamarck.*

Animal; short, carrying a solid few-whorled shell, head produced into a proboscis, which is comparatively short, tentacles contractile, slender, flexuous, margined with dark lines, having the eyes at their base on short detached stalks, foot oblong, obtuse before and behind. Branchiæ free, for the greater part consisting of a long triangular acute membrane, on the sides of which are placed parallel rounded plates. Operculum paucispiral, furnished on the under side with an apophysis which hinges on the columella.

Shell; obliquely ovate, of three very rapidly enlarging whorls, aperture transversely semilunar, columella broadly callous.

This very interesting little mollusk is the single European member of a group which is very abundant in the warmer latitudes of our hemisphere, both in the rivers and on the seashore. Of the river species, the *Neritinæ*, about a hundred and thirty are known, of which at least eighty are natives of the Philippine Islands and the islands of Polynesia. Four have been described from Madagascar, ten from Sumatra and Ceylon, four from the Ganges, six from Africa, and five from Australia. The remaining twenty are all within a limited area of the New World, enclosing the West Indies and Central America. The marine species of this family, the *Neritæ*, which are less numerous, about ninety in number, have much the same distribution, excepting that in the Eastern Hemisphere they do not come nearer to us than the latitude of Senegal and the Red Sea, while on the western coast of America they reach in a southerly direction to Peru.

The animal of *Neritina* is comparatively small and obtuse, nearly covered by its rather massively developed shell, and has only a moderately protruded proboscis. The tentacles are slender, contractile, and flexuous according to the will of the creature, and constantly in motion, and the eyes are raised on short, separately detached stalks. The foot, as may be observed in our vignette, is broadly obtusely dilated in front as far as the extremity of the proboscis. As may be gathered from the comparatively solid and overwhelming growth of the shell, *Neritina* is not a mollusk of active habits. It does not float on the surface, but adheres to stones, or crawls upon a stony or gravelly bottom. The shell is of a peculiarly oblique form, arising from its rapidly enlarging narrowly produced growth of scarcely three volutions. The operculum, which is testaceous and paucispiral, peculiarly is characterized by the presence of a hooked apophysis on the underside next the columella, on to which it slides and acts like a hinge.

Our British species of *Neritina* is:—

1. **fluviatilis.** Shell an oblique oval of three rapidly enlarging whorls, variously reticulately painted, having the columella transverse and broadly callous.

FAMILY NERITACEA.

1. **Neritina fluviatilis.** *River Neritina.*

Shell; obliquely oblong-ovate, solid, fulvous white, more or less covered with a variously reticulated net-work of blue-black or grey, spire very short, obtuse, whorls three, the last considerably produced, aperture transversely semilunar, columellar area broad, callous, slightly concave.

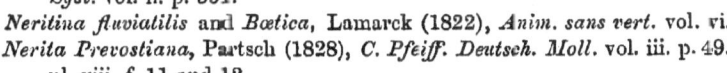

Nerita fluviatilis, Linnæus (1758), *Syst. Nat.* ed. 10. p. 777.
Theodoxus Lutetianus, Montford (1810), *Conch. Syst.* vol. ii. p. 351.
Neritina fluviatilis and *Bœtica*, Lamarck (1822), *Anim. sans vert.* vol. vi.
Nerita Prevostiana, Partsch (1828), *C. Pfeiff. Deutsch. Moll.* vol. iii. p. 49. pl. viii. f. 11 and 12.
Neritina variabilis, Hécart (1833), *Mém. Soc. Agr. Valenc.* vol. i. p. 146.
Neritina thermalis, Boubée (1833), *Bull. Hist. Nat.* p. 12.
Neritina Dalmatica, Partsch (1835), *Sow. Conch. Illus.* f. 57.
Nerita zebrina, Recluz (1841), *Rev. Zool. Soc. Cuv.* p. 341.
Nerita Mittreana, Recluz (1842), *Rev. Zool. Soc. Cuv.* p. 181 and 182.
Neritina Prevostiana, Dupuy (1851), *Hist. Moll.* vol. v. p. 593. pl. xxix. f. 2.
Neritina Bourguignati, Recluz (1852), *Journ. Conch.* p. 293.
Neritinæ Numidica, Peloponensis, meridionalis, intexta, Jordani, Sardoa, Macri, Hildreichii, trifasciata, Anatensis and *Anatolica*, auctorum.

Hab. Throughout Europe. Asia Minor. North Africa. (Adhering to stones or crawling on the gravelly bottom of rivers and streams. Also in warm springs.)

The generic features of this little river mollusk are peculiar and constant, but its specific characters are variable in the extreme, and it has been named twenty times over. In form, the shell varies from obliquely oblong to subglobose, and the painting, though generally a pattern of network, is most protean. The animal is of a pale flesh-colour, the upper parts being more or less mottled with brown, and there is generally a dark band between the tentacles, which are rather conspicuously edged with a dark line. The animal is but little protruded beyond its shell, and withdraws itself instantly on being touched. It is pretty generally diffused throughout Europe and Asia Minor, passing into North Africa. In Britain, it appears chiefly in the south-western and eastern counties of England and Ireland. Unlike the rest of our freshwater mollusks, the genus does not occur in the Middle and Northern United States of America.

Class II. ACEPHALA—WITHOUT HEAD.

The Acephala, or Headless mollusks, better known to conchologists as the Bivalves, abound in great variety in the sea, but in fresh water they are very few in number. There are scarcely more than half-a-dozen freshwater genera throughout the globe. There are no land *Acephala;* and none that respire air. All have shells. The normal type of an Acephal is a compressed bag enclosing the vital organs, including the branchiæ, or gills, to which there are two openings from behind, sometimes simple or pouted, sometimes prolonged into siphons, separate or conjoined, one for inhaling the surrounding fluid, the other for ejecting it; and there is an opening in front for the passage of a burrowing, swimming, or adhering organ, termed the foot. The Acephal having no head has no tentacles or distinct eyes, but in several of the marine genera there are pigments in certain parts of the body, or along the edge of the mantle, which have very much the appearance of eyes, and are supposed by some naturalists to be really organs of vision. Internally, on the side opposite to that of the siphonal orifices, there is an opening to the stomach, edged with two or four lips, which has been termed the mouth, but it is without jaws or tongue, or any kind of muzzle or proboscis. Externally a pair of lobes, incorporated with the integument, each secreting a shelly valve, constitutes the mantle; and the valves are connected dorsally by a cartilage, which tends to open them in opposition to the action of internal adductor muscles proceeding from the animal, which has a tendency to close them. The hinge of the valves is, in most instances, strengthened by interlocking protuberances on their dorsal margin, somewhat ambiguously designated teeth.

Our *Dreissena* is a very interesting example, high up in the series, of an Acephalous mollusk in its normal condition of a closed integument, allied to the Ascidian. In a more advanced form, as in the *Anodon* and *Unio*, there is no closed integument, but a pair of free mantle lobes, secreting, and corresponding with, the valves of the shell, and wholly enveloping the body.

Tribe I. **LAMELLIBRANCHIATA**—LAMELLA-GILLED.

The Cephalous mollusks dwelling, some in air, some in water, and some in both elements, have different modes of respiring, and their respiratory organ affords good distinguishing characters for the subdivision of the Tribes into Orders. The Acephala are all water-dwelling mollusks, and their respiratory organ is too little varied to characterize an ordinal subdivision. It is represented throughout the class by a pair of lamellæ or leaflets on each side of the viscera, quite distinct from the mantle, and they are usually clothed with extremely delicate cilia, used by the animal for producing an inflowing current from the surrounding fluid. These beautifully ciliated leaflets are exquisitely shown in Mr. Goadby's anatomical preparations.

The Lamellibranchiate Acephala are divided into two Orders, according to the mode by which the shell is affixed to the animal, whether by one central adductor muscle or by two terminal muscles.

1. **Unimusculosa.** Animal having its shell affixed by a central muscle in each valve, which is sometimes compound or furnished with one or more small auxiliary muscles.
2. **Bimusculosa.** Animal having its shell affixed by two distinct terminal muscles, one at the anterior, the other at the posterior end.

Order I. **UNIMUSCULOSA**—ONE-MUSCLED.

Animal having its shell affixed by a central muscle in each valve, which is sometimes compound, or furnished with small auxiliary muscles.

It may appear unscientific to introduce a systematic division founded on the character of a single pair of adductor muscles, when the only British mollusk inhabiting fresh water that is referred to it is possessed of more. But in viewing the entire series of *Acephala*, both freshwater and marine, a very natural subdivision is suggested by the separation into two Orders, of those in which the shell is affixed to the animal by a central muscle in each valve, as in the *Oyster*, and those in which the shell is affixed to the animal by two lateral, or rather terminal, muscles, as in the *Anodon*. No characters are permanent through any long series of natural objects, and it happens that the only member of this Order inhabiting fresh water in Europe, belongs to a family plentiful in marine species, in which the characteristic muscles of the Order undergoes a change. The adductor muscle becomes, in the *Mytilacea*, a compound muscle, and in the anterior extremity of the shell of some of them there are two small auxiliary muscles.

The only British freshwater species of *Unimusculosa* belongs to the family.

1. **Mytilacea.** Animal bearing an oblong fan-shaped shell, to which it is affixed posteriorly by a pair of strong compound adductor muscles. Beneath the umbo in each valve, there is sometimes a small auxiliary muscle affixed to a cross septum. Adhering to foreign bodies by a byssus of firm tendinous threads, proceeding from a groove in the foot.

Family I. **MYTILACEA.**

Animal bearing an oblong fan-shaped shell, to which it is affixed anteriorly by a pair of strong compound adductor muscles. Beneath the umbo in each valve, there is sometimes a small auxiliary muscle affixed to a cross septum. Adhering to foreign bodies by a byssus of firm tendinous threads, proceeding from a groove in the foot.

The *Mytilacea* are a family of about a hundred and eighty species, of which a hundred and seventy, belonging to the genera *Mytilus*, *Modiola*, *Crenella*, and *Lithodomus*, inhabit the sea, and the remaining ten are inhabitants of brackish and fresh water. Of two of these, the habitats are not known. Six are natives of Central America and the Southern United States, one is a native of Senegal, and one of Europe. The freshwater *Mytilacea* differ from those of marine habit, and from the rest of the unimusculose *Acephala*, in having the mantle closed.

The main adductor muscle of this family adheres rather near to the posterior edge of the shell, and the cicatrix indicative of its place of attachment is of a duplicated oblong form, showing it to be of a compound structure, but it is not divided into bundles of fibres in *Dreissena* as in *Mytilus*. Beneath the umbo in each valve, there is a small shelly septum, and affixed to these is a small auxiliary transverse muscle.

Our freshwater representative of this family is referred to a special genus:—

1. **Dreissena.** Animal bearing a triangular fan-shaped shell, mantle closed throughout, except for the passage of the foot and of two tubular orifices for the purposes of excretion and respiration.

Dreissena polymorpha.

Genus I. **DREISSENA**, *Van Beneden*.

Animal; subtriangular, rather compressed, bifurcated in front, bearing a subtriangular shell of which the umboes cover the bifurcations, mantle delicately fringed, closed throughout, excepting a small opening in front for the passage of the foot and two tubular orifices behind for excretion and respiration, foot ligulate spinning a copious byssus, excretory orifice developed into a small conoidal siphon, respiratory orifice larger, subtubular, armed with longitudinal rows of scale-like papillæ, of which those on the inner margin are elongated internally towards the centre.

Shell; equivalve, very inequilateral, triangular, very gibbous, umboes terminal, furnished in each valve with a cross septum for the adhesion of a small adductor muscle, cicatrix of the principal adductor muscle oblong, compound, situated within the posterior margin of each valve, hinge toothless, cartilage external, marginal.

The tranquil and gently running waters of northern and eastern Europe, between the Baltic and Caspian Seas, are inhabited by a peculiar and most prolific mussel, which has been transported among timber and on ships' bottoms to the western parts, and has become naturalized in all corners of the British Isles. It differs from the sea mussel (*Mytilus*) and from all the rest of the unimusculose Acephala, in the important feature of having the mantle closed throughout, excepting at the three orifices necessary for the passage of the foot, for excretion, and for respiration. In the *Mytili* the mantle is open, the free lobes on either side corresponding with the valves of the shell. In *Dreissena*, the shell is secreted by a closed integument, on which the mantle or calcifying portion of it is indicated by a linear fringe of papillæ.

Dreissena, named by M. Van Beneden, a Belgian naturalist, after a M. Dreissen, was first noticed more than a century ago in the river Volga, by the eminent German traveller and naturalist, Pallas, during a voyage in Russia. It was not detected in Britain until 1824, when Mr. J. De Carle Sowerby acidentally observed a gentleman fishing for perch with it, as bait, in the Commercial Docks. The mollusk had been imported from eastern or northern Europe, probably across the Baltic, either among timber or attached to the ship's bottom, and was abundantly multiplying. From the habit of *Dreissena* living in brackish as well as in fresh water, it is believed to be capable of enduring salt water for a time, as well as living for some time out of water, and thus it traverses the sea in ships or adhering to ships, and it has become naturalized in most of our docks, lakes, canals, reservoirs, and even in our water-pipes.

It will be seen on reference to our vignette, that the anterior or lower orifice is merely a simple opening for the passage of a small foot, which is comparatively short and conoid when contracted, tongue-shaped when extended. From a groove in this foot, a copious fibrous byssus is spun for the purpose of adhesion to foreign bodies. Of the two hinder or upper orifices, the one for ejecting the water is developed into a simple conoidal siphon, the other for taking it in is more complicated. The portion of the mantle in which it is situated, protrudes in a subtubular manner, and is armed with longitudinal rows of irregular scale-like papillæ. At the orifice it is contractedly pouted, and rimmed with papillæ, those on the inner edge being elongated in a more or less crowded manner towards the centre. Individuals of this mussel are wonderfully prolific, and attach themselves in groups one upon another. The shell is of a peculiar triangularly trapezoid form, the valves being affixed to the animal by an oblong compound adductor muscle just within the posterior margin, and a small auxiliary transverse muscle affixed to a cross shelf below the umbo. The hinge of the valves is toothless, the cartilage by which they are connected being marginal and external.

A few species of *Dreissena* have been discovered in the West Indies, in Central America, and in the Mississippi in the Southern United States; and one at Senegal.

1. **Dreissena polymorpha.** *Many-shaped Dreissena.*

Shell; triangularly trapezoid, very gibbous, obtusely keeled, fulvous, stained and transversely undulately waved with olive-brown, under side broadly flatly impressed, right valve sinuated at the central margin for the passage of the byssus; interior bluish-white, not pearly.

Mytilus polymorphus, pars, Pallas (1754), *Voy. Russ. App.* p. 212.
Mytilus e flavio Wolga, Chemnitz (1795), *Conch. Cab.* vol. xi. p. 256. pl. ccv. f. 2028.
Mytilus Hagenii, Bäer (1825), *Inst. Solemn. Oken's Isis*, part v. p. 525.
Mytilus Volgensis, Gray (1825), *Ann. Phil.* p. 139; *Ind. Test. Supp.* p. 3. pl. ii. f. 6.
Mytilus arca, Kickx (1834), *Desc. nouv. esp. de Moule.*
Dreissena polymorpha, Van Beneden (1834), *Bull. Acad. Sci. Brux.* vol. i. p. 105; *Ann. Sci. Nat.* pl. viii. f. 1 to 11.
Tichogonia Chemnitzii, Rossmässler (1835), *Icon.* vol. i. p. 113. f. 69.
Mytilina polymorpha, Cantraine (1837), *Ann. Sci. Nat.* vol. vii. p. 308.

Hab. Throughout Europe, chiefly in the northern and eastern parts. (In canals, rivers, docks, water-pipes, etc.)

The shell of our British *Dreissena*, compared with that of the

Senegal and Central American species, is conspicuously distinguished by its extremely gibbous trapezoidal form, and transverse zigzag colouring. In young specimens, clustering in bunches and in masses one upon another, the pretty waved pattern is particularly bright and distinct. As the specimens advance in growth, and the epidermis thickens and becomes fibrous, the pattern is nearly obliterated.

M. Moquin-Tandon and Mr. Jeffreys describe the animal as being greyish white, yellowish or fawn-colour at the posterior side, striped like the shell with zigzag marks of reddish brown. The cirrhi of the branchial orifice, arranged in concentric rows, are reddish grey, tinged at their base with brown, and the foot is grey with a delicate rosy hue.

Order II. **BIMUSCULOSA**—Two-muscled.

In this Order, which comprises the greater portion of the Acephala, marine as well as freshwater, the shell is affixed to the animal by a pair of adductor muscles, adhering to the valves of the shell at each end. All have the lobes of the mantle more or less united. We have in Britain two Families of freshwater two-muscled bivalves, the *Naiades*, including the genera *Anodonta* and *Unio*, and the *Cardiacea*, including *Pisidium* and *Cyclas*. The mantle lobes of the Naiads, the largest of all freshwater mollusks, are freely open, except where they are united behind to form the branchial and excretory siphonal orifices. In the Cockles of fresh water, which are small, some of them minute, the mantle lobes are more united on each side, and the branchial and excretory siphons are prolonged into tubes. In *Pisidium*, the tubes are blended in one, in *Cyclas* they are united for some distance and then separated. The shell of the *Naiades* is composed of firm pearly matter, covered by a thick fibrous epidermis; the shell of the *Cardiacea*, that is to say, of the freshwater genera of *Cardiacea*, is thin, covered by a horny epidermis.

Our two-muscled freshwater Acephala are comprised in two Families:—

 1. **Naiades.** Animal bearing a solid pearly shell, with the mantle lobes freely open, except behind, where they are united to form the branchial and excretory siphonal orifices, which are simply pouted. Foot large, free.

2. **Cardiacea.** Animal bearing a thin horny shell, with the mantle lobes open anteriorly for the passage of a large protruded foot, and united posteriorly to form the branchial and excretory siphons which are prolonged into tubes wholly or partially united.

Family I. NAIADES.

Animal bearing a pearly shell, with the mantle lobes freely open except behind, where they are united to form the branchial and excretory siphonal orifices, which are simply pouted. Foot large, free.

The Acephalous mollusks are less numerous in kind than the Cephals, but they include species of much larger dimensions. Among those inhabiting the sea, the giant clam (*Tridacna*) is of greatly more colossal proportions than any shell of spiral growth; and in freshwater, the *Paludina*, largest of our pond-snails, is far exceeded in size by the *Naiad*. The lakes and rivers of both hemispheres, especially the western, teem with this mollusk, a pearly bivalve of mud-dwelling habit, with a large tongue-shaped delving foot. The mantle lobes are freely open, excepting behind, where they are united to form the branchial and excretory siphonal orifices; the edges round the branchial orifice being sharply fringed, and the hinge margin of the shell is sometimes toothed, sometimes without teeth.

Such are the typical characters of the family, but there are four very distinct and remarkably local foreign genera included under this head, in which the structure of the foot and posterior siphons is greatly modified. Two, *Iridina* and *Spatha*, inhabitants of the Nile and rivers of Senegal, have the posterior siphonal orifices prolonged into short unequal tubes. The other two, *Hyria* and *Mycetopus*, natives of South America, are more different from each other. In *Hyria*, the siphonal orifices are prolonged into tubes. In *Mycetopus* there are no siphonal tubes, but the foot is singularly long and attenuated, reaching a considerable extent beyond the shell, and at the extremity it is suddenly widened into a knob. *Mycetopus* has a curiously long and narrow shell, gaping at both ends.

The shell of the *Naiades* is of an iridescent pearly substance, covered by a dark fulvous-olive, green, brown or black epidermis, and all more or less produce pearls. One or two of the North

American species of *Unio* have been observed to spin a byssus in cases of emergency. In South America M. D'Orbigny collected, at the Rio Parana, above Corrientes, an *Anodon* attached by a byssus; his genus *Byssanodonta* will not however hold, if dependence cannot be placed on this function as a generic character.

About six hundred and thirty species of Naiads, with a long array of synonyms, are now admitted, reckoning only eleven as belonging to Europe, forty-three to Asia, seventeen to Africa, and three to Australia. Of the remaining species, seventy-six belong to South America, and four hundred and eighty to North America. There are seventy-four species of four genera in the Eastern Hemisphere, and five hundred and fifty-six of six genera, in the Western, distributed as follows:—

Genera.	Europe.	Asia.	Africa.	Australia.	N. America.	S. America.
Unio	10	33	6	2	430	28
Anodonta	1	8	2	1	50	32
Dipsas	—	2	—	—	—	—
Iridina	—	—	5	—	—	—
Spatha	—	—	4	—	—	—
Hyria	—	—	—	—	—	3
Castalia	—	—	—	—	—	2
Monocondylæa	—	—	—	—	—	8
Mycetopus	—	—	—	—	—	3

The genera *Unio* and *Anodonta* are mostly associated together, the first being much the more plentiful in species, excepting in South America, where the toothless Naiads predominate over the toothed. Of the eleven European species, we have four in Britain, two confined to England, two diffused throughout the three kingdoms.

They are of the genera—

1. **Anodonta.** Shell thin and toothless.
2. **Unio.** Shell rather stout, hinge composed of interlocking erect teeth on the anterior side, and elongated marginal teeth, which are sometimes obsolete, on the posterior.

FAMILY NAIADES. 213

Anodonta cygnea.

Genus I. **ANODONTA**, *Lamarck*.

Animal; body oval, compressed, dingy greyish yellow, mantle fringed and stained with brown at the edge, gills showing a network of waved tubes, foot large, tongue-shaped, tinged with orange.

Shell; equivalve, very inequilateral, thin, rounded in front, obliquely produced and a little auriculated behind; hinge toothless, but furnished posteriorly with a long rudimentary toothlike ridge.

Anodonta, the largest of European freshwater mollusks, is provided with a largely ventricose shell of light substance, which, as its name implies, is without teeth. A strong external ligament affords hinge sufficient for its thinly calcified valves, without the support of any internal interlocking protuberances, yet a tooth of a rudimentary kind appears on the posterior side of the shell, in the form of a blunt marginal ridge. The animal has a largely developed foot of an orange-tinted yellow colour, with which it burrows into and trails along the mud, and the mantle margin is fringed and stained with brown about the edge, particularly in the parts surrounding the branchial orifice.

The lakes, canals, ponds, and gently flowing rivers throughout Europe, are all tenanted by the *Anodonta* in some form or other. It feeds on decomposed animal and vegetable matter, and the shell varies according to the quantity and character of the food, the stillness or disturbance of the water, its chemical composition, depth or shallowness, and other causes. From forty to fifty species have been made of these different phases of the European *Anodonta*, but our best authorities on the subject, Messrs. Forbes and Hanley, and Dr. Gray, are of opinion that they are all referable to one. Eleven more species have been recorded as inhabitants of the Eastern Hemisphere, eight Asiatic, chiefly Philippine, two African, and one Australian. The Western species Mr. Lea enumerates as eighty-two, of which thirty-two are South American, and fifty North American; *A. fluviatilis*, inhabiting the western and central parts of Massachusetts, being almost identical with our own species.

Dr. Gray's testimony in favour of there being only a single species of *Anodonta* in Britain is expressed in the following able manner:—
"The *Anodons* feed on decomposed animal and vegetable substances; and the size and solidity of the shell depends on the abundance of the food and the state of quietness or motion and of calcareous matter in the water in which they happen to reside. Some authors have believed them to be unisexual; but their anatomy proves that they are hermaphrodite and sufficient for themselves. Poiret supposed that of the two species he observed near Paris, one was viviparous and the other oviparous; but they all deposit eggs, which are developed in their exterior pair of gills. They have been divided into numerous species; but in ponds where there is plenty of food (and a dead dog or cat or fish affords abundance of such material), and where the water is nearly stagnant and seldom disturbed, they become of a large size, with ventricose thin shells, while in more rapid rivers with pure clear water, with very little decomposed animal or vegetable matter held in suspension, they are small, with compressed thick shells; and all intermediate forms and sizes are to be observed. After collecting many hundred specimens from various localities, I am convinced that there is only a single species found in this country."

Our British *Anodonta* is:—

1. **cygnea.** Shell a pair of thinly ventricose toothless valves, slightly auriculated on the posterior side.

FAMILY NAIADES. 215

1. **Anodonta cygnea.** *Swan Anodon.*

Shell; obliquely ovate or oblong, varying from compressed to ventricosely gibbous, anterior side rounded, a little gaping in the region of the foot, posterior side obliquely produced, slightly auriculated towards the seat of the ligament, valves roughly concentrically ridged at irregular intervals by additions of growth, upon which the epidermis lodges in fibrous wrinkles, bright verdigris green passing with age into fulvous olive, flesh-tinged towards the umboes, umboes compressed, plicately wrinkled, wrinkles few, obsolete or eroded with age, hinge toothless, in the centre, on the posterior side furnished with a rudimentary lateral tooth, in the form of a faintly developed elongated ridge.

Mytilus cygneus and *anatinus*, Linnæus (1758), *Syst. Nat.* ed. 10. p. 706.
Mytilus radiatus, Müller (1774), *Hist. Verm.* part ii. p. 209.
Mytilus Zellensis, Gmelin (1788), *Syst. Nat.* p. 3262.
Anodontites cygnæa and *anatina*, Poiret (1801), *Coq. de l'Aisne*, p. 109.
Anodonta variabilis, Draparnaud (1801), *Tabl. Moll.* p. 108.
Mytilus Avonensis, Montagu (1803), *Test. Brit.* p. 172.
Anodonta cygnæa and *anatina*, Draparnaud (1805), *Hist. Moll.* p. 133. pl. xii. f. 1, 2.

Anodonta anatina, sulcata, and *intermedia*, Lamarck (1819), *Anim. sans vert.* vol. vi. p. 85, 86.

Anodonta Cellensis, C. Pfeiffer (1821), *Deutsch. Moll.* vol. i. p. 110. pl. vi. f. 1.

Mytilus incrassatus and *maculatus*, Sheppard (1821), *Trans. Linn. Soc.* vol. xiii. p. 85. pl. v. f. 4 and 6.

Mytilus dentatus, Turton (1822), *Conch. Dict.* p. 115.

Anodon paludosus and *Avonensis*, Turton (1822), *Biv. Brit.* p. 240, 241. pl. xv. f. 1.

Anodon cygneus and *anatinus*, Turton (1822), *Conch. Brit.* pl. xlvi. p. 239.

Anodonta piscinalis, Nilsson (1822), p. 116.

Anodonta ventricosa and *ponderosa*, C. Pfeiffer (1825), *Deutsch. Moll.* vol. ii. p. 30, 31. pl. iii. and pl. iv. f. 1 to 6.

Anodonta compressa, Ziegler (1831), *Menke, Syn. Moll.* p. 106.

Symphynota cygnea, Lea (1832), *Obs. Unio*, vol. i. p. 70.

Anodonta oblonga and *minima*, Millet (1833), *Mém. Soc. Agr. Angers*, vol. i. p. 242. pl. xii. f. 1, 2.

Anodonta complanata, Ziegler (1835), *Rossm. Icon.* vol. i. p. 112. f. 68.

Anodonta Arelatensis, Jacquemin (1835), *Guide, Voy. Arles*, p. 125.

Anodonta rostrata, Rossmässler (1836), *Icon.* vol. iv. p. 25. f. 284.

Anodonta sinuosa, Manduyt (1837), *Moll. de la Vienne*, p. 15.

Anodonta Rossmassleri, Dupuy (1843), *Moll. Gers.* p. 74.

Anodonta coarctata, Potiez and Michaud (1844), *Cat. Moll. de Douai*, vol. ii. p. 142. pl. lv. f. 2.

Anodonta macilenta, Lusitana, and *ranarum*, Morelet (1845), *Moll. du Port.* p. 102 to 104. pl. xi. and xii.

Anodonta Milletii, Ray and Drouet (1848), *Rev. Zool. Soc. Cuv.* p. 225. pl. i. f. 1, 2.

Anodonta Dupuyi, Ray and Drouet (1849), *Rev. Zool. Soc. Cuv.* p. 14. pl. i. and ii.

Anodonta Grateloupiana, Gassies (1849), *Moll. de l'Agénais*, p. 193. pl. iii. and iv. f. 2.

Anodonta Rayii, Jobæ, and *subponderosa*, Dupuy (1849), *Cat. Extram. Test.* no. 25, 28, 29.

Anodonta Scaldiana and *Moulinsiana*, Dupuy (1851), *Hist. Moll.* vol. vi. p. 613 and 616. pl. xx. f. 19.

Anodonta crassiuscula and *parvula*, Drouet (1852), *Anod. de l'Aube*, p. 5, 9.

Margaron (Anodonta) cygnea, Lea (1852), *Synop. Naid.* p. 47.

Anodonta cygnea, anatina, complanata, variabilis, and *Avonensis*, Moquin-Tandon (1855), *Hist. Moll.* vol. ii. p. 557 to 562. pl. xlii. to xlvi.

Hab. Throughout Europe. Siberia. (In gently flowing rivers, brooks, canals, ponds, etc.)

The foregoing list of synonyms, though sufficiently numerous, is far from being a perfect record of the names that have been given to this protean bivalve. A mollusk of large size, dwelling amid such varied physical conditions as are presented by the places which the *Anodonta* indiscriminately inhabits, would necessarily produce a very variable shell, and some naturalists appear almost to think that each pond, canal, or river, has its peculiar species. Even Linnæus held to the opinion that we have two species in Europe, one of oblong ventricose form, *A. cygnea*, and another of oval compressed form, *A. anatina*, but these are simply the extreme varieties.

The shell of *Anodonta cygnea* is shortly rounded in front, and towards the ventral margin, where the foot is exserted, it is a little gaping. Posteriorly, where the pointed siphonal orifices protrude, the shell slopes obliquely, and is slightly auriculated. The species is very closely represented in the United States by *A. fluviatilis*.

Unio tumidus.

GENUS II. **UNIO**, *Philippson*.

Animal; elongately ovate, lobes of the mantle free, foot tongue-shaped, rather large, branchial and excretory orifices approximating, edges of the mantle around the former moderately fringed.

Shell; very inequilateral, inequivalve, anterior side short, a little gaping where the foot is exserted, posterior side more or less slopingly produced, valves rather thick, pearly, covered with a stout horny epidermis, hinge composed of strong interlocking erect teeth on the anterior side of the umbo, and elongated marginal teeth, which are sometimes obsolete, on the posterior side. Ligament external.

Of the *Unio* or pearl mollusk, we have three species in Britain, two, *U. tumidus* and *pictorum*, inhabiting ponds, lakes, and gently flowing rivers, and confined to England, chiefly the southern and eastern parts, and one, *U. margaritifer*, inhabiting the mountain rapids of all three kingdoms, in places beyond the range of the other two. The posterior teeth in this last species being reduced to mere marginal ridges, it is raised by some authors to the rank of a genus, *Alasmodon*. All three *Uniones* produce pearls, but the mountain species, *U. margaritifer*, surrounded by more turbulent elements than those that dwell in tranquil waters, is the chief contributor, although "one in a hundred might contain a pearl, and about one in a hundred of the pearls might be tolerably clear."

The animal of *Unio* merely differs from that of *Anodonta* in being of a longer subtriangular form, as indicated by the form of the shell; the foot is perhaps smaller, and the fringed edges of the posterior extremity of the mantle lobes are not quite so irregularly jagged. The shell is firmly hinged by strong, erect, interlocking teeth on the anterior side of the umboes, and elongated marginal teeth on the posterior, which, as already stated, are in *U. margaritifer* rudimentary; the want of them is compensated in that species by the presence of a longer and stronger external ligament.

The general distribution of the British species is peculiar. *U. margaritifer* abounds in the hill countries of Scandinavia, as in Britain, beyond the range of its fellows, whose central area of habitation is in France and Germany. *U. tumidus* does not range south of the Alps. *U. pictorum* is diffused throughout Europe, passing into North Africa and Russian Asia. Both species range together in Britain, but are confined to England. In the southern and eastern parts *U. tumidus* and *pictorum* are comparatively abundant, and they are found as far north as the south of Yorkshire. That they are absent from Scotland is not surprising, but it is difficult to account for their absence from the south and south-east of Ireland. "Taking the features of this distribution into consideration," says Edward Forbes in his own quaint manner, "it seems as if the *Unio margaritifer* had migrated southwards from some ancient northern centre, whilst the other *Uniones* and *Anodonta* advanced westwards and northwards, with unequal pace however, since only the last invaded Ireland."

FAMILY NAIADES.

The genus *Unio*—which name, it should be mentioned, is not the feminine noun signifying *union*, but the masculine *unio*, a pearl, in the sense used by Pliny,—has been always attributed to Retzius. Mr. Lea, in the last edition of his 'Synopsis' (p. xiii., note), states, that on a careful examination of the scarce work 'Nova Testaceorum Genera,' published in 1788, it seems to him that the genus belongs to L. M. Philippson, the author of the 'Dissertation,' Retzius being the presiding officer of the institution where this thesis was presented.

The British species of *Unio* are:—

1. **tumidus.** Shell ovately wedge-shaped, glossy verdigris green, rayed on the posterior side.
2. **pictorum.** Shell elongately oblong, fulvous olive, marked with concentric epidermic ridges.
3. **margaritifer.** Shell elongately kidney-shaped, often compressly pinched about the middle, black, posterior elongated tooth obsolete.

1. **Unio tumidus.** *Swollen Unio.*

Shell; ovately wedge-shaped, gibbously swollen, very inequilateral, anterior side short, rounded, a little gaping in the region of the foot, posterior side slopingly acuminated, valves concentrically fibrously ridged at intervals, olive green, glossy, tinged, and rayed on the posterior side with verdigris, umboes confluently undulately wrinkled, hinge composed of an erect shelving finely notched anterior tooth, and a long drawn-out posterior one in the right valve interlocking with two pairs of similar teeth in the left valve.

Unio tumidus and *ovalis*, Philippson (1788), *Nov. Test. Gen.* p. 17.
Mya ovalis, Pulteney (1799), *Cat. Shells, Dorset*. p. 27.
Mya depressa and *ovata*, Donovan (1802), *Brit. Shells*, vol. iii. and iv. pl. cx. and cxxii.
Unio rostratus, Studer (1820), *Kurz. Verz.* p. 93.
Mysca ovata and *solida*, Turton (1822), *Conch. Brit.* p. 216. pl. xvi. f. 2.
Unio inflatus, Hécart (1833), *Mém. Soc. Agr. Valenc.* vol. i. p. 145.
Unio Michaudianus, Desmoulins (1833), *Act. Soc. Linn. Bord.* vol. vi. p. 20.
Unio arcuatus, Bouchard-Chantereaux (1838), *Moll. Pas-de-Calais*, p. 91.
Margaron (Unio) tumidus, Lea (1852), *Synops. Naiad.* p. 36.
Unio (Lymnium) tumidus, Moquin-Tandon (1855), *Hist. Moll.* vol. ii. p. 577. pl. li. f. 11 to 14.

Hab. Northern and Central Europe. England, chiefly southern and eastern parts. (In gently flowing rivers, canals, and ponds.)

It will be seen by a comparison of our very characteristic figures of the shells of *U. tumidus* and *pictorum*, that the valves in the species under consideration are broader from the umbo to the margin and more wedge-shaped, the umbo being more conspicuously undulately wrinkled. Another peculiarity in the shell of *U. tumidus*, consists in the epidermis being of a glossy verdigris colour, more or less confluently painted with rays; and it may be observed that the epidermis is not disposed in such decided equidistant concentric ridges. The anterior hinge teeth are, it may be added, more elevately developed, and interlock deeper with one another. The animal has much the same colouring in both species, plain white edges to the mantle, except in the vicinity of the branchial orifice, where it is tinged with orange-brown, sometimes a little speckled. The internal nacre of *U. tumidus* is mostly of a delicate silvery-grey colour, but Dr. Gray mentions having received specimens collected in a pond in Warwickshire, in which the interior was tinged with salmon-colour. It is seldom that pearls are found in this species, secretion being feeble, and the nacre thin.

The range of *U. tumidus* in Britain is limited to England, chiefly the southern and eastern parts, extending in a northerly direction to the south of Yorkshire. It is not known in Scotland or Ireland. On the Continent it is a Northern and Central, that is to say, a Germanic, species, not having been collected south of the Alps.

2. Unio pictorum. *Painters' Unio.*

Shell; elongately oblong, anterior side moderately short, rounded, a little gaping in the region of the foot, posterior gently sloping, valves concentrically fibrously ridged at intervals, fulvous olive, ridges generally rather conspicuous, umboes but slightly wrinkled, mostly eroded; hinge similar to that of the preceding species, rather less prominently developed.

Mya pictorum, Linnæus (1758), *Syst. Nat.* ed. 10. p. 671.
Unio pictorum, Philippson (1788), *Nov. Test. Gen.* p. 17.
Unio rostrata, Lamarck (1819), *Anim. sans vert.* vol. vi. p. 77.
Mysca pictorum, Turton (1822), *Conch. Brit.* p. 245.
Unio limosus, Nilsson (1822), *Moll. Suec.* p. 110.
Unio Turtoni, Payraudeau (1826), *Moll. de Corse*, p. 65. pl. ii. f. 2, 3.
Unio Requieni and *Deshayesii*, Michaud (1831), *Comp. de Drap.* p. 106 and 107. pl. xvi. f. 24, 30.
Unio longirostris, Ziegler (1842), *Rossm. Icon.* vol. xi. p. 13. f. 738.
Unio Ardusianus, Reyniés (1843), *Lett. Moq.-Tand.* p. 5. pl. i. f. 7, 8.
Unio ponderosus, Spitzi (1844), *Rossm. Icon.* vol. xii. p. 31. f. 767.
Unio Aleronii, Companyo and Massot (1845), *Soc. Agr. Pyr.-Or.* vol. ii. p. 234. f. 2.
Unio dactylus and *mucidus*, Morelet (1845), *Hist. Moll. du Port.* p. 110, 111. pl. xiv. f. 2, 3.
Unio Philippi, Dupuy (1849), *Cat. Extram. Test.* no. 49.
Unio platyrinchoideus, *Roussii*, and *curvirostris*, Dupuy (1852), *Hist. Moll.* vol. vi. p. 649, 658. pl. xviii. f. 16, 18.
Margaron (Unio) pictorum, Lea (1852), *Synops. Naiad.* p. 36.
Unio (Lymnium) pictorum and *Requienii*, Moquin-Tandon (1855), *Hist. Moll.* vol. ii. p. 574 to 576. pl. l. f. 1 to 10. and pl. li. f. 1 to 10.

Hab. Throughout Europe. Siberia. North Africa. England, chiefly southern and eastern parts. (In gently flowing rivers, canals, and ponds.)

The *Painters' Unio*, so called from its shell having been used by painters of the last century for holding their colours, is more plentifully and much more widely diffused than the preceding species. *U. tumidus* is confined to Northern and Central Europe. *U. pictorum* ranges in a southerly direction to Sicily and North Africa, and northwards to Finland, passing over Siberia. In Britain, both species are confined to England, chiefly the southern and eastern parts, but reach sparingly to the south of Yorkshire. Neither of them appear in Scotland or Ireland. Mr. Jeffreys remarks that "the shells are still to be had of any artists'-colour-man in this country, containing a preparation of ground gold and silver leaf, for illuminating work."

The shell of *U. pictorum* may be readily recognized by its straight elongately oblong form, fulvous tone of colour, devoid of rays and concentric epidermic ridges. The umboes in this species are only slightly wrinkled, and the hinge teeth are not so prominent as in the preceding. Forbes and Hanley notice " a somewhat tortuous variety taken in the River Lea, near London, and in the northern districts of England, with the rostrum bending below the level of the incurved ventral margin, and with the hinder side either greatly produced or with its upper edge arcuated. In this form the primary teeth are apt to become rudimentary."

M. Moquin-Tandon divides this species into two, characterizing seventeen varieties of it, as follows:—*U. Requienii*; 1, shell nearly straight below; 2, shell nearly straight, cardinal teeth stronger; 3, shell large, nearly straight, of a bright green colour; 4, shell smaller, paler, slightly dilated behind; 5, shell slightly sinuous below; 6, shell elongated, sinuous below, arcuated behind; 7, shell elongated, sinuous below, a little dilated behind; 8, shell elongated, very sinuous and very arcuated. *C. pictorum*; 1, shell yellow, with green rays; 2, shell yellow, without rays; 3, shell larger, yellow or brownish, with dark bands; 4, shell rather elongated, lanceolate behind; 5, shell longer, rather narrowed posteriorly; 6, shell very large, more elongated, rather narrowed at the posterior extremity, ventricose, thick, brown; 7, shell more elongated, much narrowed behind, olive-brown; 8, shell more oval, rather truncated posteriorly, very dark brown; 9, shell smaller, shortened, a little arched, wedge-shaped posteriorly, olive.

FAMILY NAIADES.

3. **Unio margaritifer.** *Pearl-bearing Unio.*

Shell; oblong or elongately kidney-shaped, very inequilateral, rounded at both ends, anterior end a little gaping in the region of the foot, valves moderately convex, often compressly pinched about the middle, covered with a black densely fibrous epidermis, umboes mostly very much eroded, ligament long and strong, hinge composed of a blunt crested anterior tooth in the right valve, interlocking somewhat rudely with a double, similarly crested tooth in the left valve, spreading at the base in a thickened deposit towards the front adductor muscle, posterior tooth represented in each valve by an obtuse marginal ridge.

Mya margaritifera, Linnæus (1758), *Syst. Nat.* ed. 10. p. 671.
Unio margaritiferus and *crassus*, Philippson (1788), *Nov. Test. Gen.* p. 16.
Unio rugosa, Poiret (1801), *Coq. de l'Aisne*, p. 105.
Margaritana fluviatilis, Schumacher (1817), *Ess. Syst. Test.* p. 124.
Unio elongata and *sinuata*, Lamarck (1819), *Anim. sans vert.* vol. vi. part i. p. 70.
Unio elongatus and *ater*, Nilsson (1822), *Moll. Suec.* p. 106, 107.
Alasmodonta arcuata, Barnes (1823), *Sillim. Journ.* vol. vi. p. 277. pl. xi. f. 20.
Unio crassissima, Férussac (1827), *Desmoul. Moll. Girond.* p. 42.
Alasmodon margaritiferum, Fleming (1828), *Brit. Anim.* p. 417.

Unio Roissyi, Michaud (1831), *Comp. de Drap.* p. 112. pl. xvi. f. 27, 28.
Unio margaritifer, Rossmässler (1835), *Icon.* vol. i. p. 120. pl. iv.
Margaritana margaritifera, Lea (1838), *Trans. Amer. Phil. Soc.* vol. vi. p. 135.
Unio brunnea, Bonhomme (1840), *Mém. Soc. Aveyr.* vol. ii. p. 430.
Alasmodon elongatus, Thompson (1840), *Ann. Nat. Hist.* vol. vi. p. 200.
Unio tristis, Morelet (1845), *Moll. du Port.* p. 107. pl. xiii. f. 2.
Margaron (Margaritana) margaritifer and *crassus*, Lea (1852), *Synops. Naiad.* p. 39, 43.
Unio (Margaritana) margaritifer, Moquin-Tandon (1855) *Hist. Moll.* vol. ii. p. 566. pl. xlvii.
Unio (Lymnium) sinuatus, crassus, and *ater*, Moquin-Tandon (1855), *Hist. Moll.* vol. ii. p. 567 and 570, pl. xlviii. f. 1 to 3, and pl. xlix. f. 3 to 6.

Hab. Throughout Europe. (In rapids and mountain streams.)

Mankind has from a very early period delighted in adorning itself with bits of iridescent shells. There is a peculiar charm in the rainbow colours refracted by the sun's light, from their minutely undulated surface. The South Sea Islander displays the beautiful phenomena of the spectrum in a string of *Avicula* fragments. More civilized races select for their adornment the nacre deposited in superfluous globules, either on the shell's lining, or within the fleshy parts of the animal. These abnormal secretions, the produce of a surcharged or distempered gland, are the lovely pearls which grace the necklace and the diadem. All nacreous bivalves, and even bivalves like the giant clam (*Tridacna*), whose shell is as opake as marble, produce pearls, but the marble pearl is scarce, and when found not much valued. It is the iridescent pearl, especially the pearl that is tinged with rose or salmon colour, apart from its iridescence, which is the most esteemed; and of our four British Naiads, *Unio margaritifer*, with its blush-tinted interior, is the pearl-bearer *par excellence*. *U. tumidus* and *pictorum* are inhabitants of gently flowing and stagnant water; *U. margaritifer* inhabits the mountain stream, where its pearl-secreting functions are more stimulated by the turbulence of the waters. Pearls of great beauty have been procured from specimens inhabiting the Conway, North Wales. One of especial purity, presented by Sir R. Wynne to the Queen of Charles II., is now in the Crown of Queen Victoria.

U. margaritifer is diffused over the whole of Europe, appearing in the South in a robust form, known as *U. crassus*. In Britain it appears in all three kingdoms, chiefly in the mountainous parts, not visited by *U. tumidus* or *pictorum*.

The shell is stout, covered with a dense black fibrous epidermis, and it is usually tinged internally with rose or salmon colour. The hinge is more rudely developed than in the other two species. The front teeth are blunt and irregular, crested with notches, and they are supported by a thicker collateral deposit of shelly matter, reaching to the cicatrix of the anterior adductor muscle. Posteriorly the elongated tooth is reduced, as in *Anodonta*, to a blunt marginal ridge; and to compensate for this the external ligament is longer and stronger.

Family II. **CARDIACEA.**

Animal bearing a thin horny shell, with the mantle lobes open anteriorly for the passage of a large protruded foot, and united posteriorly to form the branchial and excretory siphons which are prolonged into tubes wholly or partially united.

The dykes of Holland and western Germany, and the fens of East Anglia, elicited, forty and thirty years ago, the attention of two able observers to their little bivalve mollusks, in Carl Pfeiffer, of Cassel, and the Rev. Leonard Jenyns, of Cambridge. Before their researches, the little horny cockles of our rivers, canals, and stagnant waters were associated together in one genus. Their mantle lobes are open anteriorly for the passage of a capacious extensible foot, while they are united posteriorly, as well as ventrally, to form the branchial and excretory siphons, which are prolonged into tubes. Both siphons and foot are largely extensible, and the animal has a marvellous power of withdrawing and turning them about, and often performing a whirling rotary motion by forced ejectments of the water. But of some, the siphons are united in one tube, in others, they are united for a little distance and then separated. These two forms of *Cardiacea* constitute the genera *Pisidium* and *Cyclas;* and not only are there differences in the animal, but also in the shell. In *Pisidium*, the anterior side of the shell is the longer; in *Cyclas*, the anterior side is the shorter.

The freshwater cockles are chiefly inhabitants of the temperate regions of the globe in both hemispheres. About an equal number of both kinds are known, from five-and-twenty to thirty species of each, but in the Eastern Hemisphere the number of *Pisidia* compared with *Cyclas*, is in the proportion of one to two, whilst in the Western Hemisphere it is as two to one. We have in Europe about twelve species of *Pisidium* and six of *Cyclas*; in the United States there are about six species of *Pisidium* and twelve of *Cyclas*. Nearly twice as many species are recorded by a recent monographer of the two genera, Mr. Temple Prime, but a careful examination of many of them in Mr. Cuming's collection leads me to think that probably half will have to be suppressed. Both genera appear in New Zealand and Central America, and a species of *Pisidium* has been described by Mr. Benson, from India. They are probably much more generally diffused abroad than we are at present aware of, although they are partially represented in the streams of warm, temperate, and intertropical countries by three important genera of bivalves of large size unknown in Europe—*Cyrena*, *Velorita*, and *Galathæa*.

The British freshwater *Cardiacea* belong to two genera:—

1. **Pisidium.** Animal having the siphons united in one tube. Shell thin, horny, slopingly produced in front.
2. **Cyclas.** Animal having the siphons united in one tube for some distance and then separated. Shell thin, horny, slopingly produced behind.

Pisidium amnicum. (*Slightly enlarged.*)

Genus I. **PISIDIUM**, *C. Pfeiffer*.

Animal; body oval, greyish, delicately transparent, mantle lobes united at each end, foot large, broad, attenuately elongated, siphonal tube sometimes short and obliquely subconical, sometimes elongately conical.

Shell; equivalve, inequilateral, anterior side the longer, more or less slopingly produced, thin, covered with an olive-horny epidermis, valves mostly concentrically ridged, or striated, interiorly bluish-white, umboes sometimes rather prominent, sometimes tumidly obtuse, ligament subexternal, now conspicuous, now barely discernible. Hinge composed of a small double cardinal tooth and two elevated lateral teeth in the left valve, interlocking with another double tooth and four lateral teeth of the same character, in the right valve.

The *Pisidia*, or Pea Cockles, differ essentially from the Cyclads, with which they were formerly confused, both as regard the animal and its shell. Not only do the siphonic tubes differ, as represented in our respective vignettes, the tubes in *Pisidium* being united in one, while the tubes in *Cyclas* are united for a little distance and then separated, but the shells differ also. It will be seen, that in *Pisidium*, the anterior side of the shell, the side from which the foot is protruded, is the longer, whilst in *Cyclas*, the posterior side, the side from which the siphonic tubes are exserted, is the longer. This important distinction, which appears to have escaped the notice of Mr. Jeffreys, is well defined by Forbes and Hanley, and by Baudon. The ligament may be detected, by a careful observer, to be on the shorter side of the shell in *Pisidium*, on the longer side in *Cyclas*. The valves of the shell in *Pisidium* are more depressed than in *Cyclas*, and the species are uniformly smaller, though each genus has a comparatively large species of its own, *P. amnicum* and *C. rivicola*, each larger than any representative in any other geographical province.

The European *Pisidia*, of which some thirty to forty species have been described, are probably reducible to seven or eight, of which we have five in Britain, occurring in all three kingdoms. They are not, however, so generally diffused on the Continent. *P. amnicum*, *nitidum*, and *obtusale* are found throughout Europe, but *P. amnicum* is the only species that ranges into Siberia and North Africa, while *P. obtusale* is confined chiefly to the central parts. *P. pusillum* ranges into Siberia, and is not found in the south of Europe. *P. Casertanum* and *pulchellum* are found in the south, and not in the north. The animal of *Pisidium* is extremely delicate, transparent, and almost gelatinous.

The foreign range of the genus is rather scattered, and apparently very imperfectly known. In addition to about twelve European species of *Pisidium*, there are records in Mr. Cuming's collection of the presence of a species of rather large size in Lycia, collected at an elevation of 6500 feet, one in India, *P. parvula*, Benson, collected by Dr. Bacon, and one in New Zealand, not yet described, collected by Mr. Strange.

The British species of *Pisidium* are:—
1. **amnicum.** Shell large, triangular, anteriorly slopingly produced, strongly concentrically ridged.
2. **obtusale.** Shell very small, suborbicularly ovate, nearly equilateral, moderately striated, umboes obtuse.
3. **pusillum.** Shell very small, obliquely oval, anterior side rather produced, moderately striated, umboes tumid.
4. **nitidum.** Shell very small, suborbicular, concentrically ridged and striated, glossy, umboes obtuse.
5. **Casertanum.** Shell small, triangularly orbicular, very inequilateral, anteriorly produced, finely striated, dull.
6. **pulchellum.** Shell minute, obliquely orbicular, anteriorly obliquely produced, prominently striated, dull.
7. **Henslowianum.** Shell minute, obliquely subtriangular, rhombic, finely concentrically striated, umboes calyculate.

1. **Pisidium amnicum.** *River Pisidium.*

Shell; somewhat triangularly oblique, compressed, solid, cinereous olive, glossy, strongly concentrically ridged and grooved, anterior side slopingly produced, posterior abruptly truncate; umboes comparatively pointed.

Tellina amnica, Müller (1774), *Verm. Hist.* part ii. p. 205.
Tellina striata, Schröter (1779), *Fluss-Conch.* p. 193.
Tellina rivalis, Maton (1797), *Trans. Linn. Soc.* vol. iii. p. 44. pl. xiii. f. 37, 38.
Cyclas palustris, Draparnaud (1801), *Tabl. Moll.* p. 106.
Cardium amnicum, Montagu (1803), *Test. Brit.* p. 86.
Cyclas amnica, Fleming (1814), *Edin. Encyc.* vol. vii. p. 92.
Cyclas obliqua, Lamarck (1818), *Anim. sans vert.* vol. v. p. 559.
Pisidium obliquum, C. Pfeiffer (1821), *Deutsch. Moll.* vol. i. p. 124. pl. v. f. 19, 20.
Pisidium amnicum, Jenyns (1833), *Trans. Phil. Soc. Camb.* vol. iv. p. 309. pl. xix. f. 2.
Pisidium inflatum, Mégerle (1838), *Porr. Mal. Comasc.* p. 121. pl. ii. f. 13.
Pisidium intermedium, Gassies (1849), *Desc. Pisid. Aquit.* p. 11. pl. i. f. 4.

Pisidium Grateloupianum, Normand (1354), *Coup d'œil Cycl.* p. 4.
Musculium amnicum, Adams (1858), *Gen. Rec. Moll.* vol. ii. p. 451.
Hab. Throughout Europe. Siberia. North Africa. (In gentle streams and lakes.)

This *Pisidium* is much larger than any other species, and may be recognized, not only by its size, but by its compressed oblique form and concentric sculpture. Its proportions are more than usually inequilateral, the anterior side being produced into an elongately sloping angle, rounded, while the posterior is shortly truncated. The animal is dingy white; the foot, it may be observed in our vignette, is large and extensible, broad at the base, somewhat pointed at the extremity.

Pisidium amnicum is found in all parts of the British Isles, on the muddy bottom of streams; it does not seem fond of crawling, but moves its capacious foot about in all directions, and often buries itself. It appears throughout the Continent, passing into Siberia and North Africa.

---

2. **Pisidium obtusale.**   *Obtuse Pisidium.*

Shell; suborbicularly ovate, thin, ventricose, dark yellowish green, moderately concentrically striated, nearly equilateral, both sides rounded, the anterior a little obliquely enlarged, umboes obtusely tumidly rounded and rather prominent.

Tellina minuta, Studer (1789), *Coxe, Trav. in Switz.* vol. iii. p. 439 (without characters).
Cyclas obtusalis, Lamarck (1818), *Anim. sans vert.* vol. v. p. 559.
Cyclas minima, Studer (1820), *Kurz. Verz.* p. 93.
Pisidium obtusale, C. Pfeiffer (1821), *Deutsch. Moll.* vol. i. p. 125. pl. v. f. 21, 22.
Cyclas gibba, Alder (1830), *Trans. Nat. Hist. Soc. Northumb.* vol. i. p. 41.
Pisidium fontinale var., Held (1837), *Isis,* p. 306.
Cyclas fontinalis var., Dupuy (1843), *Moll. du Gers,* p. 89.
Pisidium ventricosum, Prime (1852), *Pro. Nat. Hist. Soc. Boston, U.S.* p. 10. pl. xi.
Pisidium pusillum var., Jeffreys (1859), *Ann. Nat. Hist.* vol. iii. p. 37.
Hab. Throughout Europe, principally the central parts. United States. (In shallow pools, drains, and swamps.)

Most observers have had their attention arrested by the subglobular ventricose proportions of this species, with its obtusely swollen umboes, though some have considered it to be a variety of *T. pusillum*, Gmelin (*C. fontinalis*, Drap.). Mr. Jeffreys takes this view of the case, yet he describes it as being distinguished by more active habits, and, in his latest work, as occurring in similar situations with the typical *P. pusillum;* which is more conclusive of its distinctness than his previous statement (Ann. Nat. Hist. 3rd series, vol. iii. p. 37), that he never met with it in company with the typical form. It might, in that case, be a local variety. But the truth appears to be, that it is distributed generally throughout Europe, and Dr. Baudon is of opinion that Mr. Temple Prime's North American *P. ventricosum* is the same species.

---

3. **Pisidium pusillum.** *Little Pisidium.*

Shell; obliquely orbicularly oval, rather ventricose, cinereous-olive, but little glossy, moderately concentrically striated, striæ irregular, inequilateral, both sides rounded, the anterior rather obliquely produced; umboes moderately tumid.

Tellina pusilla, Gmelin (1788), *Syst. Nat.* p. 3231.
Cyclas fontinalis, Draparnaud (1801), *Tabl. Moll.* p. 105.
Pisidium fontinale, C. Pfeiffer (1821), *Deutsch. Moll.* vol. i. p. 125. pl. v. f. 15, 16.
Cyclas pusilla, Turton (1822), *Conch. Brit.* p. 251. pl. xi. f. 16, 17.
Pisidium pusillum, Jenyns (1833), *Trans. Phil. Soc. Camb.* vol. iv. p. 302. pl. xx. f. 4 to 6.
Hab. Northern and Central Europe. Siberia. Throughout Britain. (In ponds or among wet moss.)

In this species, the valves are less ventricose, and more strongly concentrically striated than in the preceding species, and the umboes are less tumid. Mr. Thompson, than whom we have had no more accurate observer of the specific characters and habits of the land and freshwater mollusks, says that in Ireland *P. pusillum* is the most common of the genus, and distributed universally. "It is generally to be met with," he says, "in ponds, drains, etc., but in marshy spots, both in this country and in Scotland, I have found it in company with and adhering to the same stones as land mollusca which inhabit such places, as *Vertigo palustris*, etc. In the north and south of Ireland, I have procured it among moss, which was kept moist only by the spray of the waterfall." Nilsson, the Swedish naturalist, mentions having collected living specimens beneath the bark of trees fallen in damp places. It has rather a northern range, passing into Siberia.

French conchologists characterize four varieties of *P. pusillum*, apart from two varieties of *P. obtusale*. Mr. Jenyns describes the animal as being white, with a short entire-margined siphonal tube, varying in shape from cylindrical to subconic.

4. **Pisidium nitidum.** *Shining Pisidium.*

Shell; suborbicular, compressed towards the ventral margin, ventricose towards the umboes, thin, light horn-colour, glossy, iridescent, concentrically ridged and striated, with a few deep grooves near the umboes, nearly equilateral, sides obliquely rounded, the anterior rather produced; umboes prominent, obtuse.

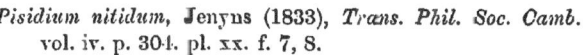

Pisidium nitidum, Jenyns (1833), *Trans. Phil. Soc. Camb.* vol. iv. p. 304. pl. xx. f. 7, 8.
Cyclas nitida, Hanley (1843), *Rec. Biv. Shells*, p. 90.
Pisidium roseum, Scholtz (1843), *Schles. Land und Wass. Moll.* p. 140.
Pisidium incertum, Normand (1854), *Coup d'œil Cycl.* p. 6.
Hab. Throughout Europe. (In stagnant waters.)

Pisidium nitidum, though not described until it was observed in the fens of Cambridgeshire and Lincolnshire in 1833, by Mr. Jenyns, is common in all the stagnant waters of Europe. It is of a rather less oblique form, more orbicular than other species, and it is characterized by a pale glossy iridescent appearance; there is also a marked peculiarity in the presence of from three to five deep grooves in the vicinity of the umboes. Mr. Jenyns notes a difference in the animal which is, perhaps, of yet greater specific importance. "The siphon," he says, "is short and funnel-shaped, with a patulous aperture, of which the margin is more or less crenated or plicated."

In the neighbourhood of Belfast, writes Mr. Thompson, *Pisidium nitidum* is found on *Utricularia vulgaris* growing in stagnant pools excavated in brickmaking.

---

5. Pisidium Casertanum. *Caserta Pisidium.*

Shell; somewhat triangularly orbicular, rather compressed, dull ash-horny, concentrically finely striated, very inequilateral, anterior side obliquely slopingly produced, posterior abruptly truncate; umboes rather obtuse.

Cardium Casertanum, Poli (1791), *Test. utr. Sicil.* vol. i. p. 65. pl. xvi. f. 1.
Pisidium cinereum, Alder (1833), *Cat. Moll. Northumb.* Supp. p. 4.

Pisidium australe, Philippi (1836), *Enum. Moll. Sicil.* vol. i. p. 39.
Cyclas lenticularis, Normand (1844), *Not. Nouv. Cycl.* p. 8. f. 7, 8.
Pisidium thermale, Dupuy (1849), *Cat. Extram. Test.* no. 238.
Pisidium lenticulare, Dupuy (1852), *Hist. Moll.* vol. vi. p. 681. pl. xxx. f. 2.
Pisidium Casertanum, Bourguignat (1853), *Voy. Mer Morte, Moll.* p. 80.
Hab. Central and Southern Europe. Syria. Throughout Britain. (In ponds and gently flowing streams.)

This species is the original *Cardium Casertanum* of Poli, described by the celebrated Neapolitan naturalist towards the close of the last century, from a specimen collected in the great aqueduct of Caserta. Excepting *P. amnicum*, it is the largest of the genus, rather compressed, finely striated, and of a dull ash-horny aspect. MM. Moquin-Tandon and Bandon refer to *P. Casertanum*, the small *P. pulchellum* of Jenyns. Forbes and Hanley believe this to be distinct, and half-a-dozen species, as will be seen by our list of synonyms, have been made of it by some authors. M. Moquin-Tandon remarks that of all the *Pisidia* this species varies the most, according to localities. *P. Baudonianum* and *rotundatum* of Cessac, and *P. sinuatum* of Gassies, he thinks may prove to be merely abnormal forms of it.

I cannot find any evidence of *P. Casertanum* ranging to the northern parts of Europe. It appears in Syria.

6. Pisidium pulchellum.   *Pretty Pisidium.*

Shell; obliquely orbicular, ventricose, sometimes rather compressed, dull horny, concentrically more or less prominently striated, very inequilateral, anterior side obliquely produced, posterior abruptly truncate; umboes sometimes obtuse, sometimes rather acute.

Cyclas fontinalis, Brown (1827), *Edin. Journ. Nat. Sci.* part i. p. 125. pl. i. f. 6, 7 (not of Draparnaud).
Pisidium pulchellum, Jenyns (1833), *Trans. Phil. Soc. Camb.* vol. iv. p. 306. pl. xxi. f. 1 to 5.
Cyclas pulchella, Hanley (1843), *Rec. Biv. Shells*, vol. i. p. 91.
Pisidium Jeannis and *Jenynsii*, Macgillivray (1843), *Hist. Moll. Aberd.* p. 248, 249.
Pisidium fontinale, Brown (1845), *Illus. Conch.* p. 94. pl. xxxix. f. 23.
Pisidium calyculatum, Gassiesianum, Iratianum, and *Normandianum,* Dupuy (1849), *Cat. Extram. Test.* p. 229 to 235.
Pisidium limosum, Gassies (1849), *Moll. de l'Agén.* p. 206. pl. ii. f. 10, 11.
Pisidium tetragonum, Normand (1854), *Coup d'œil Cycl.* p. 5.
Pisidium Casertanum var., Moquin-Tandon (1855), *Hist. Moll.* vol. ii. p. 584. pl. lii. f. 24 to 32.
Pisidium fontinale var., Jeffreys (1862), *Brit. Conch.* p. 20 (not of Draparnaud).

Hab. Central and Southern Europe. Throughout Britain. (In ditches, drains, and other stagnant water, frequently among the roots of aquatic plants.)

P. pulchellum, the smallest of our *Pisidia*, is regarded by some naturalists as a variety of *P. Casertanum*, but there are few who participate in that opinion. Though it differs little except in being smaller, and of a less triangular form, the species is easily recognized. Mr. Macgillivray divides it into as many as three species. His observations on the habits of the animal are among the best that have been made. Speaking of specimens observed in abundance in a ditch, in the neighbourhood of Aberdeen, he says, "When advancing in the water, the animal opens its valves a little, places itself erect by means of the foot, which it gradually protrudes until it considerably exceeds the shell in length. It then contracts and drags the shell quickly forward, after which it is again extended and again contracts. It is not always stretched out in a direct line, but is moved in an undulating manner, often from side to side, and appears to act as a tentacle as well as an organ of motion. The siphonal tube, which is at the same time extended, and kept so, is short, cylindrical, truncate, and undergoes little alteration, a current is seen passing out of it, and minute dark particles frequently escape. In this manner, the animal advances with considerable speed by jerks. At other times it ascends to the surface, where it proceeds in the same manner with the shell reversed, the umboes being beneath."

On the Continent *P. pulchellum* ranges with *P. Casertanum*, through the central and south parts.

---

7. **Pisidium Henslowianum.** *Henslow's Pisidium.*

Shell; rather obliquely subtriangularly orbicular, ventricose, moderately shining, yellowish horny, finely concentrically striated, inequilateral, anterior side slopingly produced, posterior arcuately rounded; umboes furnished with a lamelliform projection.

Tellina Henslowiana, Sheppard (1823), *Trans. Linn. Soc.* vol. xiv. p. 149, 150.
Cyclas appendiculata, Turton (1831), *Man.* p. 15. f. 6.
Pisidium acutum, Pfeiffer (1831), *Wiegm. Archiv.* vol. i. p. 230.
Pisidium Henslowianum, Jenyns (1833), *Trans. Phil. Soc. Camb.* vol. iv. p. 308. pl. xxi. f. 6 to 9.
Pisidium pallidum and *Jaudonianum,* Gassies (1849), *Descr. Pisid. Aquit.* pl. i. f. 10.
Pisidium Recluzianum, Bourguignat (1852), *Journ. Conch.* p. 174. pl. viii. f. 8.
Pisidium Bonnafouxiana, Cessac (1854), *Desc. Nouv. Pisid.* p. 6.
Pisidium Dupuyanum, Normand (1854), *Coup d'œil Cycl.* p. 5.

Hab. Central Europe. England and Ireland. (In ponds and ditches.)

The character upon which *P. Henslowianum* was founded as a species, is that of each umbo being surmounted by a kind of cup-like projection. These, as Dr. Gray observes, are evidently formed on the edge of very young specimens, and gradually rise to the umbo as the shell increases in size by the addition of new shelly matter to its edge. An abnormal secretion of this kind cannot be regarded as a specific character, the species is maintained by Moquin-Tandon on other general grounds, and several varieties of it are characterized, without any such appendage.

The geographical range of the species is not well determined. It has not been collected in Scotland, and its range on the Continent appears to be confined, so far as habitats are recorded, to the central parts.

Cyclas cornea.

Genus II. **CYCLAS.**

Animal; body oval, greyish or dingy white, mantle lobes open in front for the passage of a large, broad, attenuately elongated foot, closed behind to form the siphonal tubes, which are united for a little distance, and then separated, the branchial tube being the longer.

Shell; equivalve, moderately inequilateral, posterior side rather the larger, both sides rounded, sometimes smooth, sometimes concentrically ridged, thin, bluish white, covered by an olive-horny epidermis, ligament sometimes apparent, sometimes barely discernible, hinge composed of a sublamellar double cardinal tooth, and two elevated lateral teeth in the right valve interlocking with another double tooth, and four lateral teeth, of similar character, in the left valve.

We have in Britain five species of *Cyclas*, of which two, *C. cornea* and *lacustris*, are more or less plentiful throughout, the first especially so, in almost endless variety, and both range over Siberia. Of the remaining three, *C. pallida* and *Pisidioides* appear in England only, while *C. rivicola* appears in England chiefly, the only other recorded habitat being the neighbourhood of Dublin. The last-named species, *C. rivicola*, is one of much interest. It is larger than any other *Cyclas* in any part of the world, and it is a native chiefly of our metropolitan rivers. There are no more Cyclads in Europe than are to be found in Britain, and only two additional species have been collected in any other part of our hemisphere, one in Borneo of a compressly grooved type, and one in

New Zealand, which is almost identical, but smaller, with the European *C. rivicola*. In the Western Hemisphere the Cyclads are more numerous, with, however, very little variation of type. There are five well-established species in the Northern United States, and four in the Southern. Mr. Temple Prime has described several more. Three to four species inhabit Jamaica, and Mr. Cuming possesses one Cyclad from Tobasco, Mexico, and another (*C. maculata*, Morelet) collected by himself in Panama.

The animal of *Cyclas*, as already noticed, has the siphonic tubes separated after a little distance, the branchial tube being the longer, and is active in its habits, making vigorous use of its foot, progressing by sudden jerks, performing a rotary motion in the water by forced ejectments of it, floating by its foot in contact with the under surface of the water, and even suspending itself by a few byssus-like threads. The animal is also able to sustain life for some time out of water. Living specimens of *C. maculata* were collected by Mr. Cuming while searching among the roots of trees for land-shells in the garden of a ruined convent in Old Panama, where they had been left after the rainy season, and would doubtless be preserved until its recurrence.

The shell of *Cyclas* is a thin bluish-white substance, covered by a firm greenish-olive or fulvous horny epidermis. The hinge is composed of rather conspicuous elevated lamellar lateral teeth and sublamellar cardinal teeth, the cardinal teeth being more developed in *Cyclas* than in *Pisidium*. The ligament is small and linear, sometimes scarcely apparent on the outside.

The British species of *Cyclas* are:—
1. **rivicola.** Shell large, oval-globose, slightly lunuled, nearly equilateral, densely ridge-striated, ligament apparent.
2. **cornea.** Shell small, suborbicular, nearly equilateral, almost smooth, ligament scarcely apparent.
3. **Pisidioides.** Shell moderate, oblong-oval, posteriorly broadly subtriangularly sloped, concentrically finely ridged.
4. **ovalis.** Shell moderate, oblong-oval, compressed, nearly equilateral, smooth, umboes nearly central.
5. **lacustris.** Shell small, squarely orbicular, rhombic, compressly expanded at the sides, smooth, umboes calyculate.

1. **Cyclas rivicola.** *Brook Cyclas.*

Shell; oval-globose, thin, ventricose, compressed towards the ventral margin, densely concentrically ridge-striated, bluish white, covered with a dark olive-green glossy epidermis, nearly equilateral, anterior side rounded, with a faintly impressed lunule beneath the umboes, posterior rather produced, slightly truncate, umboes obtuse; ligament apparent.

Tellina cornea var., Schröter (1779), *Gesch. Fluss-Conch.* p. 189. pl. iv. f. 4.
Cyclas cornea var., Draparnaud (1801), *Tabl. Moll.* p. 105.
Cardium corneum var., Montagu (1803) *Test. Brit.* p. 86.
Cyclas rivicola, Leach (1818), *Lam. Anim. sans vert.* vol. v. p. 558.
Cyclas sabulicola, Krynicki (1831), *Bourg. Monog. Sphær.* p. 13.
Cyclas æquata, Sheppard (1840), *Gray, Turt. Man.* p. 280.
Sphærium rivicola, Bourguignat (1853), *Rev. Soc. Cuv.* p. 345.

Hab. Central Europe. Metropolitan, Midland, and Northern Counties of England. Neighbourhood of Dublin. (In rivers and canals.)

This species, which is more abundant in our metropolitan rivers, the Thames, the Lea, and the New River, than in any other part of the kingdom, is much the largest species of the genus. Its shell is oval-globose, densely concentrically ridge-striated, of a clear bluish white internally, covered externally by a rich glossy olive-green epidermis. The animal is of a fawn-grey colour, the siphonal tubes being tinged with rose or tawny, while the foot is white.

C. rivicola inhabits many parts of the Continent, including Holland, Belgium, Germany, and France, but it is much less widely diffused than *C. cornea*. It is not found in Scotland, and it is only lately that an Irish habitat, neighbourhood of Dublin, has been attributed to it. It has a very near representative in New Zealand, in a species of rather smaller size not yet described, collected by Mr. Strange; and in the United States in Mr. Say's *C. similis*, inhabiting the muddy banks of the Connecticut.

The *Cyrenæ*, which inhabit the rivers and estuaries of the intertropical parts of both hemispheres, are very closely connected with the *Cycladæ* through this species. They are especially abundant in the Malayan islands and peninsulas, where they live imbedded in the mud of mangrove swamps.

2. Cyclas cornea. *Horny Cyclas*.

Shell; suborbicular, thin, more or less globosely ventricose, very finely concentrically striated, almost smooth, bluish white, covered with a rather dull yellowish horny or olive epidermis, inequilateral, anterior side short, rounded, posterior obliquely rotundately produced; umboes obtuse, ligament scarcely apparent.

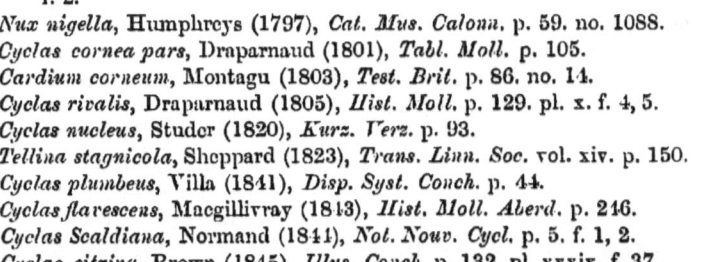

Musculus exiguus, Lister (1678), *Hist. Anim. Angl.* p. 150. pl. ii. f. 31.
Tellina cornea, Linnæus (1758), *Syst. Nat.* ed. 10. p. 678.
Tellina rivalis, Müller (1774), *Verm. Hist.* part ii. p. 202.
Sphærium corneum, Scopoli (1777), *Intr. Nat. Hist.* p. 397. no. 88.
Cardium nux, Da Costa (1778) *Test. Brit.* p. 173. pl. xiii. f. 2.
Nux nigella, Humphreys (1797), *Cat. Mus. Calonn.* p. 59. no. 1088.
Cyclas cornea pars, Draparnaud (1801), *Tabl. Moll.* p. 105.
Cardium corneum, Montagu (1803), *Test. Brit.* p. 86. no. 14.
Cyclas rivalis, Draparnaud (1805), *Hist. Moll.* p. 129. pl. x. f. 4, 5.
Cyclas nucleus, Studer (1820), *Kurz. Verz.* p. 93.
Tellina stagnicola, Sheppard (1823), *Trans. Linn. Soc.* vol. xiv. p. 150.
Cyclas plumbeus, Villa (1841), *Disp. Syst. Conch.* p. 44.
Cyclas flavescens, Macgillivray (1843), *Hist. Moll. Aberd.* p. 246.
Cyclas Scaldiana, Normand (1844), *Not. Nouv. Cycl.* p. 5. f. 1, 2.
Cyclas citrina, Brown (1845), *Illus. Conch.* p. 132. pl. xxxix. f. 37.
Sphærium citrinum and *Scaldianum*, Normand (1854), *Coup d'œil Cycl.* p. 1.

Hab. Throughout Europe. Siberia. (Both in flowing and stagnant water.)

This well-known species abounds in rivers, brooks, lakes, ponds, ditches, and even in drains throughout Europe, from Lapland to the islands of the Mediterranean, and over a large portion of Russian Asia. It is also closely represented in the United States by Mr. Say's *C. partumeia*. Compared with *C. rivicola*, the shell is very much smaller and more globose, but it is extremely variable. Sometimes additions are made to the margin of the valves, generally of a paler colour than the rest, throwing the valves into a convexity, which makes the shell almost as broad as it is long, while sometimes, as in the specimen represented in our animal-vignette, the valves are compressed with the umboes almost beaked. The concentric sculpture of the shell is generally fine, sometimes almost obsolete; the epidermis is mostly rather dull.

The animal of *C. cornea* is of a dingy white, with the siphonal orifices tinged with flesh-colour. In a drawing of Mr. Berkeley's now before me, they are quite pink, while the foot is white. It has the habit of floating on the water by means of the under surface of the foot like *Physa*.

M. Moquin-Tandon describes the variety named by M. Normand *Scaldiana* as being larger and ruder, with the umboes more inflated, and offering a passage towards *C. rivicola*, especially in having the ligament partially external.

3. **Cyclas Pisidioides.** *Pisidium-like Cyclas.*

Shell; oblong-oval, subtriangular, thin, ventricose, concentrically finely ridged, bluish-white, covered with a glossy yellowish-olive epidermis, inequilateral, rounded in front, produced into a broad subangular slope behind, umboes moderately convex.

Sphærium Pisidioides, Gray (1856), *Ann. Nat. Hist.* 2nd series, vol. xviii. p. 25.
Cyclas Pisidioides, Gray (1857), *Turt. Man.* p. 255.
Sphærium corneum var, Jeffreys (1862), *Brit. Conch.* p. 6.

Hab. Paddington Canal.

I give this species *quantum valeat*. If, as Mr. Jeffreys states, it is a variety of *C. cornea*, in which the posterior side is more than usually produced into a slope, the concentric striæ coarser, and the ligament slightly perceptible on the outside, it is a very extreme variety. These are features partaking more of the character of *C. rivicola*, and my example of *C. Pisidioides*, here figured from the collection of Dr. Battersby, has more the appearance of that species than of *C. cornea*. It was described in 1856 from specimens collected in the Paddington Canal.

Dr. Gray describes the animal of *Cyclas Pisidioides* as having the siphons united nearly to the end, the upper shorter, subconic; orifices circular, simple, the lower rather large, about twice the length of the upper when expanded.

4. **Cyclas ovalis.** *Oval Cyclas.*

Shell; oblong-oval, thin, rather compressed, smooth, marked concentrically at intervals with striæ of growth, bluish white, covered with a pale yellowish-ash epidermis, nearly equilateral, sides expandedly rounded, the posterior being the larger, umboes rather swollen, nearly central, ligament slightly apparent.

Cyclas lacustris, Draparnaud (1805), *Hist. Moll.* p. 130, pl. x. f. 6, 7 (not *Tellina lacustris* of Müller).
Cyclas ovalis, Férussac (1807), *Ess. Méth. Conch.* 2nd ed. p. 128 and 136.
Cyclas consobrina, Férussac (1818), *Blainv. Dict. Sci. Nat.* vol. xii. p. 279.
Sphærium Deshayesianum, Bourguignat (1853), *Rev. Soc. Cuv.* p. 345.
Sphærium ovale, Bourguignat (1854), *Mém. Soc. Phys. Bord.* vol. i.
Sphærium pallidum, Gray (1856), *Ann. Nat. Hist.* 2nd ser. vol. xvii. p. 465.
Cyclas pallida, Gray (1857), *Turt. Man.* p. 254. f. 61.
Hab. Central Europe. England. (In canals and ponds.)

We are indebted to M. Bourguignat for having pointed out that this species is the *Cyclas lacustris* of Draparnaud, which is not the *Tellina lacustris* of Müller, a species more generally known as *C. calyculata,* Draparnaud. Férussac made this observation within two years of the publication of Draparnaud's mistake, and gave the species the name, soon after abandoned, which is now restored to it. The shell is of a rather compressed oval-oblong form, nearly equilateral, and almost smooth, the surface being marked at distant intervals with rather prominent lines of growth.

Among the chief characters of the species is its pale drab colour. Six years since, some specimens of *C. ovalis* brought to Dr. Gray from the Grand Junction Canal, near Kensal Green, were observed to be distinct from any other species described in his 'Manual,' and were named by him *C. pallida.* The animal he described as having the foot large, and the siphons united nearly to the tip with the apices conical. Mr. Daniel also collected specimens in the Surrey Canal. "A living specimen," says Mr. Jeffreys, "taken early in February, and kept in a vessel by itself, gave birth about three weeks afterwards to some young ones at intervals of two or three days. Immediately on being excluded they were very active, and used their long foot as an organ of progression, by extending it to its full length; and after attaching its point to the bottom of the vessel like a leech, they drew up their shell to it."

5. Cyclas lacustris. *Lake Cyclas.*

Shell; somewhat squarely orbicular or rhombic, very thin, but little ventricose, light ash-horny substance, nearly equilateral, sides somewhat compressly expanded at the margin, posterior side rather the larger; umboes nearly central, rather prominent, capped with the calyculate shell of the young fry.

Tellina lacustris, Müller (1774), *Verm. Hist.* pars ii. p. 204 (not of Draparnaud).
Cardium lacustre, Montagu (1803), *Test. Brit.* p. 89.
Cyclas calyculata, Draparnaud (1805), *Hist. Moll.* p. 130. pl. x. f. 14, 15 (not of C. B. Adams).
Cyclas lacustris, Férussac (1807), *Ess. Méth. Conch.* p. 128. no. 4 (not of Draparnaud).
Tellina tuberculata, Alten (1812), *Syst. Erd. und Fluss-Conch.* p. 4. pl. i. f. 1.
Tellina tenera, Schrank (1814), *Gärtn. Ann. Méth.* p. 316. no. 2.
Cyclas Ryckholtii, Normand (1844), *Nat. Nouv. Cycl.* p. 7. f. 5.
Cyclas Terveriana, Dupuy (1849), *Cat. Extram. Test.* no. 87.
Sphærium lacustre, *Ryckholtii*, and *Terverianum*, Bourguignat (1853), *Rev. Soc. Cuv.* p. 345.
Sphærium Brochonianum, Bourguignat (1854), *Mém. Soc. Sci. Bord.* vol. i.
Sphærium Creplini and *Jeannotii*, Normand (1854), *Coup d'œil Cycl.* p. 2.
Hab. Throughout Europe. Siberia. Kamtchatka. North Africa. England and Ireland. (In lakes, canals, ponds, etc.)

Cyclas lacustris is distinguished from the rest of the Cyclads by the square or rather rhombic form of the shell, arising from its thinly compressed expansion at the sides. It is also characterized more than any other species of the genus by a cup-like capping of the umboes, which is the embedded shell of the young fry. In colour it is a semi-transparent ash-tinged horn, smooth, but marked at intervals, commonly rather distant, with lines of growth. The animal is greyish-white, sometimes tinged with rose.

DISTRIBUTION

Name.	In Britain.	In Europe.
Achatina acicula	England and Ireland	Throughout
Acme lineata	Throughout	Throughout
Ancylus fluviatilis	Throughout	Throughout
—— lacustris	Throughout	Throughout
Anodonta cygnea	Throughout	Throughout
Arion ater	Throughout	Throughout
—— hortensis	Nearly throughout	Nearly throughout
Assiminea Grayana	Greenwich and Woolwich	
Azeca tridens	England	Central
Balea perversa	Throughout	Throughout
Bulimus acutus	South and west of England and Ireland	South and West
—— montanus	South and west of England and Ireland	Central
—— obscurus	Throughout	Throughout
Bythinia Leachii	Throughout	Throughout
—— similis	England	Central and South
—— tentaculata	Throughout	Throughout
Carychium minimum	Throughout	Throughout
Clausilia biplicata	Central and south of England	Central
—— laminata	England, chiefly south. Ireland, rare	Nearly throughout
—— perversa	Throughout	Throughout
—— Rolphii	Middle and south of England	Central
Conovulus bidentatus	North, south, and west of England and Ireland	South and West
—— denticulatus	South and south-west of England	South and West
—— myosotis	South and south-west of England	Central and South
Cyclas cornea	Throughout	Throughout
—— lacustris	England and Ireland	Throughout
—— ovalis	England	Central
—— Pisidioides	South of England	. . . ?
—— rivicola	Metropolitan, middle, and north of England. Ireland, Dublin	Central
Cyclostoma elegans	England, south of Yorkshire	Central and South
Dreissena polymorpha	Throughout	Throughout

OF SPECIES.

In other Parts of the Globe.	Habits.
Syria. Asia Minor. North Africa. Madeira.	Of secluded habits, buried under stones or in the soil.
Siberia	In damp places, under stones and among moss.
North Africa. Madeira .	Adhering to stones and plants in shallow water.
. . . .	Adhering to stems and leaves of plants in shallow water.
Siberia	In gently flowing rivers, brooks, canals, ponds, etc.
. . . .	In woods and gardens in shady places, about wells or pumps.
. . . .	In gardens and roadside meadows, in damp places.
. . . .	On muddy banks and inlets of the Thames.
. . . .	Widely sparingly distributed in wooded districts.
Azores. Madeira . . .	Among moss and lichens in crevices of walls.
. . . .	Abundant in chalky or sandy places near the sea.
. . . .	In wooded districts, among decayed leaves.
Siberia	On old walls, under stones, or among trees, moss, etc.
North Africa	In gentle streams and still water.
Siberia	In brackish ditches near the Thames at Greenwich.
. . . .	In gentle streams and still waters.
Siberia. North Africa .	Among moss or at the roots of grass in wet places.
. . . .	In woods and hedges, among the roots of shrubs.
. . . .	Mostly in beech woods, about the trunks of trees.
. . . .	In crevices of walls, rocks, and trees, or under stones.
.	Among dead leaves or beneath the bark of trees.
. . . .	In crevices of rocks and timber, among wet moss, or under stones on the banks of rivers near the sea.
. . . .	In mud or among the roots of plants near the sea.
. .	In mud or under stones about the mouths of tidal rivers.
Siberia	In all waters, both flowing and stagnant.
Siberia. North Africa .	In lakes, canals, ponds, etc.
. . . .	In canals and ponds.
. . . .	In canals.
.	In rivers and canals.
Canary Islands	Under stones and about the roots of shrubs.
. . . .	In canals, rivers, docks, aqueducts, water-pipes, etc.

Name.	In Britain.	In Europe.
Geomalacus maculosus	Ireland, co. Kerry	
Helix aculeata	Throughout	Throughout
—— arbustorum	Throughout	Throughout
—— aspersa	Throughout	Central and South
—— Cantiana	England, chiefly south-east	Central and South
—— Carthusiana	England, chiefly south-east	Central and South
—— cricetorum	Nearly throughout	Central and South
—— fasciolata	Throughout, but local	Central and South
—— fulva	Throughout	Throughout
—— fusca	Throughout	West of France, near the sea.
—— hispida	Throughout	Central and South
—— lamellata	North Britain	North of Germany, Sweden
—— lapicida	Central and south of England	Northern and Central
—— nemoralis	Throughout	Throughout
—— obvoluta	England, Hants	Central
—— Pisana	West of England. South-east of Ireland. Jersey	Central and South
—— pomatia	South of England	Central, but local
—— pulchella	Throughout	Throughout
—— pygmæa	Throughout	Throughout
—— revelata	South-west of England. Channel Islands	France
—— rotundata	Throughout	Throughout
—— rufescens	Central and south of England and Ireland	Central and South
—— rupestris	Throughout	Central and South
—— sericea	England, chiefly west and south	Throughout, but local
—— virgata	Nearly throughout	Central and South
Limax agrestis	Throughout	Throughout
—— brunneus	North-east of England	West of France
—— cinereus	Throughout	Throughout
—— flavus	Throughout	Throughout
—— gagates	Throughout, but local	Central, South, and West
—— marginatus	Throughout	Central
—— Sowerbyi	England, chiefly south	Central
—— tenellus	North of England	South of France
Lymnæa auricularia	Middle and south of England and Ireland	Nearly throughout, local
—— glabra	England	North and Central
—— glutinosa	England	Sweden to the Pyrenees
—— involuta	Ireland, Killarney	

In other Parts of the Globe.	Habits.
. . . .	Among moss at the shady base of moist rocks.
Azores	Under dead leaves, moss, and stones, in damp places.
. . . .	In woods and gardens, or under rocks, in damp places.
North Africa. Azores.	
United States. Brazil .	In gardens and woods, in crevices of old walls, etc.
. . . .	Among hedges, chiefly in chalk districts.
. . . .	Among grass, chiefly on the chalk downs.
. . . .	Chiefly on heaths and downs inland or near the sea.
North Africa	Chiefly in sand-banks and under stones near the sea.
Siberia. Azores . . .	Under stones, moss, or leaves, in damp shady places.
. . . .	In woods and bushy places, under leaves or brambles.
Siberia	Under stones, fallen trees, decaying leaves, etc.
. . . .	In woods, among dead leaves
. . . .	In chalky districts, chiefly among rocks.
United States	In gardens, woods, fields, chalk cliffs, quarries, etc.
. . . .	Among moss at the roots of trees.
North Africa. Azores.	
Canaries	In fields and on sand-hills near the sea.
. . . .	In woods and pathway-hedges.
Azores. Madeira. Siberia.	
Thibet. United States .	Under stones or wood, or on walls, wet or dry.
Azores. Siberia . . .	Under stones or leaves in moist woods or in grass.
. . . .	Among grass or at the roots of shrubs.
Azores	Under stones and ruins, or about the bark of trees.
North Africa. Madeira .	In fields, gardens, and hedges, in damp places.
Madeira	Chiefly among rocks in mountainous places.
Irkutsk, Siberia. Caucasus	Vicinity of damp mossy banks or under stones.
North Africa	Chiefly in sandy and chalky districts near the sea.
Madeira. United States .	Abundant in fields, woods, and gardens.
. . . .	In damp woods or on the banks of streams.
Syria. Madeira	In gardens, about outhouses, and in cellars.
Syria. Madeira	In caves in woods, about outhouses, cellars, wells.
Madeira	By the roadside or at the base of old walls.
. . . .	Mostly on trees, especially beech, ash, and walnut.
. . . .	In gardens.
. . . .	In woods.
Siberia. Cashmere . .	In ponds, lakes, and marshes, or on aquatic plants.
. . . .	Partially diffused in drains, ditches, and shallow pools.
Syria	In stagnant pools and ditches, local.
. . . .	In a small lake on Cromaylaun mountain.

Name.	In Britain.	In Europe.
Lymnæa limosa	Throughout	Throughout
—— palustris	Throughout	Throughout
—— stagnalis	England, chiefly middle and south. Ireland.	Nearly throughout
—— truncatula	Throughout	Throughout
Neritina fluviatilis	Chiefly south-west and east of England and Ireland	Throughout
Paludina contecta	England, chiefly middle and south	Chiefly Central
—— vivipara	England, chiefly middle and south	Chiefly Central
Physa hypnorum	Throughout	North and Central
—— fontinalis	Throughout	Throughout
Pisidium amnicum	Throughout	Throughout
—— Casertanum	Throughout	Central and South
—— Henslowianum	England and Ireland	Central
—— obtusale	Throughout	Throughout
—— nitidum	Throughout	Throughout
—— pulchellum	Throughout	Central and South
—— pusillum	Throughout	North and Central
Planorbis albus	Throughout	Throughout
—— carinatus	England and Ireland	Throughout, but partial
—— complanatus	Throughout	Throughout
—— contortus	Throughout	Throughout
—— corneus	Middle and south-east of England and Ireland	Nearly throughout
—— crista	Throughout	Throughout
—— fontanus	Throughout	Throughout
—— glaber	Throughout	Throughout
—— nitidus	Throughout, chiefly south	Chiefly Central
—— spirorbis	Throughout	Throughout
—— vortex	Throughout	Throughout
Pupa Anglica	Throughout	Portugal
—— cylindracea	Throughout	Throughout
—— muscorum	Throughout	Throughout
—— secale	England, chiefly south and west	Central and South
Succinea elegans	Throughout	Throughout
—— oblonga	Throughout, rare and local	Throughout
—— putris	Throughout	Throughout
Testacella haliotidea	England and Ireland, chiefly south	Central and South
—— Maugei	Central and south of England and Ireland	Central and South
Unio margaritifer	Throughout	Throughout

In other Parts of the Globe.	Habits.
Siberia. Thibet. Afghanistan	In ponds, lakes, ditches, springs, at various elevations.
Siberia. North Africa	In shallow muddy waters.
Siberia. Cashmere	In slow streams, canals, ponds, marshes.
Siberia. North Africa. Afghanistan	On the muddy margins of stagnant and slow waters.
Asia Minor. North Africa	On the gravelly bottom of rivers, in warm springs, etc.
. . . .	In canals, ponds, and gently running waters.
United States	In canals, ponds, and gently running waters.
Arctic Siberia	In ponds and ditches, on blades of grass and other plants.
. . . .	On aquatic plants, both in stagnant and running water.
Siberia. North Africa	In gently flowing streams and lakes.
Syria	In ponds and gently flowing streams.
. . . .	In ponds and ditches.
United States	In shallow ponds, drains, and swamps.
. . . .	In ditches, drains, and other stagnant water.
. . . .	In ditches, drains, and other stagnant water.
Siberia	In ponds, ditches, swamps, or among wet grass.
Siberia. North Africa	Everywhere common on water plants.
Siberia	In stagnant marshes, ponds, ditches, etc.
Siberia. North Africa	Everywhere in marshes, ponds, canals, ditches, etc.
Siberia	In ponds, ditches, stagnant marshes, etc.
Siberia. North Africa	In muddy ponds and ditches.
North Africa	On water flags in ponds and ditches.
Siberia	On water plants in ponds and clear streams.
North Africa. Madeira	On water plants in ponds, lakes, and marshes.
. . . .	On duckweed and other plants in ponds and ditches.
Siberia. North Africa	On plants in shallow stagnant water.
Siberia. North Africa	On plants in shallow stagnant water.
North Africa	Under stones, and among dead leaves and moss.
North Africa	Under stones and among dead leaves and moss.
Siberia	Under stones and among dead leaves and moss.
. . . .	Under stones, among rocks, trees, moss, chiefly in chalk-districts.
. . . .	On mud and water flags in wet places.
. . . .	On plants in sandy places near the sea.
. . . .	On mud and flags in wet places.
North Africa. Canaries	In gardens, burrowing deep into the ground.
Canaries	In gardens, burrowing deep into the ground.
. . . .	In rapids and mountain streams.

Name.	In Britain.	In Europe.
Unio pictorum	England, chiefly south and east	Throughout
—— tumidus	England, chiefly south and east	North and Central
Valvata cristata	Throughout	Throughout
—— piscinalis	Throughout	Throughout
Vertigo alpestris	North of England	North and Central
—— antivertigo	England and Ireland	Throughout
—— edentula	Throughout	North and Central
—— minuta	Throughout, rare	Throughout
—— Moulinsiana	West of Ireland?	Central
—— pusilla	England and Ireland, chiefly south and east	North and Central
—— pygmæa	Throughout	Throughout
—— striata	Throughout	Central
—— vertigo	West of England and Ireland	Central
Vitrina pellucida	Throughout	North and Central
Zonites alliaria	Throughout	Throughout
—— cellarius	Throughout	Throughout
—— crystallinus	Throughout	Throughout
—— excavatus	Throughout, but local	North and Central
—— nitidulus	Throughout	Throughout
—— nitidus	Throughout	Central and South
—— purus	Throughout	North and Central
—— radiatulus	Throughout	Central
Zua subcylindrica	Throughout	Throughout

TOTAL.

Species inhabiting Britain throughout 76
Species inhabiting England and Ireland, only 20
Species inhabiting England only 29
Species inhabiting Ireland only 3

128

Of our 128 British species, there are, therefore, in
    Scotland 76
    England 125
    Ireland 99

In other Parts of the Globe.	Habits.
Siberia. North Africa	In gently flowing rivers, canals, and ponds.
	In gently flowing rivers, canals, and ponds.
Siberia	Adhering to submerged stones and sticks in gentle water.
Siberia. Asia Minor	Adhering to submerged stones and sticks in gentle water.
. . . .	Under stones among dead leaves on the hills.
. . . .	In marshy places, on water flags and other plants.
Siberia	Under leaves or about the roots of grass in wet places.
. . . .	Under stones in damp shady places on hills.
. . . .	Under stones in marshy places.
.	
. . . .	Among wet moss or under stones in woods.
Siberia. Azores . . .	Under logs of wood or stones in wet places.
. . . .	Under stones or leaves or at the roots of grass in wet places.
. . . .	Among the roots of grass in wet places.
. . . .	Under stones, leaves, or moss.
. . . .	In gardens, among leaves and under stones.
Madeira. United States.	
South Africa. New Zealand	Chiefly in cellars, drains, and under loose bricks or stones.
. . . .	Under stones and among moss.
. . . .	Under decaying timber, etc.
. . . .	Among moss and under stones in sheltered places.
Thibet. United States	Under stones in shady places in pine-beds and orchid-houses.
. . . .	In woods, under decaying trees, leaves, etc.
United States	Among moss and under stones.
Siberia. Cashmere. Thibet. North Africa. Madeira.	
United States	Under stones, logs of wood, and leaves, wet and dry.

TOTAL.

Of which 43 are inhabitants of land, and 30 of fresh water.
Of which 12 are inhabitants of land, and 8 of fresh water.
Of which 19 are inhabitants of land, and 10 of fresh water.
Of which 2 are inhabitants of land, and 1 of fresh water.

79   49

Of which 43 are inhabitants of land, and 30 of fresh water.
Of which 77 are inhabitants of land, and 48 of fresh water.
Of which 60 are inhabitants of land, and 39 of fresh water.

Eastern Hemisphere.

Name.	Countries in which there are other Species.	
	Of European type.	Of other types.
Achatina	Ceylon, Hindostan	West Africa.
Ancylus	Siberia, Teneriffe, Natal, Bengal, New Zealand.	Tasmania.
Anodonta	Siberia.	Various.
Arion	......?	......?
Assiminea		
Azeca	North Africa, Madeira, Canaries.	
Balea	Azores, Madeira, Tristan d'Acunha	
Bulimus	Asia Minor	Various.
Bythinia	Asia, North Africa.	
Carychium	Siberia, India, North Africa.	
Clausilia	Asia Minor, Madeira, Egypt, India, China, Japan	Cambojia. Siam.
Conovulus		
Cyclas	New Zealand	Borneo.
Cyclostoma	North Africa	Madagascar.
Dreissena		Senegal.
Geomalacus		
Helix	Various	Various.
Limax	New Zealand, Mauritius, Ascension Island, South Africa.	
Lymnæa	Western Asia, Thibet	India, Malayan Islands, Australia.
Neritina	Asia Minor, Egypt	Philippines, Madagascar, Ceylon, Hindostan, Africa, Australia.
Paludina	Siberia, Japan, Hindostan, Philippine Islands, Central Africa	Siberia, Japan, Siam, Bengal, China, Philippine Islands.
Physa	Syria, China, Singapore, Philippine Islands	Egypt, Natal, Australia, New Zealand.
Pisidium	Lycia, India, New Zealand	......?
Planorbis	Ceylon, India, Burmah, South Africa, Australia.	
Pupa	Asia Minor, Madeira, Canaries, India, Ceylon	Asia Minor, Madagascar, South Africa, Australia, New Zealand.
Succinea	India, China, South and West Africa, Australia, Tasmania.	
Testacella	Canaries.	
Unio	Various	Various.
Valvata	Siberia.	
Vertigo	......?	......?
Vitrina	Madeira, Philippines, Ceylon, India, Siam, Malayan Islands, Australia, New Caledonia.	
Zonites	......?	......?
Zua	Madeira.	

OF GENERA.

Western Hemisphere.

Name.	Countries in which there are other Species.	
	Of European type.	Of other types.
Achatina	Central America, West Indies	Central America, West Indies.
Ancylus	Bahia, Chili, Central America, West Indies, United States, Oregon, Newfoundland.	
Anodonta.	United States.	Various.
Arion	Puget Sound.	
Assiminea.		
Azeca		
Balea		
Bulimus	Various	Various.
Bythinia	North America.	
Carychium.		
Clausilia	West Indies and Central America.	
Conovulus.		
Cyclas	United States, Jamaica, Panama, Mexico.	
Cyclostoma	West Indies and Central America, Florida	West Indies and Central America.
Dreissena.		West Indies, Central America, South United States.
Geomalacus.		
Helix	Various	Various.
Limax	North and South America.	
Lymnæa	United States, Central America.	
Neritina	West Indies	Polynesia, West Indies, Central America.
Paludina	United States.	United States.
Physa	Bahia, West Indies, Central America, Mazatlan	Bahia, West Indies, Central America, Mazatlan.
Pisidium	Greenland, United States, Honduras, Valparaiso, Coquimbo.	
Planorbis	Bahia, West Indies, United States.	Chili, Bahia, West Indies, United States.
Pupa	West Indies, United States	West Indies, United States, Bolivia.
Succinea	Polynesia, West Indies	Polynesia, West Indies, Bolivia, Mexico.
Testacella	West Indies.	
Unio	Various	Various.
Valvata	West Indies, United States	Guatemala, United States.
Vertigo	. . . . . ?	. . . . . ?
Vitrina	United States.	
Zonites	. . . . . ?	. . . . . ?
Zua	. . . . . ?	

DISTRIBUTION AND ORIGIN OF SPECIES.

THE geographical distribution of animals over the terrestrial portion of the globe, including its lakes, rivers, and stagnant pools, is pretty well known as regards mammals, birds, reptiles, and fishes, but of mollusks the record is as yet imperfect. Mr. Woodward has given in his 'Manual' a sketch of the general distribution of molluscan life over the globe in geographical provinces and regions, illustrated by lists of genera; and I had the pleasure of communicating, in 1851, to the 'Annals and Magazine of Natural History,' an outline of the division of the terrestrial portion in provinces, founded on a detailed comparison of species. Objections were, I believe, raised to my sketch of a system of geographical provinces, on the ground that it was based on the distribution of a single genus. That genus, however, *Bulimus*, numbering eight hundred species, is almost world-wide in its geographical range; and I found on the completion of my monograph of the great genus *Helix*, which is cosmopolitan, and numbers after very considerable reduction fifteen hundred species, that the distribution of the two genera coincide, with the exception that *Helix* has an important additional province in North America. Inland mollusks are distributed throughout the globe in geographical provinces, over which the species appear to disperse through migration or transport, from that point or centre of each province in which the species are most numerous, and most highly developed. Islands that have been upheaved from beneath the bed of the ocean have a molluscan fauna of their own. Madeira and the Galapagos Islands, for example, possess an aboriginal offspring, small and of insular character, quite distinct from the few species that have become colonized in them by transmission from other lands.

The British Isles, geologists tell us, once formed a portion of the European Continent. The character and distribution of their land and freshwater mollusks, which are Continental species in diminished number and variety, go to support that view. The great Caucasian province, of which they are outlying fragments, embraces the whole of Europe and Russian Asia, with a portion of Japan, the whole of North Africa, Asia Minor, Arabia, Persia, Thibet, and a portion of India. The specific centre of its molluscan population is about the region of Hungary and the Caucasus. In India and beyond Thibet the species commingle with those of another province, the Malayan, which embraces Siam, Cambojia, Malacca and the Indian Archipelago, China, probably, in part, for we know little or nothing as yet of either the fauna or flora of that country, and Corea, with the remaining portion of Japan, and whose specific centre is about the region of Borneo, and the Molucca and Philippine groups. A partial subdivision of the Caucasian province of distribution is indicated in Southern Europe by the chain of the Pyrenees, Alps, and Carpathians; and the districts north and south of that mountain-barrier have been designated the Germanic and Lusitanian regions. It will, however, be seen that the land and freshwater mollusks on either side of this chain are of the same ideal type.

Edward Forbes, as all naturalists are aware, propounded a theory, supported in a most comprehensive manner by established geological theories, to the effect that the British Isles were peopled with animals and plants from the European continent, partly by transmission on floating masses of ice, chiefly by transmission through migration before the land became isolated. His theory was based on the general and traditional belief of mankind that all the individuals of a species have descended from a single progenitor (or pair); and the point in the geographical province at which it might be assumed there was originally the most numerous and highly developed assemblage of progenitors, he called its specific centre. During the sixteen years which have elapsed since this theory was broached, and more especially since the loss of its illustrious founder, great progress has been made in the accession and collation of materials indispensable for its consideration. The desire which it has elicited among naturalists to inquire into the origin of species, has stimulated a more careful registration of habitats, and more detailed comparisons of local faunas and floras; and a vast amount of additional information has been collected on the general distribution over the globe of mollusks. Only sixteen of the British land mollusks are adduced in evidence by Edward Forbes, in his celebrated memoir (Mem. Geol. Surv., vol. i. p. 336), while the freshwater species are not alluded to in any way, and the idea of a distribution of individuals of Old World species in the New, and *vice versâ*, is not allowed to either kind. "Species of opposite hemispheres," says the author "placed under similar conditions, are representatives, and not identical." The distribution of species over the globe, so far as I am able to gather from the land and freshwater mollusks, appears to require that we should take for granted the doctrine of a plurality of progenitors for each species; and the term 'specific centre' to indicate merely that point of a geographical province, in which the species are most numerous and come nearest to our notion of an ideal type of the group.

The doctrine of more than one point of origin for a species, considered with reference to the typical character and distribution of land and freshwater mollusks, rests mainly on the following propositions:—

1. Land species with greater facilities of migration than freshwater species, are less widely and evenly diffused.
2. Land and freshwater species of opposite hemispheres are not always representatives, but are sometimes identical.
3. The range of land and freshwater species over areas (zoological provinces) indicated by uniformity of type, is not arrested by the intervention of sea.

§ 1.

Land species with greater facilities of migration than freshwater species are less widely and evenly diffused.

The doctrine of the migration of all the individuals of a species from a single parent (or pair), involves the conclusion that species, permanent, as I think, in their character, and immutable. diminish in number in their march from the specific centre of a province towards its confines. Out of many hundreds of land mollusks inhabiting the Caucasian province at its specific centre, only ninety have reached the British Isles, of which thirty-five stop short of Scotland, and nineteen of Ireland. Their progress northwards, it may be argued, is arrested to a great extent by change of climate, and in all directions by foes, by mountain barriers, by rivers, and by other physical and unknown causes. It will readily be conceded that land species have greater facilities of locomotion than freshwater species, particularly species inhabiting stagnant ponds and ditches; and it should follow, according to the doctrine of migration, that the further off freshwater species are from the specific centre of a province, the more diminished in number than land species they would be. The very contrary is the fact. Out of five hundred and sixty species of *Helix* inhabiting the Caucasian province, a very large proportion of which are assembled at its specific centre, we have but twenty-four in Britain, of which only eleven range throughout. The disproportion in number of the species of *Clausilia* is larger still. This genus is especially populous at its specific centre. Between two and three hundred species inhabit Austria and Hungary, yet we have but four in Britain, of which only one ranges throughout.

Let us now turn to the sluggish mud-dwelling *Lymnæacea* of the ponds and ditches of the province. There are not six species, it may be safely stated, in all Europe more than there are in Britain. They have no particular centre of creation. There is no evidence to show whether the alleged primogenitors of our British species were created in Siberia, Hungary, or Thibet. There is scarcely any variation either in the form or number of the species in those remote localities. Of *Planorbis* scarcely more than fifteen species inhabit the whole Caucasian province, and we have eleven of them in Britain, all ranging throughout, with the exception that two, *P. carinatus* and *corneus*, partial in their distribution in Europe stop short of Scotland. Of *Physa* and *Lymnæa* it is extremely doubtful whether there are any species throughout the province more than we have in Britain. Neither of *Ancylus*, which lives attached, limpet-like, to sticks and stones, and has very limited facilities of migration, are there any species throughout the province more than we have in Britain. The species of these genera described by recent Continental authors as new, are worthless.

Edward Forbes mentioned, as a strong point in his argument, that the great mass of the flora and fauna of the British Isles, migrated during the post-pliocene epoch, over the elevated bed of the glacial sea, from the Germanic region. He urges in support of this view, that "every plant universally distributed in these islands is Germanic," and "that the great mass of our pulmoniferous mollusca have come from the same quarter." It does not appear to me necessary to bring forward any geological phenomena to account for the universal distribution in the British Isles, of plants and animals which were and are distributed universally on the Continent in the same latitude. The Germanic portion of the Continent surrounds all the Continental side of Britain.

§ 2.

Land and freshwater species of opposite hemispheres are not always representatives, but are sometimes identical.

Species of one province are not unfrequently simulated at distant points of another province by other species. These are representatives. Some of the British *Lymnææ*, for example, are represented with a remarkable degree of parallelism in North America, in the midst of a fauna which in other respects is as remarkably distinct. *L. limosa* has a very near representative in *L. catascopium*, *L. auricularia* in *L. macrostoma*, *L. stagnalis* in *L. jugularis*, *L. palustris* in *L. elodes*, *L. desidiosa* in *L. truncatula*. The evidence of the existence of indigenous land and freshwater species in North America, which are identical with indigenous British species, is very limited, but it is unimpeachable. Out of our hundred and twenty-eight species, ten inhabit the United States, but six of them, *Limax agrestis*, *Zonites cellarius*, *nitidus*, and *radiatulus*, and *Helix aspersa* and *nemoralis*, are not indigenous. They have been conveyed accidentally in casks and other packages, and have become acclimatized chiefly among the outlying islands, and in the vicinity of the maritime cities, one species, *Z. cellarius*, having spread to some distance inland. The remaining four species, which I take to be indigenous in the United States as well as in Britain, are *Pisidium obtusale*, *Paludina vivipara*, *Helix pulchella*, and *Zua subcylindrica*. The *Pisidium* is considered by Mr. Temple Prime, who describes it under the name *P. ventricosum*, to be a distinct species, but Dr. Baudon, our latest authority on that genus, asserts that it is identical with *P. obtusale*. I will only speak then of the range of the remaining three. *Paludina vivipara* lives abundantly along the south shore of Lake Michigan, in Indiana, Illinois, Missouri, Arkansas, Alabama, Georgia, South Carolina, and Florida. *Helix pulchella*, says Dr.

Binney, "inhabits all the Atlantic States, from Maine to South Carolina, and from Vermont to Council Bluffs on the Missouri." Of *Zua subcylindrica*, he says: "It is distributed over a vast expanse of country; it has been noticed in the North-western Territory, near the Lake of the Woods and Lake Winnipeg, in Ohio, in all the Middle States, and in every State of New England." Now it is a curious fact that the two last-named species are quite abnormal forms in the Caucasian molluscan fauna. Not one of the other twenty-three British *Helices* have any typical affinity with *Helix pulchella*; and there are no other species of *Zua* in Britain. Concomitant with this, they have the most extended distribution both in space and elevation of all mollusks in the Eastern Hemisphere,—both species ranging throughout Europe, and in Asia from Northern Siberia to Thibet.

§ 3.

The range of land and freshwater species over areas (zoological provinces) indicated by uniformity of type, is not influenced by the intervention of sea.

It was remarked by Kirby, that physical conditions are not the primary causes of the zoological provinces, which he regarded as fixed by the Will of the Creator. The intervention of sea appears to have no influence on the type or diffusion of species. The boundaries of two distinct provinces meet across a continent, as in West Africa, or across an island, as in Japan, while parts of a province may be widely separate by an intervening sea, as in the case of the Mediterranean. There are, it is true, fewer Caucasian species in North Africa than in Southern Europe, so would there be if the intermediate space were dry land. The North African species are not members of another fauna; they are outlying southern species of the Caucasian type, just as the British species are outlying western species of that type. North Africa has, however, from fifty to sixty species of the Caucasian type, chiefly in Algeria and Egypt, of its own; while Britain has only three species of its own, *Assiminia Grayana*, *Geomalacus maculosus*, and *Lymnæa involuta*. The sea around the islands of the Asiatic Archipelago, which enclose the specific centre of the Malayan province of distribution, predominates in marked typical force in Burmah, Siam, and Cambojia, and in India as far northward and westward as Sikkim-Himalaya. Two remarkable land shells, amongst a characteristic assemblage of others, collected lately in Cambojia, are conspicuous examples of species of the same typical character, separated by an intervening sea. *Helix Cambojiensis* is a species of which the only other example of similar type, *H. Brookei*, occurs in Borneo; and *Bulimus Cambojiensis* is of a type which appears in Burmah in *B. atricallosus*, while both belong to a particular specific type, *B. citrinus*, centred in the Moluccas, and not found in the intervening Philippine Islands. In India Malayan forms appear at the north-western confines of the province, among the Nilgherry and Khasiah Hills, in the richly coloured *Cyclophori*, intermingled with small Caucasian forms.

One of the most remarkable instances of the types of one province mingling with the types of another, appears in Siberia. It was shown in 'Johnston's Physical Atlas,' published in 1852, that the tiger ranges from India, as high up as the 50th parallel of latitude, where the river is frozen over for at least six months in the year. Mr. T. W. Atkinson during his travels in the Valley of the Amoor, met with the tiger, lynx, and panther, in the Middle region, almost as plentiful as in the jungles of Hindostan, mingling in their distribution with British species, such as the fox and stag. Every night, while encamping, fires were lighted to frighten away the tigers. Having been led by this to examine the species of land and freshwater mollusks of the Amoor, described by Gerstfeldt, I was impressed with the same startling phenomenon. *Paludina Ussuriensis* and *Melania Amurensis*, two singularly characteristic species, of undoubted Malayan type, mingle with Caucasian species (many of them British) in the isothermal latitude of Iceland, within five degrees of the limit of permanent ground frost.

LIST OF WORKS

REFERRED TO IN THE SYNONYMY.

ADAMS, H. AND A   The Genera of Recent Mollusca. London, 1858.
ALBERS, J. C.   Malacographia Maderensis, sive enumeratio Molluscorum quæ in insulis Maderæ et Porto Sancti reperiuntur. Berlin, 1854.
ALDER, J.   Transactions of the Natural History Society of Northumberland, Durham, and Newcastle. Newcastle, 1830.
ALDER, J.   Magazine of Zoology and Botany, vol. ii. London, 1837.
ALERON.   Bulletin de la Société Philomathique de Perpignan. Perpignan, 1837.
ALLMAN, G. J.   Annals and Magazine of Natural History, vol. xvii. London, 1846.
ALTEN, J. W.   Systematische Abhandlung über die Erd- und Fluss-Conchylien, welche um Augsburg, etc. Augsburg, 1812.
ANTON, H. C.   Verzeichniss der Conchylien welche sich in der Sammlung befinden. Halle, 1839.
BAER, K. E. VON.   Oken's Isis, Encyclopedische Zeitung, part v. Jena, 1825.
BARNES, D. W.   Silliman's American Journal of Science. Newhaven, 1823.
BAUDON, A.   Catalogue des Mollusques du département de l'Oise. Beauvais, 1853.
BAUDON, A.   Essai monographique sur les Pisidies Françaises. Paris, 1857.
BEAN, W.   Annals and Magazine of Natural History, vol. vii. London, 1834.
BECK, H.   Index Molluscorum præsentis ævi Musæi principis augustissimi Christiani Frederici. Copenhagen, 1837.
BENEDEN, VAN.   Bulletin de l'Académie des Sciences de Bruxelles, vol. i. Brussels, 1834.
BINNEY, A.   Journal of the Natural History Society of Boston, United States, vol. i. Boston, 1837.
BINNEY, A.   The Terrestrial Air-breathing Mollusks of the United States and the adjacent Territories of North America. Boston, U. S., 1851.

BONHOMME, J. Notice sur les Mollusques bivalves fluviatiles observées jusqu'à ce jour aux environs de Rodez. 1840.

BOUBÉE, N. Bulletin d'Histoire naturelle de France; Mollusques. Paris, 1833.

BOUCHARD-CHANTEREAUX. Catalogues des Mollusques terrestres et fluviatiles dans le département de Pas-de-Calais. Boulogne, 1838.

BOUILLET, J. B. Catalogue des espèces et variétés de Mollusques terrestres et fluviatiles dans la haute et basse Auvergne, etc. Clermont-Ferrand, 1836.

BOURGUIGNAT, J. R. Voyage autour de la Mer Morte, exécuté par F. de Saulcy : Mollusques. Paris, 1853.

BOURGUIGNAT, J. R. Revue et Magasin zoologique de la Société Cuviérienne. Paris, 1853 to 1855.

BOURGUIGNAT, J. R. Mémoires de la Société des Sciences physiques et naturelles de Bordeaux. Bordeaux, 1854.

BOURGUIGNAT, J. R. Étude synonymique sur les Mollusques des Alpes maritimes, publié par A. Risso en 1826. Paris, 1861.

BRARD. C. P. Histoire des Coquilles terrestres et fluviatiles qui vivent aux environs de Paris. Paris, 1815.

BROWN, T. Account of the Irish Testacea, in Memoirs of the Wernerian Society of Edinburgh. 1818.

BROWN, T. Edinburgh Journal of Natural History, vol. i. Edinburgh, 1827.

BROWN, T. Illustrations of the Recent Conchology of Great Britain and Ireland. London, 1845.

BRUGUIÈRE, J. G. Encyclopédie méthodique, vol. vi., article Vers. Paris, 1789.

BRUMATI, L. Catalogo sistematico delle Conchiglie terrestri e fluviatili osservate nel territorio di Monfalcone. Goritz, 1838.

CALCARA, P. Monographia dei generi Clausilia e Bulimus, con l'aggiunta di alcune, etc. Palermo, 1840.

CANTRAINE, F. Bulletin de l'Académie de Bruxelles. Brussels, 1836.

CANTRAINE, F. Annales des Sciences naturelles de Bruxelles. Brussels, 1837.

CESSAC, P. DE. Bulletin de la Société naturelle de Creuse. Creuse, 1854.

CHARPENTIER, J. DE. Catalogue des Mollusques terrestres et fluviatiles de la Suisse. Neufchatel, 1837.

CHARPENTIER, J. DE. Zeitschrift für Malocologie. 1847.

CHEMNITZ, J. H. Neues systematisches Conchylien-Cabinet, vol. iv. to xi. Nuremberg, 1780 to 1786.

CLARKE, B. J. Annals and Magazine of Natural History, vol. xii. and xvi. London, 1843 and 1855.

COMPANYO. Rapport des Mollusques terrestres et fluviatiles du département des Pyrénées-Orientales. Perpignan, 1837.

COSTA, E. M. DA. Historia naturalis Testaceorum Britanniæ. London, 1778.

CRISTOFORI AND JAN. Catalogus in iv. sectiones divisus Rerum Naturalium, etc. Milan, 1832.

CUVIER, G. Leçons d'Anatomie comparée: Tableau. Paris, 1800.

DESHAYES, G. P. Histoire naturelle des Animaux sans vertèbres, de Lamarck. 1835 to 1845.

DESHAYES, G. P. Expédition scientifique de Morée. Paris, 1836.

DESMAREST, A. G. Bulletin de la Société Philomathique. Paris, 1814.

DESMOULINS, C. Catalogue des espèces de Mollusques testacés dans le département de la Gironde. Bordeaux, 1827.

DESMOULINS, C. Actes de la Société Linnéenne de Bordeaux. Bordeaux, 1833.

DILLWYN, L. W. A Descriptive Catalogue of Recent Shells, etc. London, 1817.

DONOVAN, E. The Natural History of British Shells. London, 1800 to 1805.

DRAPARNAUD, J. P. R. Tableau des Mollusques terrestres et fluviatiles de la France. Montpellier, 1801.

DRAPARNAUD, J. P. R. Histoire naturelle des Mollusques terrestres et fluviatiles de la France. Montpellier, 1805.

DROUET, H. Revue et Magasin zoologique de la Société Cuviérienne. Paris, 1852 to 1854.

DUMONT, F. Bulletin de la Société d'Histoire naturelle de Savoie. Geneva, 1850 to 1853.

DUPUY, D. Essai sur les Mollusques terrestres et fluviatiles du département du Gers. Paris, 1843.

DUPUY, D. Histoire naturelle des Mollusques terrestres et d'eau douce qui vivent en France. Paris, 1847 to 1852.

DUPUY, D. Catalogus extramarinorum Galliæ Testaceorum, etc. Paris, 1849.

DUVAL. Revue zoologique de la Société Cuviérienne. Paris, 1345.

FARINES, J. Annales des Sciences Naturelles. Second series, vol. ii. Paris, 1834.

FÉRUSSAC, J. J. and A. Histoire générale et particulière des Mollusques terrestres et fluviatiles, etc. Paris, 1819 to 1832.

FÉRUSSAC, A. E. Concordance systématique pour les Mollusques terrestres et fluviatiles de la Grande-Bretagne. Paris, 1820.

FÉRUSSAC, A. E. Tableau systématique des Animaux Mollusques, classés en familles, etc. Paris, 1822.

FITZINGER, L. Systematische Verzeichniss der im Erzherzogthum Oesterreich, etc. Vienna, 1833.

FLEMING, J. Brewster's Edinburgh Encyclopædia, art. Conchology. Edinburgh, 1814.

FLEMING, J. A History of British Animals. Edinburgh, 1828.

FORBES, E. Annals and Magazine of Natural History, vol. ii. London, 1838.

FORBES AND HANLEY. A History of British Mollusca and their Shells. London, 1853.
FORSTER, F. Ideen über die Gebilde der Clausilien. Berlin, 1842.
GÆRTNER, G. Versuch einer systematischen Beschreibung der im Wetterau bisher entdeckten Konchylien. Hainau, 1813.
GASSIES, G. B. Tableau des Mollusques terrestres et d'eau douce de l'Agénais. Paris, 1849.
GASSIES, G. B. Description des Pisidies de l'Aquitaine. Paris, 1849.
GASSIES AND FISCHER. Monographie du genre Testacelle. Paris, 1856.
GMELIN, J. F. Caroli a Linné Systema Naturæ per regna tria, etc., ed. xiii. Leipzig, 1788 to 1790.
GOULD, A. A. Report of the Invertebrata of Massachusetts. Cambridge, U. S., 1841.
GRAELLS, M. P. Catalogo de los Moluscos terrestres y de agua dulce observados en España. Madrid, 1846.
GRAS, A. Description des Mollusques fluviatiles et terrestres du département de l'Isère. Grenoble, 1840.
GRATELOUP, J. Distribution Géographique de la famille des Limaciens. Bordeaux, 1855.
GRAY, J. E. London Medical Repository, vol. xv. London, 1821.
GRAY, J. E. Annals of Philosophy. London, 1824 and 1825.
GRAY, J. E. Turton's Manual of the Land and Freshwater Shells of the British Isles, 2nd and 3rd editions. London, 1840 and 1857.
GRAY, J. E. Proceedings of the Zoological Society of London. London, 1847.
GRAY, J. E. Catalogues of Shells in the British Museum. London, 1855.
GRAY, J. E. Annals and Magazine of Natural History, Second Series, vol. xvii. and xviii. London, 1856.
HANLEY, S. Catalogue of Recent Bivalve Shells. London, 1842 to 1860.
HARTMANN, J. D. W. System der Erd- und Flussmollusken, etc., in Steinmüller Neue Alpina Wintherthur. Nürnberg, 1821.
HARTMANN, J. D. W. System der Erd und Süsswasser Gasteropoden, etc. Nürnberg, 1821.
HARVEY, W. H. Transactions of the Linnean Society of London, vol. xvii. London, 1834.
HÉCART. Catalogue des Coquilles terrestres et fluviatiles des environs d Valenciennes. Valenciennes, 1833.
HELD, F. Aufzahlung der in Bayern lebenden Mollusken, in Isis, vol. iv. 1836 and 1837.
HOY, T. Transactions of the Linnean Society, vol. i. London, 1790.
HUMPHREYS, G. Museum Calonnianum. Catalogue of the Calonne Museum. London, 1797.
JACQUEMIN, E. Guide du Voyageur à Arles. Arles, 1835.
JEFFREYS, J. G. Transactions of the Linnean Society of London, vol. xvi. London, 1830.

JEFFREYS, J. G.   Annals and Magazine of Natural History.   Second Series, vol. xvi., to Third Series, vol. iv.   London, 1855 to 1860.

JEFFREYS, J. G.   British Conchology; or, an Account of the Mollusca, etc.   Vol. i., Land and Freshwater Shells.   London, 1862.

JENYNS, L.   Transactions of the Philosophical Society of Cambridge, vol. iv.   Cambridge, 1833.

JOHNSTON, G.   Edinburgh New Philosophical Journal, vol. v.   Edinburgh, 1828.

JURINE, L.   Helvetischer Almanach.   Verzeichniss der Mollusken, etc. 1817.

KICKX, J.   Synopsis Molluscorum Brabantiæ australi indigenorum.   Louvain, 1830.

KICKX, J.   Description d'une nouvelle espèce fluviatile du genre Mytilus.   Brussels, 1838.

KLEES, J. G.   Dissertatio et descriptiones Testaceorum circa Tubingam indigenorum.   Tübingen, 1818.

KRYNICKI, J.   Bulletin of the Natural History Society of Moscow.   Moscow, 1832 to 1837.

LAMARCK, J. B. M.   Annales du Muséum, vol. vi.   Paris, 1805.

LAMARCK, J. B. M.   Histoire naturelle des Animaux sans vertèbres.   Paris, 1815 to 1822.

LAFON-DU-CUJULA.   Description des Mollusques terrestres et fluviatiles du Lot-et-Garonne.   Paris, 1806.

LEA, J.   Observations on the genus Unio.   Philadelphia, 1829 to 1862.

LEA, J.   A Synopsis of the Family of the Naiades.   Philadelphia, 1852.

LEACH, W. E.   A Synopsis of the Mollusca of Great Britain, 1820.   London, 1852.

LIGHTFOOT, J.   Philosophical Transactions of the Royal Society, vol. lxxvi.   London, 1786.

LINNÆUS, C.   Fauna Succica, sistens animalia Sueciæ regni.   First and second editions; Stockholm, 1746 and 1767.

LINNÆUS, C.   Systema Naturæ, per regna tria naturæ, etc., 10th and 12th editions.   Stockholm, 1758 and 1767.

LISTER, M.   Historia animalium Angliæ, tres tractatus.   London, 1678.

LOWE, R. T.   Proceedings of the Zoological Society of London.   London, 1854.

MACGILLIVRAY, W.   A History of the Molluscous Animals of the Counties of Aberdeen, Kincardine, and Banff.   London, 1843.

MATON, W. G.   Transactions of the Linnean Society of London, vol. iii.   London, 1797.

MATON AND RACKETT.   Transactions of the Linnean Society of London, vol. viii.   London, 1807.

MAUDUYT, L.   Tableau des Mollusques terrestres et fluviatiles du département de la Vienne.   Poitiers, 1839.

MENKE, K. T.   Synopsis methodica Molluscorum generum, etc., quæ in museo Menkeano adservantur.   Pyrmont, 1830.

MERMET, C. Histoire des Mollusques terrestres et fluviatiles vivant dans les Pyrénées-Occidentales. Pau, 1843.

MICHAUD, A. L. G. Complément de l'Histoire naturelle des Mollusques terrestres et fluviatiles de Draparnaud. Verdun, 1831.

MILLER, J. S. Annals of Philosophy, Second Series, vol. vii. London, 1822.

MILLET, P. A. Guérin's Magasin de Zoologie. Paris, 1843.

MILLET, P. A. Mollusques de Maine-et-Loire, etc. Angers, 1854.

MITTRE, M. H. Revue et Magasin de Zoologie de la Société Cuviérienne. Paris, 1841.

MONTAGU, G. Testacea Britannica, or Natural History of British Shells. London, 1803.

MONTFORT, D. DE. Conchyliologie systématique et classification méthodique des Coquilles. Paris, 1810.

MOQUIN-TANDON, A. Actes de la Société Linnéenne de Bordeaux, vol. xv. Bordeaux, 1849.

MOQUIN-TANDON, A. Histoire naturelle des Mollusques terrestres et fluviatiles de France, etc. Paris, 1855.

MORELET, A. Description des Mollusques terrestres et fluviatiles du Portugal. Paris, 1845.

MULLER, O. F. Vermium terrestrium et fluviatilium Historia, etc. Copenhagen, 1774.

MULLER, A. Wiegmann's Archives. Berlin, 1838.

NILSSON, S. Historia Molluscorum Sueciæ terrestrium et fluviatilium, etc. Lund, 1822.

NORMAND, N. A. J. Notice sur plusieurs espèces de Cyclades découvertes dans les environs de Valenciennes. Valenciennes, 1844.

NORMAND, N. A. J. Description de six Limacés nouvelles observées aux environs de Valenciennes. Valenciennes, 1852.

NORMAND, N. A. J. Coup d'œil sur la famille des Cyclades dans le département du Nord. Valenciennes, 1854.

OKEN, L. Lehrbuch der Zoologie. Leipzig, 1815.

OLIVI, A. G. Zoologia Adriatica, ossia Catalogo ragionato degli animali del golfo e delle lagune di Venezia, etc. Bassano, 1792.

PALLAS, P. S. Voyage en Russie, nouvelle édition, revue par Lamarck et Langlès. Paris, 1794.

PAYRAUDEAU, B. C. Catalogue descriptif et méthodique des Annélides et des Mollusques de l'Ile de Corse. Paris, 1826.

PENNANT, T. British Zoology, illustrated by plates and brief explanation. London, 1777.

PFEIFFER, C. Naturgeschichte Deutscher Land und Süsswasser Mollusken, etc. Cassel, 1821.

PFEIFFER, L. Symbolæ ad historiam Heliciorum. Cassel, 1841 to 1846.

PFEIFFER, L. Monographia Heliciorum viventium, sistens descriptiones systematicas, etc. Cassel, 1847 to 1853.

PHILIPPI, R. A. Enumeratio Molluscorum Siciliæ, tum viventium, tum tellure tertiaria fossilium, etc. Berlin, 1836 and 1844.

PHILIPPSON, L. M. Dissertatio historico-naturalis, sistens nova Testaceorum Genera. Lund, 1788.

POIRET, J. L. M. Coquilles fluviatiles et terrestres observées dans le département de l'Aisne et aux environs de Paris; Prodrome. Paris, 1801.

POLI, J. X. Testacea utriusque Siciliæ eorumque historia et anatome, tabulis æneis illustrata. Parma, 1791.

PORRO, C. Malacologia terrestre e fluviale della provincia Comasco. Milan 1838.

POTIEZ AND MICHAUD. Galerie des Mollusques et Coquilles du Muséum de Douai. Paris, 1838 and 1844.

PRIDEAUX. Zoological Journal, vol. i. London, 1824.

PRIME, T. Proceedings of the Natural History Society of Boston, United States. Boston, U. S., 1852.

PULTENEY, R. Catalogue of the Birds, Shells, and some of the most rare Plants of Dorsetshire. London, 1799.

PUTON, E. Essai sur les Mollusques terrestres et fluviatiles des Vosges. Epinal, 1847.

RAY AND DROUET. Revue et Magasin zoologique de la Société Cuviérienne. Paris, 1848 and 1849.

RAZOUMOWSKY, J. Histoire naturelle de mont Jorat, etc. Lausanne, 1789.

RECLUZ, C. A. Revue et Magasin zoologique de la Société Cuviérienne. Paris, 1841 and 1842.

REEVE, L. Proceedings of the Zoological Society of London, part xviii. London, 1850.

RISSO, A. Histoire naturelle des principales productions de l'Europe méridionale, vol. iv. Paris, 1826.

ROISSY, F. Histoire naturelle des Mollusques. Suites à Buffon de Sonnini. Paris, 1801 to 1805.

ROSSMÄSSLER, E. A. Iconographie der Land und Süsswasser Mollusken, etc. Dresden, 1835 to 1854.

SAY, T. Nicholson's American Encyclopædia, vol. iv. Philadelphia, 1817.

SAY, T. American Conchology, or Descriptions of the Shells of North America. New Harmony, 1830 to 1832.

SCACCHI, A. Osservazioni zoologiche (Testacei). Naples, 1833.

SCHOLTZ, H. Schlesien Land und Süsswasser Mollusken. Breslau, 1843 and 1853.

SCHRÖTER, J. S. Die Geschichte der Flussconchylien mit vorzüglicher, etc. Halle, 1779.

SCHUMACHER, C. F. Essai d'un nouveau système des habitations des Vers testacés. Copenhagen, 1817.

SCOPOLI, J. A. Introductio ad Historiam naturalem, sistens genera lapidum, plantarum et animalium, etc. Prague, 1777.

SHEPPARD R. Transactions of the Linnean Society of London, vol. xiv. London, 1825.

SHUTTLEWORTH, R. J. Ueber Land und Süsswasser Mollusken von Corsica. Berne, 1843.

SOWERBY, G. B. The Genera of Recent and Fossil Shells. London, 1820 to 1824.

STARK, J. Elements of Natural History. Edinburgh, 1828.

STEWART, C. Elements of Natural History. Edinburgh, 1817.

STUDER. Fauna Helvetica in Coxe's Travels in Switzerland. London, 1789.

STUDER. Kurzes Verzeichniss der bis jezt in unserm Vaterlande entdeckten Conchylien. Berne, 1820.

STURM, J. Deutschland Fauna in Abbildungen nach der Natur, mit Beschreibungen. Nuremberg, 1803 to 1809.

SWAINSON, W. A Treatise on Malacology, or the natural classification of Shell-fish. London, 1840.

TAPPING, T. The Zoologist, vol. xiv. London, 1856.

THOMPSON, W. Annals and Magazine of Natural History, vol. vi. London, 1840.

THOMPSON and GOODSIR. Annals and Magazine of Natural History vol. v. London, 1840.

TURTON, W. A Conchological Dictionary of the British Isles. London, 1819.

TURTON, W. The Bivalve Shells of the British Isles systematically arranged. London, 1830.

TURTON, W. Manual of the Land and Freshwater Shells of the British Isles. London, 1831.

VALLOT. Exercice sur l'Histoire naturelle. Catalogue descriptif de 62 mollusques terrestres de la Côte-d'Or. Dijon, 1801.

VESTEND. Bulletin de l'Académie de Bruxelles, vol. iii. Brussels, 1835.

VILLA, A. and J. B. Dispositio systematica Conchyliarum terrestrium et fluviatilium quæ observantur. Milan, 1841.

WALKER, G. Testacea minuta rariora, nuperrime detecta in arena littoris Sandvicensis. London, 1784.

WOOD, W. Index Testaceologicus, or a Catalogue of Shells, British and Foreign; Supplement. London, 1828.

INDEX.

	Page
Abida	
secale	109
Achatina	
acicula	97
acuta	97
Algira	97
Goodallii	95
lubrica	93
subcylindrica	93
tridens	95
Acicula	
acicula	97
eburnea	97
fusca	179
lineata	179
Acme	
fusca	179
lineata	179
Moutoni	179
spectabilis	180
Acteon	
bidentatus	131
denticulatus	129
Acroloxus	
lacustris	172
Alasmodon	
elongatus	224
margaritiferum	223
Alasmodonta	
arcuata	223
Alæa	
antivertigo	116
cylindrica	123
edentula	122
marginata	110
minutissima	123
nitida	122
palustris	116
pygmæa	118
revoluta	122
substriata	119
vulgaris	118
Alexia	
denticulata	130
Amulia	19
Ameria	150
Amnicola	183
Amphibina	
putris	42

	Page
Amphibulina	
oblonga	44
succinea	42
Amphipeplea	
glutinosa	167
involuta	168
lacustris	157
Amplexus	
crenellus	83
paludosus	83
Ampullaria	192
Ancylastrum	
fluviatilis	171
Ancylus	
Baconi	170
Barilensis	170
capuloides	171
cyclostoma	171
deperditus	171
Deshayesianus	171
fluviatilis	171
gibbosus	171
Haldemanni	170
Janii	171
lacustris	172
Moquinianus	172
obtusus	171
riparius	171
Sibiricus	172
simplex	171
strictus	171
Verreauxi	170
vitraceus	171
Anodon	
anatinus	216
Avonensis	216
cygneus	216
paludosus	216
Anodonta	
anatina	215
Arelatensis	216
Cellensis	216
coarctata	216
complanata	216
compressa	216
crassiuscula	216
cygnea	215
Dupuyi	216
Grateloupiana	216

	Page
Anodonta	
intermedia	216
Jobæ	216
Lusitana	216
macilenta	216
Milletii	216
minima	216
Moulinsiana	216
oblonga	216
parvula	216
piscinalis	216
ponderosa	216
ranarum	216
Rayii	216
Rossmassleri	216
rostrata	216
Scaldiana	216
sinuosa	216
subponderosa	216
sulcata	216
variabilis	215
ventricosa	216
Anodontites	
anatina	215
cygnea	215
Aplexa	
hypnorum	153
Aplostoma	47
Arianta	
arbustorum	63
Arion	
albus	10
ater	9
circumscriptus	11
empiricorum	10
flavus	11
foliolatus	11
fuscatus	10
fuscus	11
hortensis	11
melanocephalus	10
subfuscus	10
rufus	10
Assiminea	
Grayana	183
Auricella	
Carychium	127
lineata	179
myosotis	131

INDEX.

	Page
Auricula	
alba	131
oidentata	131
Carychium	127
denticulata	129
dubia	132
erosa	131
gracilis	127
lineata	179
Micheli	132
minima	127
myosotis	130
personata	129
tenella	129
Azeca	
Matoni	95
Nouletiana	95
trideus	95
Balea	
fragilis	106
perversa	106
Sarsi	105
Tristensis	105
ventricosa	105
Bathyomphalus	140
Bradybœna	
Brunonensis	66
cœlata	75
Cantiana	66
Carthusiana	68
circinata	75
hispida	76
plebeia	76
rufescens	75
Buccinum	
acicula	97
auricula	159
fossarum	164
fragile	161
glabrum	165
glutinosum	167
medium	157
palustre	162
peregrum	157
rivale	157
roseo-labiatum	161
stagnale	161
terrestre	97
truncatulum	164
Bulimnus	
Lackhamensis	90
obscurus	91
Bulimus	
acicula	97
acutus	88
articulatus	89
Asticrianus	91
utricallosus	258
auricularius	159
bidens	100
Cambojiensis	258
citrinus	104
Clausilioides	105
elongatus	89
exiguus	127
fontinalis	151
fragilis	161

	Page
Bulimus	
Funcki	105
glaber	165
glutinosus	167
hordeaceus	91
Humberti	91
hypnorum	153
Lackhamensis	90
leucostoma	105
limosus	157
lineatus	179
litoralis	89
lubricus	93
Meukeanus	95
mininus	127
Montacuti	90
montanus	90
muscorum	110
obscurus	164
obscurus	91
palustris	162
pereger	157
perla	151
perversus	103
perversus	106
stagnalis	161
succineus	42
tentaculatus	189
truncatus	164
turritella	89
unidentatus	111
variabilis	89
ventricosus	89
viviparus	196
Bulla	
fontinalis	151
fluviatilis	151
hypnorum	153
turrita	153
Byssanodonta	212
Bythinella	158
Bythinia	
jaculator	189
Kickxii	190
Leachii	190
Michaudi	190
Moutonii	188
similis	188
tentaculata	189
ventricosa	190
Cœcilianella	
acicula	97
anglica	97
Brandelii	98
Liesvillei	97
nanodea	98
nyctelia	98
raphidia	98
subsaxana	98
Syriaca	98
tumulorum	98
Cœcilioides	
acicula	97
Calcarina	45
Cardium	
amnicum	228
Casertanum	232

	Page
Cardium	
corueum	238
lacustre	231
nux	241
Carocolla	
lapicida	74
maculata	68
Carychium	
cochlea	179
denticulatum	130
elongatum	127
fuscum	179
lineatum	179
Menkeanum	95
minimum	127
myosotis	131
nanum	127
personatum	129
politum	95
Rayiunum	127
striolatum	127
tridentatum	127
Castalia	212
Cepæa	
hortensis	64
nemoralis	64
Chilina	155
Chilostoma	
pulchella	83
Chilotrema	
lapicida	74
Chondropoma	176
Chondrus	
secale	109
Cingulifera	
arbustorum	63
Cionella	
acicula	97
lubrica	93
Circinnaria	
pulchella	83
Clausia	
parvula	106
Clausilia	
abietina	103
ampla	100
bidens	100
biplicata	101
cordata	102
cruciata	103
derugata	100
dubia	103
Everettii	103
fragilis	106
infulæformis	102
lamellata	100
laminata	100
lucida	100
Montagui	101
Mortilletii	102
nigricans	103
obtusa	103
papillaris	101
parvula	103
perversa	103
perversa	106
plicata	101

	Page		Page		Page
Clausilia		Cyclas		Fruticicola	
plicatula	102	ovalis	240	circinata	75
quadrata	102	pallida	240	hispida	76
radicans	102	palustris	228	sericea	77
Reboudii	103	Pisidioides	239	Fusulus	
Rolphii	102	plumbeus	238	fragilis	106
roscida	103	pulchella	233	Geomalacus	
rostrata	102	pusilla	230	maculosus	13
rugosa	103	rivalis	238	Glandina	
similis	101	riricola	237	acicula	97
uniplicata	106	Ryckboltii	241	Gonostoma	
ventricosa	101	sabulicola	237	obvoluta	73
vivipara	101	Scaldiana	238	Granaria	
vulnerata	102	Terveriana	241	secale	109
Cochlea		Cyclophorus	258	Gulnaria	
fasciata	64	Cyclostoma		ampla	159
ericetorum	71	achatinum	196	auricularia	159
pomatia	61	contectum	194	Hartmanni	159
unifasciata	63	costulatum	176	Monnardii	159
virgata	69	dentatum	176	ovata	157
vivipara	194	elegans	177	peregra	157
vulgaris	59	fuscum	179	Gyraulus	
Cochlicella		impurum	189	hispidus	159
meridionalis	69	jaculator	189	Gyrorbis	192
Cochlicellus		lineatum	179	Helicella	
acutus	89	marmoreum	177	alliaria	48
Maroccanus	89	obtusum	199	cellaria	47
Cochlicopa		simile	188	crystallina	53
lubrica	93	sulcatum	176	ericetorum	72
tridens	95	viviparum	194	fasciolata	71
Cochlodonta	109	Daudebardia	39	glabra	48
Cochlohydra	82	Delomphalus		hispida	76
Cœnatoria		rotundatus	84	nitida	51
adspersa	60	rupestris	85	nitidula	49
pomatia	61	saxatilis	85	nitidiosa	50
Columna		Dipsas	212	revelata	78
lubricus	93	Discus		rupestris	84
Conovulus		crystallinus	53	succinea	51
albus	132	pygmæus	86	variabilis	70
bidentatus	131	rotundatus	84	Helicigona	
denticulatus	129	Dreissena		lapicida	74
myosotis	130	polymorpha	209	Helicogena	
Conulus		Elisma		hortensis	64
fulvus	80	fasciata	89	hybrida	64
Coretus	188	Elona	189	libellula	64
Corneola		Ena		nemoralis	64
pulchella	83	montana	90	pomatia	61
Crenella	206	obscura	91	Helicodonta	
Cryptomphalus	60	Epistylia	46	obvoluta	73
Cyclas		Ericia	177	Helicolimax	
æquata	227	Eruca		pellucida	30
amnica	228	fragilis	106	Heliomaue	60
appendiculata	224	muscorum	123	Helix	
calyculata	241	umbilicata	111	acicula	97
citrina	225	Eulimax	20	aculeata	81
consobrina	240	Euparypha		acuta	74
cornea	225	rhodostoma	69	acuta	83
flavescens	226	Euryomphala		affinis	74
fontinalis	230	pygmæa	86	albella	63
gibba	229	rotundata	84	albella	71
lacustris	241	rupestris	84	albula	77
lenticularis	231	Ferussacia		Algira	45
minima	229	subcylindrica	93	aliena	85
nitida	231	Fruticicola		alliacea	43
nucleus	228	aculeata	81	alliaria	43
obliqua	228	Carthusiana	66	alpestris	63
obtusalis	229	Carthusianella	68	alternata	72

Helix	Page	Helix	Page	Helix	Page
angigyra	73	fœtida	48	neglecta	71
angusta	43	fontana	147	nemoralis	64
arbustorum	63	fontinalis	162	nitens	47
arenaria	67	fossaria	164	nitens	48
arenosa	72	fragilis	161	nitida	51
aspersa	59	fulva	80	nitida	67
auricularia	159	fusca	64	nitida	148
bidens	100	fusca	79	nitidiosa	49
bifasciata	89	Galloprovincialis	66	nitidula	49
bilabiata	73	Gibsii	67	nitidula	80
bimarginata	67	glabra	48	nivea	71
Brookei	258	glabra	165	obscura	91
buccinata	90	glaphyra	47	obvoluta	73
Bullaoides	151	globularis	77	occidentalis	78
cœlata	75	glutinosa	167	octanfracta	165
Cambojiensis	258	Goodalli	95	octona	165
Canigouensis	63	Gothica	63	Ofltonensis	122
Cantiana	66	Granatelli	81	olivetorum	45
caperata	70	granulata	77	pallida	66
carinata	142	grisea	61	paludosa	82
Carthusiana	67	Gussoniana	61	palustris	162
Carthusianella	67	Helmii	49	parmula	74
Carychium	127	hispida	76	pellucida	39
cellaria	47	holoserica	73	peregra	157
chersina	56	hortensis	59	perversa	101
cincta	64	hortensis	64	perversa	103
cingenda	68	hybrida	64	perversa	104
circinata	75	hydatina	53	perversa	106
clandestina	75	inflata	157	petholata	68
cochlea	143	Itala	71	petronella	52
cochlea	179	Kirbii	86	piligera	77
compactilis	196	Lackhamensis	90	Pisana	68
complanata	143	lacustris	143	piscinalis	198
concinna	76	lamellata	18	planata	142
contorta	146	lapicida	74	planorbis	142
cornu-arietis	138	lenticula	50	planorbis	144
cornea	138	lenticularis	147	planospira	63
corrugata	75	ligata	61	plebeium	76
corrugata	79	limacoides	39	pomatia	61
corvus	162	limbata	67	Poneutina	78
costata	82	limbata	142	pulchella	82
crassa	146	limosa	42	pura	49
crassa	162	limosa	157	pusilla	84
crenella	82	lineata	148	putris	42
crenulata	70	Listeri	74	putris	157
cretacea	88	littorea	183	pygmæa	86
cristata	200	lubrica	93	quinquefasciata	64
crystallina	53	lucida	47	radiata	83
Cumberlandiana	72	lucida	51	radiatula	50
cylindrica	118	lucifuga	64	revelata	78
derugata	100	lucorum	60	rhodostoma	68
diaphana	39	lurida	76	rhombea	143
depilata	76	lutea	157	rotundata	83
disjuncta	70	lutescens	61	rudis	76
Draparnaudi	139	marmorata	153	rufescens	75
eburnea	53	Mazzullii	60	rufilabris	67
elegans	69	minuta	82	rufina	75
elongata	44	minuta	86	rupestris	84
elliptica	39	monilifera	70	saxicola	56
erica	71	montana	75	saxatilis	84
ericetorum	71	montana	90	Scarburgensis	81
excavata	52	Mortoni	80	scale	109
exigua	122	muscorum	110	seminulum	81
fasciata	196	mutabilis	64	sericea	77
fascicularis	198	nana	138	spinulosa	81
fasciolata	70	nautilea	141	spirorbis	145

INDEX. 271

	Page
elix	
spirula	85
stagnalis	161
stagnorum	91
striata	70
striata	71
striatula	50
striutula	162
strigata	68
strigata	70
striolata	75
subalbida	69
subcylindrica	93
subglobosa	64
subrufescens	79
subterranea	30
subviricis	78
succinea	42
succinea	51
sylvestris	90
Taurica	61
tentaculata	189
terebra	143
teres	157
Terverii	70
trigonophora	73
trochiformis	80
trochulus	80
truncatula	164
turgidula	63
Turtoni	83
turturum	64
umbilicata	84
umbilicata	111
umbilicata	146
variabilis	69
variegata	60
ventricosa	196
vertigo	120
virgata	69
viridula	52
vitrea	53
vitrina	52
vivipara	195
vortex	144
Wittmanni	63
Xartarti	63
zonaria	68
zonaria	69
Hemithalamus	
lacustris	148
Hippeutis	
lenticularis	147
Hyalina	
pellucida	39
Hydrobia	
Angarensis	188
similis	188
ventrosa	186
Hydrocena	183
Hygromane	85
Hyria	212
Iphigenia	102
Iridina	212
Isthmia	112
Jaminea	
denticulata	130

	Page
Jaminea	
edentula	122
marginata	110
muscorum	111
quinquedens	130
secale	109
Latomus	
lapicida	74
Lenticula	
lapicida	74
Leptolimnea	
elongata	165
Leuconia	132
Limacella	
unguicula	24
parma	26
Limacina	
pellucida	39
Limax	
agrestis	20
albus	10
Alpinus	26
antiquorum	26
arboreus	21
arborum	21
ater	10
aureus	12
bilobatus	20
brunneus	22
carinatus	17
cinereus	25
Cyrenæus	26
fasciatus	11
filans	20
flavus	11
flavus	24
gagates	19
glaucus	21
hortensis	21
intermedius	11
luteus	10
maculatus	26
marginatus	17
marginatus	21
maximus	25
rufus	10
salicetum	21
salicium	20
scaudens	21
Sowerbyi	17
subfuscus	10
succineus	10
sylvaticus	20
tenellus	23
tunicata	20
Valentianus	26
variegatus	24
virescens	10
Limnea	
Burnetti	157
canalis	159
Doublieri	164
fontinalis	151
glacialis	157
intermedia	157
lineata	157
marginata	157

	Page
Limnea	
microstoma	164
Nouletiana	157
raphidia	161
thermalis	157
Trencaleonis	157
turrita	153
variabilis	165
Limneus	
acronicus	159
acutus	159
ampullaceus	159
auricularius	159
communis	162
elongatus	165
fontinalis	157
glaber	165
glutinosus	167
Grayanus	183
Hartmanni	159
involutus	168
major	161
minutus	164
ovatus	157
palustris	162
pereger	157
subulatus	165
tinctus	162
truncatulus	164
vulgaris	157
Limnophysa	
minuta	164
palustris	162
truncatula	164
Lithodomus	206
Littorina	
anatina	188
Lochea	10
Lucena	
pulchella	83
putris	42
Lucilla	47
Lymnæa	
auricularia	159
Boissii	158
catascopium	155
corvus	162
desidiosa	155
disjuncta	162
elodes	155
elongata	165
fasciata	89
fontinalis	199
fossaria	164
fragilis	161
fuscus	162
glabra	165
glutinosa	167
Hookeri	157
involuta	168
jugularis	155
Laokhameusis	90
lacustris	157
leucostoma	165
limosa	157
lubrica	93
macrostoma	155

INDEX.

	Page
Lymnæa	
oblonga	164
obscura	91
octanfracta	165
palustris	162
peregrina	157
putris	157
reflexa	163
stagnalis	160
tentaculata	159
teres	158
truncatula	164
umbrosa	163
Vahlii	158
vivipara	194
Vogesiaca	162
Lymnium	220
Lymnus	161
Margaritana	
fluviatilis	223
margaritifera	224
Margaron	
crassus	224
cygnea	216
margaritifer	224
pictorum	221
tumidus	220
Marpessa	100
Melania	
Amurensis	258
setosa	82
Melantho	193
Merdigera	
montana	90
obscura	91
Milax	
gagates	19
Sowerbii	17
Modiola	206
Monacha	
Carthusianella	67
sericea	77
Monocondylæa	213
Musculius	
amnicus	229
Musculus	
exiguus	238
Mya	
depressa	220
margaritifera	223
ovalis	220
ovata	220
pictorum	221
Mycetopus	212
Mysca	
ovata	220
pictorum	221
solida	220
Mytilina	
polymorpha	209
Mytilus	
arca	209
anatinus	215
Avonensis	215
cygueus	215
dentatus	216
Hagenii	209

	Page
Mytilus	
incrassatus	216
maculatus	216
polymorphus	209
radiatus	215
Volgensis	209
Zellensis	215
Myxas	
Mülleri	167
Nanina	52
Natica	
vivipara	194
Nauta	
hypnorum	153
Nautilus	
crista	141
lacustris	148
Nerita	
elegans	177
fasciata	195
fluviatilis	203
hepatica	183
jaculator	189
Mittreana	203
obtusa	199
piscinalis	198
Prevostiana	203
pusilla	198
valvata	200
vivipara	194
zebrina	203
Neritina	
Anatensis	203
Anatolica	203
Bœtica	203
Bourguignati	203
Dalmatica	203
fluviatilis	203
Hildreichii	204
intexta	203
Jordani	203
Macri	203
meridionalis	203
Numidica	203
Peloponensis	203
Prevostiana	203
Sardoa	203
thermalis	203
trifasciata	203
variabilis	203
Nux	
nigella	238
Odostomia	
biplicata	101
Carychium	127
juniperi	109
laminata	100
muscorum	110
nigricans	103
perversa	106
sexdentata	116
vertigo	120
Omphiscola	
glabra	165
Ouchidium	7
Optidoceros	183
Ovatella	129

	Page
Oxychilus	
cellarius	47
ericetorum	74
lucidus	51
nitidulus	49
Paludina	
achatina	196
acuta	190
ampullacea	195
contecta	194
crystallina	194
decipiens	190
fasciata	169
gigantea	193
Grayana	183
impura	189
jaculator	189
Kickxii	190
Listeri	194
Michaudi	190
obvolutus	73
similis	188
tentaculata	189
Ussuriensis	193
ventricosa	190
vivipara	195
vulgaris	196
Paludinella	183
Paludomus	193
Parmacella	5
Patella	
cornea	171
fluviatilis	171
lacustris	172
oblonga	172
Patula	
pygmæa	86
rupestris	84
rotundata	84
Petasia	
trochiformis	80
Physa	
acuta	152
aurantia	150
cornea	153
fontinalis	151
hypnorum	153
influviata	150
Maugeræ	150
turrita	153
Physopsis	151
Phytia	131
Pisidium	
acutum	234
amnicum	228
australe	232
Bounafouxianum	234
calyculatum	233
Casertanum	232
cinereum	232
Dupuyanum	234
fontinule	233
Gassiesianum	233
Grateloupianum	229
Henslowianum	234
incertum	231
inflatum	228

INDEX.   273

	Page
Pisidium	
intermedium	228
Iratianum	233
Jaudonanum	234
Jenynsii	233
Joannis	233
lenticulare	232
limosum	233
nitidum	231
Normandianum	233
obliquum	228
obtusale	229
pallidum	234
pusillum	230
pulchellum	233
Recluzianum	234
roseum	231
tetragonum	233
thermale	233
ventricosum	229
Planorbis	
aclopus	138
acutus	142
adelosius	138
albus	139
anthracus	138
Banaticus	138
bicarinatus	136
bulla	151
carinatus	142
clausulatus	148
complanatus	143
complanatus	147
complanatus	148
compressus	144
contortus	146
cornens	138
cornu	140
crista	141
cristatus	141
disciformis	142
Draparnaldi	139
Dufouri	138
elophilus	138
Etruscus	138
fontanus	147
glaber	140
fragilis	145
gyrorbis	140
hispidus	139
imbricatus	141
lævis	140
lenticularis	147
lentus	136
leucostoma	145
lutescens	142
marginatus	143
Metidjensis	138
nautileus	141
nautilus	148
nitidus	148
Nordenskioldi	138
obvolutus	73
Perezii	145
planatus	142
purpura	138
reticulatus	139

	Page
Planorbis	
rhombeus	143
Rossmassleri	140
rotundatus	145
septemgyratus	145
Sheppardi	143
similis	138
spirorbis	145
submarginatus	144
tenellus	144
trivolvis	136
turritus	153
villosus	139
vortex	144
umbilicatus	142
umbilicatus	143
Platyla	179
Polita	
cellaria	47
crystallina	53
fulva	80
glabra	48
lucida	51
uitidiosa	50
nitidula	49
Polyphemus	
acicula	97
Pomatia	
adspersa	60
pomatia	61
antiquorum	61
Pomatias	
elegans	177
Prolepis	11
Pupa	
Anglica	112
antivertigo	116
bideus	100
bigranata	110
Charpentieri	117
cylindracea	111
edentula	122
elatior	105
exigua	127
fragilis	106
Goodallii	95
juniperi	109
marginata	110
Menkeana	95
minuta	123
minutissima	123
Moulinsiana	117
muscorum	110
muscorum	123
nana	121
obtusa	123
octodentata	116
perversa	106
pygmæa	118
ringens	112
rugosa	103
secale	109
Sempronii	111
sexdentata	116
Shuttleworthiana	119
substriata	119
tridens	95

	Page
Pupa	
umbilicata	111
Venetzii	121
vertigo	120
Pupilla	
Draparnaudii	111
marginata	110
muscorum	110
umbilicata	111
Pupula	
lineata	179
Pyramidula	
rupestris	84
Pythia	
denticulata	129
mysotis	131
Radix	
auriculatus	159
Rissoa	
anatina	168
ventrosa	186
Segmentaria	
lacustris	148
Segmentina	
lineata	148
nitida	148
Seraphia	
tridentata	127
Simpulopsis	38
Spatha	212
Sphærium	
Brochonianum	241
citrinum	238
corneum	238
Creplini	241
Deshayesianum	240
Jeannotii	241
lacustre	241
ovale	240
pallidum	240
Pisidioides	239
rivicola	237
Ryckholtii	241
Scaldianum	238
Terverianum	241
Stagnicola	
communis	162
elegans	161
minuta	164
octanfracta	165
vulgaris	161
Stomodonta	
antivertigo	116
edentula	123
fragilis	106
marginata	110
muscorum	123
pygmæa	118
rugosa	103
secale	109
umbilicata	111
Styloides	
acicula	97
lubricus	93
Succinea	
abbreviata	44
acuta	43

T

INDEX.

	Page
Succinea	
amphibia	42
arenaria	44
Corsica	43
elegans	43
gracilis	43
Levantina	43
Mülleri	42
oblonga	44
ovalis	41
Pfeifferi	43
putris	42
Symphynota	
cygnea	216
Syncera	183
Tachea	
hortensis	64
nemoralis	64
Tanalia	193
Tanychlamys	52
Tapada	
oblonga	44
putris	42
succinea	43
Teba	
caperata	71
Cantiana	66
cingenda	68
rufescens	75
spinulosa	81
vagata	70
Tebennephorus	6
Tellina	
amnica	228
cornea	238
Heuslowiana	234
lacustris	241
minuta	229
pusilla	230
rivalis	226
rivalis	238
stagnicola	235
striata	225
tenera	241
tuberculata	241
Testacella	
Antillarum	29
bisulcata	30
Burdigalensis	32
Canariensis	32
Cauigonensis	30
Companyonii	30
episcia	32
Europœa	30
Galliæ	30
Galloprovincialis	30
haliotidea	30
Matheronii	29
Maugei	32
Medii-Templi	30
Oceanica	32
scutulum	30
Thea	
Terverii	70
virgata	70
Theba	
Carthusiana	66

	Page
Theba	
Carthusianella	67
Charpentieri	67
ericetorum	72
fulva	80
Pisana	68
Theodoxus	
Lutetianus	203
Tichogonia	
Chemnitzii	209
Tornatellina	95
Torquatella	
muscorum	110
Torquilla	
secale	100
Tricula	183
Tridacna	224
Trigonostoma	
obvolutum	73
Trochulus	
hispidus	76
Trochus	
cristatus	200
terrestris	80
Truncatella	
Grayana	174
lineata	179
polita	179
Turbo	
achatinus	196
adversus	151
Anglicus	112
bidens	100
biplicatus	101
Carychium	127
chrysalis	110
cristatus	200
cylindraceus	111
edentulus	122
elegans	177
fasciatus	88
fontinalis	199
fuscus	179
glaber	93
helicinus	82
janitor	189
juniperi	109
laminatus	100
Leachii	190
marginatus	110
Maroccana	89
muscorum	110
Myrmecidis	84
nautileus	141
nigricans	103
nucleus	189
paludosus	82
patulus	159
perversus	103
perversus	106
reflexus	177
rupium	91
sexdentatus	110
stagnalis	161
striatus	177
tentaculatus	189
thermalis	199

	Page
Turbo	
triaufractus	42
tridens	95
tumidus	177
vertigo	121
Unio	
Aleronii	221
arcuatus	220
Ardusianus	221
ater	223
brunnea	221
crassissima	223
crassus	223
curvirostris	221
dactylus	221
Deshayesii	221
elongatus	223
inflatus	220
limosus	221
longirostris	221
margaritifer	223
margaritiferus	223
Michaudianus	220
mucidus	221
ovalis	220
Philippi	221
pictorum	221
platyrinchoideus	221
ponderosus	221
Requieni	221
Roissyi	224
rostrata	221
rostratus	220
Roussii	221
rugosa	223
sinuata	223
tristis	224
tumidus	219
Turtoni	221
Vaginulus	6
Vallonia	
rosalia	83
Valvata	
alpestris	197
Amooreusis	197
Baicalensis	197
branchialis	200
contorta	197
cristata	200
depressa	199
humeralis	197
minuta	199
Moquiniana	199
obtusa	199
pygmæa	198
piscinalis	198
planorbis	200
prasina	197
pulchella	200
pupoidea	197
similis	188
sincera	197
spirorbis	200
tricarinata	197
trochlea	197
Velletia	
lacustris	172

INDEX.

	Page
Verticillus	45
Vertigo	
alpestris	119
Anglica	112
angustior	121
antivertigo	116
curta	119
cylindrica	123
edentula	122
hamata	121
heterostropha	120
lepidula	122
minuta	123
minutissima	123
Moulinsiana	117
muscorum	123
nana	121
nitida	122
palustris	116
plicata	121
pupula	123
pusilla	120
pygmœa	118
quinquedentata	118
secale	109
septemdentata	116
sexdentata	116
similis	118
striata	119
substriata	119
Venetzii	121

	Page
Vertigo	
vertigo	121
vulgaris	118
Vertilla	120
Vitrina	
beryllina	39
Dillwynii	39
margaritacea	79
membranacea	79
Mülleri	39
pellucida	39
Vivipara	
communis	194
fasciata	196
fluviorum	196
Volvaria	
bidentata	131
Voluta	
alba	131
bidentata	131
denticulata	129
reflexa	129
ringens	129
Vortex	
cellaria	47
lapicida	74
obvoluta	73
Xerophila	
Pisana	68
ericetorum	72

	Page
Xerophila	
striata	71
Terverii	70
variabilis	70
Zenobia	66
Zonites	
alliaria	48
cellarius	47
crystallinus	53
cricetorum	72
excavatus	52
fulvus	80
fuscus	79
glaber	48
lucidus	47
lucidus	51
nitidus	51
nitidulus	49
purus	49
pygmæus	86
radiatus	84
radiatulus	50
rotundatus	84
rupestris	84
striatula	50
Zospeum	126
Zua	
Boissii	93
lubrica	93
subcylindrica	93

JOHN EDWARD TAYLOR, PRINTER,
LITTLE QUEEN STREET, LINCOLN'S INN FIELDS.

LOVELL REEVE & CO.'S

PUBLICATIONS IN

Natural History, Science, Travels,

ANTIQUITIES, ETC.

> "None can express Thy works but he that knows them;
> And none can know Thy works, which are so many
> And so complete, but only he that owes them."
> — *George Herbert.*

LONDON:
LOVELL REEVE AND CO., 5, HENRIETTA STREET, COVENT GARDEN.
1863.

CONTENTS.

	PAGE
FLOWERING PLANTS	3
COLONIAL AND FOREIGN FLORAS	5
FERNS	6
MOSSES	6
SEAWEEDS	6
FUNGI	7
ZOOLOGY	7
INSECTS	8
MOLLUSKS AND SHELLS	8
GEOLOGY	10
CHEMISTRY	10
GEOGRAPHY	10
VOYAGES AND TRAVELS	10
ANTIQUITIES	11
PHOTOGRAPHY	11
MISCELLANEOUS	13
SERIALS	14
NEW WORKS	15
FORTHCOMING WORKS	16

LOVELL REEVE & CO.'S PUBLICATIONS.

FLOWERING PLANTS.

Genera Plantarum

Ad Exemplaria imprimis in Herbariis Kewensibus servata definita; auctoribus G. BENTHAM et J. D. HOOKER. Vol. I., Pars I., sistens Dicotyledonum Polypetalarum Ordines LVI: Ranunculaceas—Connaraccas.

Royal 8vo, 21s.

The Field Botanist's Companion;

Being a Familiar Account, in the Four Seasons, of the Flowering Plants most common to the British Isles. By THOMAS MOORE, F.L.S., F.R.H.S. With coloured Figures and Dissections by W. Fitch of 110 species.

In One Volume, 24 coloured plates, 21s.

Handbook of the British Flora;

A Description of the Flowering Plants and Ferns Indigenous to, or Naturalized in, the British Isles. For the use of Beginners and Amateurs. By GEORGE BENTHAM, F.L.S.

In One Volume, 680 pages, 12s.

Illustrations of the Nueva Quinologia of Pavon.

With Observations on the Barks described, by JOHN ELIOT HOWARD, F.L.S., and 30 elaborately Coloured Plates by W. FITCH.

Imperial folio, half-morocco, gilt edges, £6. 6s.

Select Orchidaceous Plants.

By ROBERT WARNER, F.R.H.S. With Notes on Culture by B. S. WILLIAMS, Author of 'The Orchid-Grower's Manual,' and 'Hints on the Cultivation of Ferns;' assisted by some of the best Growers.

To be completed in Ten Quarterly Parts, imperial folio, each, with four coloured plates, 10s. 6d.

Curtis' Botanical Magazine;

Comprising the Plants of the Royal Gardens of Kew, and of other Botanical Establishments in Great Britain, with suitable Descriptions. By Sir W. J. HOOKER, D.C.L., F.R.S., Director of the Royal Gardens of Kew. Third Series.

Royal 8vo, 6 coloured plates by Fitch, 3s. 6d. monthly.

Vols. I. to XVIII., each, with 72 coloured plates, 42s.

A Complete Set of the Second Series, in 17 vols., new, £35. 14s. The only copy remaining.

Vols. I. to XLV of the First Series, bound in 24 vols., whole calf (except the last vol., which is in Numbers), £18.

The Floral Magazine.

Comprising Figures and Descriptions of New Popular Garden Flowers. By the Rev. H. H. DOMBRAIN.

Imperial 8vo, 4 coloured plates by ANDREWS, 2s. 6d. Monthly.
Vols. I. and II., each, with 64 coloured plates, 42s.

The Rhododendrons of Sikkim-Himalaya;

Being an Account of the Rhododendrons recently discovered in the Mountains of Eastern Himalaya. By J. D. HOOKER, M.D., F.R.S.

Imperial folio, 30 coloured plates by Fitch, £3. 16s.

Illustrations of Sikkim-Himalayan Plants.

Chiefly selected from Drawings made in Sikkim under the superintendence of the late J. F. Cathcart, Esq., Bengal Civil Service. The Botanical Descriptions and Analyses by J. D. HOOKER, M.D., F.R.S.

Folio, 24 coloured plates and illuminated title-page by Fitch, £5. 5s.

The Victoria Regia.

By Sir W. J. HOOKER, F.R.S.

Elephant folio, 4 coloured plates by Fitch, 21s.

Pescatorea.

Figures of Orchidaceous Plants, chiefly from the Collection of M. PESCATORE. Edited by M. LINDEN, with the assistance of MM. G. LUDDEMAN, J. E. PLANCHON, and M. G. REICHENBACH.

Folio, Parts I. to XII., each, 4 coloured plates, 7s.

The Tourist's Flora.

A Descriptive Catalogue of the Flowering Plants and Ferns of the British Islands, France, Germany, Switzerland, and Italy. By JOSEPH WOODS.

8vo, 18s.

Journal of Botany and Kew Miscellany.

Original Papers by eminent Botanists, Communications from Botanical Travellers, etc. Edited by W. J. HOOKER, D.C.L., F.R.S.

Vols. IV. to IX., each, 12 plates, some coloured, £1. 4s.
A complete set in 9 vols, half calf, £10. 16s.

The London Journal of Botany.

Edited by Sir W. J. HOOKER, D.C.L., F.R.S., Director of the Royal Gardens of Kew.

Vol. VII., completing the Series, 23 plates, plain, 30s.

Icones Plantarum.

Figures of new and rare Plants,. By Sir W. J. HOOKER, D.C.L., F.R.S. New series.

Vol. V., 100 plates. 31s. 6d.

COLONIAL AND FOREIGN FLORAS.

Flora Hongkongensis;
A Description of the Flowering Plants and Ferns of the Island of Hongkong. By GEORGE BENTHAM, V.P.L.S. With a Map of the Island.
In One Volume, 550 pages, 16s.

Flora of the British West Indian Islands.
By A. H. R. GRISEBACH, M.D., Professor of Botany in the University of Göttingen.
Parts I. to V., 5s. each. To be completed in 7 Parts.

The Botany of the Antarctic Voyage
of H.M.SS. "Erebus" and "Terror" in the years 1839-1843, under the command of Captain Sir J. C. ROSS, R.N., F.R.S. By JOSEPH DALTON HOOKER, M.D. F.R.S.

1. *Flora of Lord Auckland and Campbell's Islands, and of Fuegia, the Falkland Islands, etc.*
   In 2 vols., 200 plates, £10. 15s. coloured; £7. 10s. plain.

2. *Flora of New Zealand.*
   In 2 vols., 130 plates, £13. 2s. 6d. coloured; £9. 5s. plain.

3. *Flora of Tasmania.*
   In 2 vols., 200 plates, £17. 10s. coloured; £12. 10s. plain.

Cryptogamia Antarctica;
Or, Cryptogamic Plants of the Antarctic Islands. Issued separately.
In One Volume, quarto, £4. 4s. coloured; £2. 17s. plain.

On the Flora of Australia,
Its Origin, Affinities, and Distribution; being an Introductory Essay to the 'Flora of Tasmania.' By JOSEPH DALTON HOOKER, M.D., F.R.S.
128 pages, quarto, 10s.

On the Flora of New Zealand;
Its Origin, Affinities, and Geographical Distribution; being an Introductory Essay to the 'Flora of New Zealand.' By JOSEPH DALTON HOOKER, M.D., F.R.S.
40 pages, quarto, 2s.

Outlines of Elementary Botany,
As Introductory to Local Floras. By GEORGE BENTHAM, V.P.L.S.
45 pages, stitched, 2s. 6d.

Botany of the Voyage of H.M.S. Herald.
Under the command of Captain Kellett, R.N., C.B., during the Years 1845-51. By Dr. BERTHOLD SEEMANN, F.L.S. Published under the authority of the Lords Commissioners of the Admiralty.
Royal 4to, 100 plates, £5. 10s.

FERNS.

The British Ferns.

Coloured Figures and Descriptions, with Analyses of the Fructification and Venation, of the Ferns of the British Isles, Systematically Arranged. By Sir W. J. HOOKER, K.H., D.C.L., etc.

Royal 8vo, 66 coloured plates by Fitch, £2. 2s.

Garden Ferns.

Coloured Figures and Descriptions, with Analyses of the Fructification and Venation, of the Ferns best adapted for Cultivation in the Garden, Hothouse, and Conservatory. By Sir W. J. HOOKER, K.H., D.C.L., etc.

Royal 8vo, 64 coloured plates by Fitch, £2. 2s.

Filices Exoticæ.

Century of Exotic Ferns, particularly of such as are most deserving of Cultivation. By Sir W. J. HOOKER, K.H., D.C.L.

Royal 4to, 100 coloured plates by Fitch, £6. 11s.

Ferny Combes.

A Ramble after Ferns in the Glens and Valleys of Devonshire. By CHARLOTTE CHANTER. *Second Edition.*

Fcp. 8vo, 8 coloured plates by Fitch, and a Map of the County, 5s.

MOSSES.

Handbook of the British Mosses;

Being a Description, with Coloured Figures and Dissections, of all the Mosses inhabiting the British Isles. By the Rev. M. J. BERKELEY, M.A., F.L.S.

[*In the press.*

SEAWEEDS.

Synopsis of British Seaweeds.

Descriptions, with Critical Remarks, of all the known Species, abridged from Professor Harvey's 'PHYCOLOGIA BRITANNICA.'

A pocket volume, 220 pages, 5s.

Phycologia Britannica.

A History of the British Seaweeds; containing coloured Figures and Descriptions of all the Species of Algæ inhabiting the Shores of the British Islands. By WILLIAM HENRY HARVEY, M.D., F.R.S., Professor of Botany to the Dublin Society.

In 4 vols. royal 8vo, 360 coloured plates, £6. 6s.

Phycologia Australica.

A History of Australian Seaweeds, containing Coloured Figures and Descriptions uniform with the 'Phycologia Britannica.' By WILLIAM HENRY HARVEY, M.D., F.R.S.

Vols. I. to IV., each, containing 60 coloured plates, 30s.
To be completed in Five Volumes.

Nereis Australis.

Figures and Descriptions of Marine Plants collected on the Shores of the Cape of Good Hope, the extra-tropical Australian Colonies, Tasmania, New Zealand, and the Antarctic Regions. By Professor HARVEY, M.D., F.R.S.

Imperial 8vo, Two Parts, each, containing 25 coloured plates, £1. 1s.

FUNGI.

Outlines of British Fungology,

Containing Characters of above a Thousand Species of Fungi, and a Complete List of all that have been described as Natives of the British Isles. By the Rev. M. J. BERKELEY, M.A., F.L.S. With Coloured Figures and Dissections of 170 Species by FITCH.

8vo, 24 coloured plates, 30s.

The Esculent Funguses of England.

An Account of their Classical History, Uses, Characters, Development, Nutritious Properties, Modes of Cooking, etc. By the Rev. Dr. BADHAM. With 12 coloured plates. [*In the press.*

Illustrations of British Mycology.

Figures and Descriptions of the Funguses of interest and novelty indigenous to Britain. By Mrs. HUSSEY.

Royal 4to; First Series, 90 coloured plates, £7. 12s. 6d.;
Second Series, 50 plates, £4. 10s.

ZOOLOGY.

Zoology of the Voyage of H.M.S. Samarang,

Under the command of Captain Sir Edward Belcher, C.B., F.R.A.S., during the Years 1843-46. Edited by ARTHUR ADAMS, F.L.S.

The Vertebrata, with 8 plates, by John Edward Gray, F.R.S.; the Fishes, with 10 plates, by Sir John Richardson, F.R.S.; the Mollusca, with 24 plates, by Arthur Adams, F.L.S., and Lovell Reeve, F.L.S.; the Crustacea, with 13 plates, by Arthur Adams, F.L.S., and Adam White, F.L.S.

Royal 4to, 55 coloured plates, £3. 10s.

Zoology of the Voyage of H.M.S. Herald,

Under the command of Captain Kellett, R.N., during the Years 1845-51. By Sir J. RICHARDSON. Edited by Professor Edward Forbes, F.R.S. Published under the authority of the Lords Commissioners of the Admiralty.

Part I. Fossil Mammals, 15 double plates. Royal 4to, 21s.
Part II. Fossil Mammals, 10 plates. Royal 4to, 10s. 6d.
Part III. Reptiles and Fishes, 10 plates. Royal 4to, 10s. 6d.

INSECTS.

Curtis' British Entomology.

Illustrations and Descriptions of the Genera of Insects found in Great Britain and Ireland, containing coloured figures, from nature, of the most rare and beautiful species, and, in many instances, of the plants upon which they are found.

Complete in 8 vols., 8vo, 770 coloured copper plates, £16. 16s.

Curtis' British Entomology in Monographs.

Orders.	Plates.	£.	s.	d.	Orders.	Plates.	£.	s.	d.
APHANIPTERA	2	0	1	6	HYMENOPTERA	125	3	3	0
COLEOPTERA	256	6	8	0	LEPIDOPTERA	193	4	16	0
DERMAPTERA	1	0	1	0	NEUROPTERA	13	0	7	0
DICTYOPTERA	1	0	1	0	OMALOPTERA	6	0	3	6
DIPTERA	103	2	12	0	ORTHOPTERA	5	0	3	0
HEMIPTERA	32	0	16	6	STREPSIPTERA	3	0	2	0
HOMOPTERA	21	0	11	0	TRICHOPTERA	9	0	5	0

A Reissue of the Orders Coleoptera, Diptera, Hymenoptera, and Lepidoptera, in Monthly Parts, each containing 5 plates, with text, price 2s. 6d., commences January 1st, 1863.

Insecta Britannica;

Vols. II. and III., Diptera. By FRANCIS WALKER, F.L.S.

8vo, each, with 10 plates, 25s.

MOLLUSKS AND SHELLS.

The Land and Freshwater Mollusks

Indigenous to and Naturalized in the British Isles. By LOVELL REEVE, F.L.S. With finely-executed Wood-Engravings of the Shell of each Species by G. B. SOWERBY, and of the Living Animal of each Genus by O. JEWITT.

8vo, 10s. 6d.

Elements of Conchology;

An Introduction to the Natural History of Shells, and of the Animals which form them. By LOVELL REEVE, F.L.S.

2 vols., 62 coloured plates, £2. 16s.

Conchologia Systematica.

A Complete System of Conchology; in which the Lepades and Conchiferous Mollusca are described and classified according to their Natural Organization and Habits. By LOVELL REEVE, F.L.S.

2 vols. 4to, 300 coloured plates, £8. 8s.

Conchologia Iconica.

Figures and Descriptions of the Shells of the Mollusca, with Remarks on their Affinities, Synonymy, and Geographical Distribution. By LOVELL REEVE, F.L.S., F.G.S. The Drawings by G. B. SOWERBY, F.L.S. Monthly. In Parts, demy 4to, each, containing 8 coloured plates, 10s. Part 223 just published.

CONCHOLOGIA ICONICA IN MONOGRAPHS.

Genera.	Plates	£. s. d.	Genera.	Plates.	£. s. d.
Achatina	23	1 9 0	Mangelia	8	0 10 6
Achatinella	6	0 8 0	Melania	59	3 14 6
Amphidesma	7	0 9 0	Melanopsis	3	0 4 0
Ampullaria	28	1 15 6	Melatoma	3	0 4 0
Anatina	4	0 5 6	Mesalia & Eglisia	1	0 1 6
Anculotus	6	0 8 0	Mesodesma	4	0 5 6
Anomia	8	0 10 6	Meta	1	0 1 6
Arca	17	1 1 6	Mitra	39	2 10 0
Argonauta	4	0 5 6	Modiola	11	0 14 0
Artemis	10	0 13 0	Monoceros	4	0 5 6
Aspergillum	4	0 5 6	Murex	36	2 5 6
Avicula	18	1 3 0	Myadora	1	0 1 6
Buccinum	14	0 18 0	Myochama	1	0 1 6
Bulimus	89	5 12 0	Mytilus	11	0 14 0
Bullia	4	0 5 6	Nassa	29	1 17 0
Calyptræa	8	0 10 6	Natica	30	1 18 0
Cancellaria	18	1 3 0	Nautilus	6	0 8 0
Capsa	1	0 1 6	Navicella	8	0 10 6
Capsella	2	0 3 0	Nerita	19	1 4 6
Cardita	9	0 11 6	Neritina	37	2 7 0
Cardium	22	1 8 0	Oliva	30	1 18 0
Cassidaria	1	0 1 6	Oniscia	1	0 1 6
Cassis	12	0 15 6	Paludomus	3	0 4 0
Chama	9	0 11 6	Partula	4	0 5 6
Chiton	33	2 2 0	Patella	42	2 13 0
Chitonellus	1	0 1 6	Pecten	35	2 4 0
Columbella	37	2 7 0	Pectunculus	9	0 11 6
Conus	47	3 11 0	Pedum	1	0 1 6
Corbula	5	0 6 6	Perna	6	0 8 0
Crania	1	0 1 6	Phasianella	6	0 8 0
Crassatella	3	0 4 0	Phorus	3	0 4 0
Crenatula	2	0 3 0	Pinna	34	2 3 0
Crepidula	5	0 6 6	Pirena	2	0 3 0
Crucibulum	7	0 9 0	Placunanomia	3	0 4 0
Cyclophorus	20	1 5 6	Pleurotoma	40	2 10 0
Cyclostoma	23	1 9 0	Psammobia	8	0 10 6
Cymbium	26	1 13 0	Psammotella	1	0 1 6
Cypræa	27	1 14 0	Pterocera	6	0 8 0
Cypricardia	2	0 3 0	Purpura	13	0 17 0
Delphinula	5	0 6 6	Pyrula	9	0 11 6
Dolium	8	0 10 6	Ranella	8	0 10 6
Donax	9	0 12 6	Ricinula	6	0 8 0
Eburna	1	0 1 6	Rostellaria	3	0 4 6
Fasciolaria	7	0 9 0	Sanguinolaria	1	0 1 6
Ficula	1	0 1 6	Scarabus	3	0 4 0
Fissurella	15	1 0 6	Simpulopsis	2	0 3 0
Fusus	21	1 6 6	Siphonaria	7	0 9 6
Glauconome	1	0 1 6	Soletellina	4	0 5 6
Haliotis	17	1 1 0	Spondylus	18	1 3 6
Harpa	4	0 5 6	Strombus	19	1 4 6
Helix	210	13 5 0	Struthiolaria	1	0 1 6
Hemipecten	1	0 1 6	Terebra	27	1 14 0
Hemisinus	6	0 8 0	Terebratula	11	0 14 0
Hinnites	1	0 1 6	Thracia	3	0 4 0
Hippopus	1	0 1 6	Tridacna	8	0 10 6
Ianthina	5	0 6 6	Trigonia	1	0 1 6
Io	3	0 4 0	Triton	20	1 5 6
Isocardia	1	0 1 6	Trochita	3	0 4 6
Leptopoma	8	0 10 6	Trochus	16	1 0 6
Lingula	2	0 3 0	Turbinella	13	0 17 0
Lithodomus	5	0 6 6	Turbo	13	0 17 0
Littorina	18	1 3 0	Turritella	11	0 14 6
Lucina	12	0 14 0	Umbrella	1	0 1 6
Lutraria	5	0 7 0	Voluta	22	1 8 0
Mactra	21	1 6 6	Vitrina	10	0 13 0
Malleus	3	0 4 0	Vulsella	2	0 3 0

"This great work is intended to embrace a complete description and illustration of the Shells of Molluscous Animals; and so far as we have seen, it is not such as to disappoint the large expectations that have been formed respecting it. The figures of the Shells are all of full size: in the descriptions a careful analysis is given of the labours of others; and the author has apparently spared no pains to make the work a standard authority on the subject of which it treats."

ATHENÆUM.

GEOLOGY.

The Geologist.

A Magazine of Geology, Palæontology, and Mineralogy. Illustrated with highly finished Wood Engravings. Edited by S. J. Mackie, F.G.S., F.S.A. Published Monthly. Price 1s. 6d.

Vol. V., numerous Wood Engravings, 18s.

CHEMISTRY.

Chemical Analysis, Qualitative and Quantitative.

By Dr. Henry M. Noad, F.R.S. [*In the Press.*

Part 1., 'Qualitative,' in the Spring.

GEOGRAPHY.

A Survey of the Early Geography of Western Europe,

as connected with the First Inhabitants of Britain, their Origin, Language, Religious Rites, and Edifices. By Henry Lawes Long, Esq.

8vo, 6s.

VOYAGES AND TRAVELS.

Three Cities in Russia.

By Professor C. Piazzi Smyth, F.R.SS.L. & E., Astronomer Royal for Scotland, Author of 'Teneriffe, an Astronomer's Experiment,' etc.

2 Vols., post 8vo, Maps and Wood-Engravings, 26s.

Travels on the Amazon and Rio Negro,

With an Account of the Native Tribes, and Observations on the Climate, Geology, and Natural History of the Amazon Valley. By Alfred R. Wallace, Esq. With Remarks on the Vocabularies of Amazonian Languages, by R. G. Latham, M.D., F.R.S.

Royal 8vo, 6 plates and maps, 18s.

Western Himalaya and Tibet;

Narrative of a Journey through the Mountains of Northern India. By Thomas Thomson, M.D., Assistant-Surgeon, Bengal Army.

8vo, Tinted Lithographs and Map, 15s.

LOVELL REEVE AND CO.'S PUBLICATIONS. 11

Travels in the Interior of Brazil,

Principally through the Northern Provinces and the Gold and Diamond Districts. By GEORGE GARDNER, M.D., F.L.S. *Second Edition.*

8vo, Plate and Map, 12s.

Narrative of a Walking Tour in Brittany.

By JOHN MOUNTENEY JEPHSON, B.A., F.S.A. Accompanied by Notes of a Photographic Expedition by LOVELL REEVE, F.L.S.

Royal 8vo, with Map by Arrowsmith, and Stereoscopic Frontispiece, 12s.; or with 90 Photographic Vignettes, £2. 2s.

*** Issued separately are 90 stereographs, mounted on cards for use in the Stereoscope, in box with lock and key, £4. 4s.

The Conway.

Narrative of a Walking Tour in North Wales; accompanied by Descriptive and Historical Notes. By J. B. DAVIDSON, Esq., M.A.

Extra gilt, 20 stereographs of Welsh Scenery, 21s.

ANTIQUITIES.

Manual of British Archæology.

By the Rev. CHARLES BOUTELL, M.A.

CONTENTS.—Chap. 1. Architecture.—2. Architectural Accessories.—3. Sepulchral Monuments.—4. Heraldry.—5. Seals.—6. Coins.—7. Palæography, Illuminations and Inscriptions.—8. Arms and Armour.—9. Costumes and Personal Ornaments.—10. Pottery, Porcelain and Glass.—11. Miscellaneous Subjects.

Royal 16mo, 20 coloured plates, 10s. 6d.

Horæ Ferales.

Studies in the Archæology of the Northern Nations. By the late JOHN M. KEMBLE, M.A. Edited by R. G. LATHAM, M.D., F.R.S., and AUGUSTUS W. FRANKS, F.S.A. [*In the press.*

4to, 30 plates, some coloured, £3. 3s.

PHOTOGRAPHY.

Sketches in India,

Taken at Hyderabad and Secunderabad, in the Madras Presidency. By CAPTAIN ALLAN N. SCOTT, Madras Artillery. Edited by C. R. WELD.

100 Photographic Vignettes, £3. 3s.

Mounted as Slides for the Stereoscope, £5. 5s.

The Stereoscopic Magazine.

A Gallery for the Stereoscope of Landscape Scenery, Architecture, Antiquities, Natural History, Rustic Character, etc. With Descriptions. 2s. 6d. Monthly.

3 vols., each complete in itself and containing 50 Stereographs, £2. 2s.

The Stereoscopic Cabinet.

Monthly Packet of Slides for the Stereoscope.

3 slides, 2s. 6d. Monthly.

The Foreign Stereoscopic Cabinet.

Monthly Packet of Foreign Slides for the Stereoscope.

3 slides, 2s. 6d. Monthly.

Stonyhurst College and its Environs;

Photographed by ROGER FENTON, Esq.

15 slides, 15s.

The Conway Stereographs.

20 slides, 20s.

The Brittany Stereographs.

90 slides, in box with lock and key, £4. 4s.

The Isle of Wight Stereographs.

22 slides, 21s.

Interior of the British Museum.

By ROGER FENTON, Esq.

25 slides, 25s.

English Castles and Abbeys.

14 slides, 14s.

Foreign Castles and Abbeys.

24 slides, 24s.

English Cathedrals and Churches.

21 slides, 21s.

Foreign Cathedrals and Churches.

38 slides, 38s.

Calvaries and Crosses.

10 slides, 10s.

Druidical Remains.
6 slides, 6s.

⁎⁎* Any of the above Stereographs may be had singly at 1s. each; also a List of 440 subjects.

Folding Stereoscopes in cases, 3s. 6d. each.

MISCELLANEOUS.

The Gate of the Pacific.
By Captain BEDFORD PIM, R.N., F.R.G.S.   [*In January.*

Demy 8vo, 8 Chromo-Lithographs, Maps and Cuts.

Phosphorescence;
Or, the Emission of Light by Minerals, Plants, and Animals. By Dr. T. L. PHIPSON, F.C.S., Member of the Chemical Society of Paris, etc. etc.

Numerous Illustrations, 5s.

Shakespeare's Sonnets,
Reproduced in Facsimile from the First Printed edition of 1609, by the New Process of Photo-Zincography in use at her Majesty's Ordnance Survey Office. From the unrivalled Copy in the Library of Bridgewater House, by permission of the Right Hon. the Earl of Ellesmere.

10s. 6d.

Literary Papers on Scientific Subjects.
By the late Professor EDWARD FORBES, F.R.S., selected from his Writings in the 'Literary Gazette.' With a Portrait and Memoir.

Small 8vo, 6s.

A Treatise on the Growth and Future Treatment
of *Timber Trees.* By G. W. NEWTON, of Ollersett, J.P.

Half-bound calf, 10s. 6d.

Parks and Pleasure Grounds;
Or, Practical Notes on Country Residences, Villas, Public Parks, and Gardens. By CHARLES H. J. SMITH, Landscape Gardener.

Crown 8vo, 6s.

The Planetary and Stellar Universe.
A Series of Lectures. With Illustrations. By R. J. MANN.

12mo, 5s.

The Artificial Production of Fish.
By PISCARIUS. *Third Edition.* 1s.

NEW SERIES OF BRITISH NATURAL HISTORIES.

British Flora.
By G. BENTHAM. 12s. [*Ready*.

British Fungology.
By the Rev. M. J. BERKELEY. 30s. [*Ready*.

British Ferns.
By Sir W. HOOKER. 42s. [*Ready*.

British Field Plants.
(The Field Botanist's Companion.) By THOMAS MOORE. 21s. [*Ready*.

British Land and Freshwater Mollusks.
By LOVELL REEVE, F.L.S. 10s. 6d. [*Ready*.

British Mosses.
By the Rev. M. J. BERKELEY. [*In the Press*.

SERIALS.

Illustrated Handbook of the British Flora.
By G. BENTHAM, F.R.S., President of the Linnean Society. Wood Engravings, with Dissections, of every Species, from original Drawings by W. Fitch.
2s. 6d. Monthly.

Botanical Magazine.
New and Rare Plants. With Descriptions by Sir W. HOOKER, D.C.L., F.R.S., and 6 coloured plates, with Dissections, by W. Fitch.
3s. 6d. Monthly.

Floral Magazine:
New Popular Garden Flowers. With Descriptions by the Rev. H. H. DOMBRAIN, and 4 coloured plates by Andrews.
2s. 6d. Monthly.

Select Orchidaceous Plants.
By R. WARNER, F.R.H.S.
Folio, 4 superbly coloured plates, 10s. 6d. Quarterly.

Phycologia Australica.
By Dr. HARVEY, F.R.S.
6 coloured plates, 3s. Monthly.

Conchologia Iconica.
By LOVELL REEVE, F.L.S.
4to, 8 coloured plates, 10s. Monthly.

Geologist.
  Illustrated Magazine of Geology, Palæontology, and Mineralogy.
  Wood engravings, 1s. 6d. Monthly.

Curtis' British Coleoptera.
  5 coloured copper-plates, 2s. 6d. Monthly.

Curtis' British Lepidoptera.
  5 coloured copper-plates, 2s. 6d. Monthly.

Curtis' British Hymenoptera.
  5 coloured copper-plates, 2s. 6d. Monthly.

Curtis' British Diptera.
  5 coloured copper-plates, 2s. 6d. Monthly.

Stereoscopic Magazine.
  3 stereographs, with Descriptions, 2s. 6d. Monthly.

Stereoscopic Cabinet.
  3 stereoscopic slides, 2s. 6d. Monthly.

Foreign Stereoscopic Cabinet.
  3 Foreign stereoscopic slides, 2s. 6d. Monthly.

RECENT WORKS.

Reeve's British Land and Freshwater Mollusks.
  Wood Engravings. 10s. 6d.

Dr. Phipson's Phosphorescence.
  Coloured Frontispiece and Wood Engravings, 5s.

Moore's Field Botanist's Companion.
  24 coloured plates, 21s.

Professor Smyth's Three Cities in Russia.
  2 vols., Maps and Illustrations, 26s.

Shakespeare's Sonnets. Facsimile of the First Edition.
  10s. 6d.

Howard's Nueva Quinologia.
  30 coloured plates, £6. 6s.

Hooker's British Ferns.
  66 coloured plates, 42s.

Hooker's Garden Ferns.
64 coloured plates, 42s.

Berkeley's Fungology.
24 coloured plates, 30s.

Captain Scott's Sketches in India.
100 photographs, with Descriptions, £3. 3s.

Grisebach's Flora of the West Indies.
Parts I. to V., each 5s.

Harvey's Phycologia Australica.
Vols. I. to IV., each 30s.

FORTHCOMING WORKS.

The Gate of the Pacific.
By Captain BEDFORD PIM, R.N., F.R.G.S. [*In January.*

Chemical Analysis, Qualitative and Quantitative.
By Dr. HENRY M. NOAD, F.R.S. [Part I., QUALITATIVE, *in the Spring.*

Bentham's Illustrated Handbook of the British Flora.
[*Publishing Monthly.*

Curtis' British Entomology in Orders.
        COLEOPTERA.   [*Publishing Monthly.*
        LEPIDOPTERA.   "
        HYMENOPTERA.   "
        DIPTERA.   "

Badham's Esculent Funguses.
New Edition.

Berkeley's Mosses.

Horæ Ferales.
By the late JOHN M. KEMBLE, M.A. Edited by Dr. LATHAM and A. W. FRANKS.

Seemann's Flora Vitiensis.

www.ingramcontent.com/pod-product-compliance
Lightning Source LLC
Chambersburg PA
CBHW022045230426
43672CB00008B/1074